Learning With Spheres

This book provides, for the very first time, a critical edition and an English translation (accompanied by critical notes and technical analyses) of the chapter on spheres (*golādhyāya*) from Nityānanda's *Sarvasiddhāntarāja*, a Sanskrit astronomical text written in seventeenth-century Mughal India.

Readers will learn how terrestrial and celestial phenomena were understood by early modern Sanskrit astronomers using spherical geometry. The technical discussions in this book, supported by the critically edited Sanskrit text and geometric diagrams, offer an opportunity for historians of the astral sciences to understand developments in astronomy in seventeenth-century Mughal India from a more nuanced perspective. These are supplemented through explorations of modernity, mathematics, and mythology and how they thrived within Sanskrit astronomical discourse at the courts of the Mughal emperors.

This book will be of interest to historians and philosophers of science, in particular those interested in the history of non-Western astral sciences. The book will be a valuable resource for scholars studying the general history of Sanskrit astronomy in the Indian subcontinent as well as those interested in the technical aspects of Sanskrit and Indo-Persian astronomy in Mughal India.

Anuj Misra is a Gerda Henkel Fellow at the University of Copenhagen, Denmark. His research focuses on medieval and early modern exchanges in Sanskrit astral sciences and includes articles and book chapters on the influence of Islamicate thought in the Sanskrit astronomy of Mughal India.

Scientific Writings from the Ancient and Medieval World
Series editor: John Steele, Brown University, USA

Scientific texts provide our main source for understanding the history of science in the ancient and medieval world. The aim of this series is to provide clear and accurate English translations of key scientific texts accompanied by up-to-date commentaries dealing with both textual and scientific aspects of the works and accessible contextual introductions setting the works within the broader history of ancient science. In doing so, the series makes these works accessible to scholars and students in a variety of disciplines including history of science, the sciences, and history (including Classics, Assyriology, East Asian Studies, Near Eastern Studies and Indology).

Texts will be included from all branches of early science including astronomy, mathematics, medicine, biology, and physics, and which are written in a range of languages including Akkadian, Arabic, Chinese, Greek, Latin, and Sanskrit.

The Foundations of Celestial Reckoning
Three Ancient Chinese Astronomical Systems
Christopher Cullen

The Babylonian Astronomical Compendium MUL.APIN
Hermann Hunger and John Steele

The Gaṇitatilaka and its Commentary
Two Medieval Sanskrit Mathematical Texts
Alessandra Petrocchi

The Medicina Plinii
Latin Text, Translation, and Commentary
Yvette Hunt

Learning With Spheres
The *golādhyāya* in Nityānanda's *Sarvasiddhāntarāja*
Anuj Misra

https://www.routledge.com/classicalstudies/series/SWAMW

Learning With Spheres

The *golādhyāya* in Nityānanda's *Sarvasiddhāntarāja*

Anuj Misra

Routledge
Taylor & Francis Group
LONDON AND NEW YORK

First published 2023
by Routledge
4 Park Square, Milton Park, Abingdon, Oxon OX14 4RN

and by Routledge
605 Third Avenue, New York, NY 10158

Routledge is an imprint of the Taylor & Francis Group, an informa business

British Library Cataloguing-in-Publication Data
A catalogue record for this book is available from the British Library

ISBN: 978-1-138-58357-3 (hbk)
ISBN: 978-1-032-31636-9 (pbk)
ISBN: 978-0-429-50668-0 (ebk)

DOI: 10.4324/9780429506680

Typeset in TeX-Gyre-Pagella
by codeMantra

**To those who taught me to think,
to those who sat with me thinking,
and to those who loved me as I thought.**

तं वन्दे परनिण्यदां कवितमं सिद्धान्तवादेषु य-
मानन्दस्थितिवर्तनं सुमुदितं नित्यं हि भाष्यं विना ।
साध्येयं ह्यपि वर्णनामिति मतिं मत्वा श्रयेत्कं पुन-
र्यत्नित्योर्मिरुदेत्यविज्ञमनसा भाष्ये ऽनुजे शाम्यतु ॥

taṃ vande paraniṇyadāṃ kavitamaṃ siddhāntavādeṣu yam-
ānandasthitivartanaṃ sumuditaṃ nityaṃ hi bhāṣyaṃ vinā |
sādhyeyaṃ hy api varṇanām iti matiṃ matvāśrayet kaṃ punar-
yatnityormir udety avijñamanasā bhāṣye 'nuje śāmyatu ||

I praise him, the giver of great mystery, who is the wisest among those who talk about the *siddhānta*s, [the one who remains] abiding in a state of bliss forever delighted without ever speaking. But who should one appeal to thinking that one might get an explanation? May this perpetual wave of despair that arises in an ignorant mind be quelled in a later commentary [that follows].

Contents

Figures

Tables

1 Introduction

The astral sciences were among the first organised systems of investigating the heavens in a long tradition of human inquiry about the natural world. The study of astral sciences grew out of the cradle of theology, mythology, and prophecy, and over the course of its history, it itself became the grounds where the methods of arithmetic, geometry, and abstraction thrived.[1] By medieval times, astronomers began determining the movements of the heavens with increasingly sophisticated numerical, geometrical, and graphical techniques. Their findings not only gave them insights into the inner workings of nature, but also invested them with the power to explore this knowledge for social benefit. Spherical geometry now explained the motions of celestial objects and catarchic astrology interpreted their destined impact. This meant that practising astronomers could accrue fame in their societies as scholars in service of imperial, noble, or papal courts.

For Sanskrit astronomers, the royal patronage of the Mughal courts (1526–1857)[2] brought them in direct contact with their Persianate counterparts as they both jostled for imperial sponsorship.[3] The Mughal emperors instituted translation projects that offered the opportunity of employment or endorsement to a newly emergent class of professional astronomers and bilingual interpreters. It is in this competing cosmopolitan world of Mughal India that Sanskrit astronomy truly begins to engage with Islamicate (Arabic and Persian) ideas.[4] The complex discourses that followed were shaped by the power struggles of language, culture, and identity as medieval Islamicate astronomy was now cast into the language of Sanskrit.

At the court of Mughal Emperor Shāh Jahān (r. 1628–1658), the Hindu Pandit Nityānanda (fl. 1630/1650) worked alongside the Muslim scholar Mullā Farīd al-Dīn Dihlavī (d. ca. 1629/1632) to translate into Sanskrit the latter's Persian *zīj* (a handbook of astronomical tables), the *Zīj-i Shāh Jahānī* (ca. 1629/1630), itself based upon the famous *Zij-i Jadīd-i Sulṭānī* (ca. 1438/1439) of Mirzā Ulugh Beg. Nityānanda's *Siddhāntasindhu* 'The Ocean of *siddhānta*s' (ca. early 1630s), like the

DOI: 10.4324/9780429506680-1

Zīj-i Shāh Jahānī, is an enormous work that includes theoretical discussions on hermeneutics, logic, metaphysics, mathematics, and astronomy, along with a large number of astronomical, calendrical, and geographical tables. It is the largest (and among the earliest) Sanskritic presentation of Islamicate astronomy composed in the style of an Islamicate *zīj*.

With the *Sarvasiddhāntarāja* 'The King of all *siddhāntas*' (1639), a later adaptation of his *Siddhāntasindhu*, Nityānanda reformulated his ideas to follow the material and the metrical standards of a traditional canonical treatise (*siddhānta*) in Sanskrit astral sciences (*jyotiḥśāstra*).[5] The contents of the *Sarvasiddhāntarāja* are arranged in three main chapters (*adhyāyas*), namely, the *gaṇitādhyāya* 'chapter on computations', the *golādhyāya* 'chapter on spheres', and the *yantrādhyāya* 'chapter on instruments'. The *golādhyāya* provides the geometrical basis to understand the computational methods for planetary motion described in the *gaṇitādhyāya*, while the *yantrādhyāya* describes the methods of construction of various astronomical instruments. The work is entirely composed in Sanskrit metrical poetry with verses of various lengths in an assortment of complex metres. By integrating Islamicate views (*yāvanika-mata*) with traditional Sanskrit canonical opinions (*saiddhāntika-mata*) and mythohistorical narratives (*paurāṇika kathā*), Nityānanda demonstrates a creativity, originality, and innovation hitherto rarely seen in any Sanskrit text on astronomy.[6] In fact, the transformation of Islamicate astronomy to suit the linguistic and ontological paradigms of traditional Sanskrit science makes Nityānanda's *Sarvasiddhāntarāja* a unique experiment in syncretic epistemology where mathematics, mythology, and modernity contend and cooperate simultaneously.

The present book provides, for the very first time, a critical edition and an English translation (accompanied by critical notes and technical analyses) of the contents of the 'chapter on spheres' (henceforth, called the *golādhyāya*) from Nityānanda's *Sarvasiddhāntarāja*. In the *golādhyāya*, Nityānanda describes (i) the different types of spheres, namely, the terrestrial sphere or the sphere of the Earth (*bhūgola*), the sphere of asterisms (*bhagola*), the oblique celestial sphere (*khagola*), and the visible celestial sphere (*dṛggola*); (ii) the great and small circles on these spheres that correspond to various terrestrial, chronometric, and astronomical measurements; and (iii) the mythohistorical accounts of geography and cosmography from Purāṇic sources.

Several topics discussed in this chapter are of Greco-Islamicate (Ptolemaic) origin; for example, the latitudinal division of the inhabited land (oecumene) on Earth into seven climes or the distribution of 48 constellations in relation to the ecliptic. Their inclusion in a seventeenth-century Sanskrit astronomical *siddhānta* is itself novel and unique for its time. There are some instances where Nityānanda rejects traditional world-views (based on the Purāṇas lit. the ancient word) in favour of more contemporary ones, whereas in other instances, he accepts both the mythohistorical and mathematical points of view as valid despite the inherent contradictions between them. In many ways, Nityānanda's attempts to reconcile heterogeneous opinions are a reflection of the changing episteme of early modern Sanskrit astronomy, and by examining the intricacies of his thoughts on spheres, this book hopes to bring to the fore some of the ways in which traditional Sanskrit astronomy functioned and flourished in early modern Mughal India.

1.1 Indian astronomy: a brief overview

Within India, the study of *jyotiḥśāstra* has enjoyed a rich history of scholarship: the various types of technical texts (e.g., *siddhānta*s, *karaṇa*s, *koṣṭhaka*s, *tantra*s), commentaries (*bhāṣya*s), and annotations (*vārttika*s) illustrate its capaciousness to include a diverse range of opinions.[7] However, within a global discussion on the history of science, the Indian contribution remains remarkably under-represented, especially when compared to the extent of similar studies across Babylonian, Hellenistic, and Islamicate cultures of science.[8] The lack of the availability of translations of technical Sanskrit literature (e.g., texts in mathematical astronomy) in modern-day languages has meant that the Indian contribution to the global history of science continues to remain peripheral and secluded. However, studies like Pingree (1963, 1973b, 1976b, 1978a, 1978b), Mak (2013), Mercier (2018), Gansten (2019, 2020) and Misra (2021, 2022) have demonstrated the interconnectedness of the study of astral sciences in India and the world. The study of Indian astronomy offers a perspective to examine the history of science in a society where the power struggles of sociocultural identities, linguistic hegemonies, and philosophical predispositions shape scientific discourses in unique ways.[9] Moreover, the changing geopolitical landscape of India has meant that there were extended periods of its history where traditional ideas were confronted with foreign theories. The ensuing arguments in accepting, accommodating, or repudiating these theories demonstrate the vitality of thought among Indian astronomers (see, e.g., Pingree 1996a and Minkowski 2002).

In contrast to the Greco-Islamicate practice of using kinematic celestial models, the study of astronomy in India focused on developing and improving numerical algorithms to calculate planetary motion. Its emphasis on computations allowed it to maintain a versatile structure without committing itself to any immutable ideals or geometrical forms. In fact, this approach was in line with the emphasis on memory and orality in Indian sciences. Technical texts were often composed in versified (metrical) Sanskrit for mnemonic purposes, and in mathematical astronomy, mastery was demonstrated by compositions that reflected both poetic elegance and numerical sophistication. This, however, is not to deny the importance of geometry in Indian mathematical astronomy: on the contrary, epicyclic and eccentric planetary models from non-Ptolemaic traditions made their way into Sanskrit *siddhānta*s as early as the fifth century (Pingree 1971). These models were developed and modified over time to match the computational advancement, in particular, in the works of the Kerala (Niḷā) school of astronomers and mathematicians (see, e.g., Shastri 1969, Shukla 1973, Ramasubramanian, Srinivas and Sriram 1994 and Ramasubramanian 1998).

Table 1.1 sketches an approximate chronology of the different periods in the history of Indian astronomy, as described in S. B. Dikshit (1969), Pingree (1978a), S. B. Dikshit (1981), Pingree (1981b), Subbarayappa and Sarma (1985), Rao (2000), Sen and Shukla ([1985] 2000), Plofker (2009) and Divakaran (2018) all of which, taken together, survey the field fairly exhaustively. A significant aspect of Indian astronomy is the origin, development, competition, and coexistence of different schools of thoughts called *pakṣa*s in the medieval India. The differences between the *pakṣa*s were based on their choice of astronomical parameters, e.g., the integer number of complete rotations indicating the mean motions of the planets (over very large cyclical periods of time called *yuga*s). The proponents of these *pakṣa*s wrote foundational and canonical treatises (*siddhānta*s) that differed on specific computational nuances, e.g., on the effect of double-epicyclic forces acting on the mean positions of planets and causing their longitude to dislocate towards their true positions (Pingree 1971).[10] By the middle of the second millennium, the corpus of technical astronomical texts in Sanskrit grew to include commentaries (*bhāṣya*s), technical expositions (*tantra*s), disquisitions (*pradīpikā*s), annotations (*vārttika*s), compendiums (*sāra*s or *saṅgraha*s), book on rationales (*yukti*s), logic (*nyāya*s), explanatory guides (*vivaraṇa*s), and mnemonic aphorisms (*sūtra*)—all of these based on commenting or critiquing the canonical texts.

Along with these exegetical texts, shorter (and more concise) texts called *karaṇa*s were written with a stronger emphasis on practical rules of computations. The production of these *karaṇa*s led to a profusion of simpler astronomical table-texts (*sāraṇī*s or *koṣṭhaka*s) written for the benefit of practical astronomy (see, e.g., Montelle and Plofker 2018). These table-texts became standard reference books for fortune-tellers, priests, and astrologers who, in turn, composed additional calendrical, astrological, and divinatory texts for practical purposes, namely, almanacs and ephemerides (*pañcāṅga*s); books of omens (*saṃhitā*s); and texts on genethlialogy (*jātaka-paddhati*), catarchic astrology (*muhūrta-paddhati*), and interrogations (*praśna-vicāra*).

Towards the middle of the second millennium, Islamicate astronomy from the Marāgha and Samarqand schools began circulating in the centres of learning in northern India. Sanskrit astronomers, particularly those under the direct or indirect patronage of the Mughal courts (more on this in Section 1.1.3), debated the validity and utility of this foreign astronomy in their works.[11] It is here that we find seventeenth-century Hindu astronomers like Nityānanda (fl. 1630/1650), Munīśvara

Table 1.1 An approximate chronology of the periods in the history of Indian astronomy

Date	*Period*	*References*
BCE		
ca. *ante* 2000	Indus Valley Civilisation	Ashfaque (1977) and Vahia and Menon (2011)
ca. 1500–1000	Ṛg Vedic Period	
ca. 1000–500	Late Vedic Period	S. B. Dikshit (1969) and Ôhashi (1993)
ca. 500–300	Beginnings of *Vedāṅga* Astronomy	
CE		
ca. 100–200	Introduction of Greek Horoscopy	Pingree (1978d)
ca. 200–400	*Vedāṅga* Astronomy	Ôhashi (2002)
ca. 400	Introduction of Greek Mathematical Astronomy	Pingree (1978a)
ca. 500–1200	Canonical (*saidhhāntika*) Astronomy	S. B. Dikshit (1981)
ca. 1200/1300–1700/1800	Indian and Islamicate Astronomy	Ansari (1995), Rahman (1998), and Ansari (2005, 2016)
ca. post 1700	Indian and European Astronomy	Ansari (1985), Kochhar and Narlikar (1995) and Sen and Shukla ([1985] 2000)

(fl. 1646), and Kamalākara (fl. 1658) composing their own canonical works that engage with Islamicate ideas in different ways. The separation or syncretism of Indian and Islamicate astronomy seen in their writings offers a liminal perceptive on the linguistic, cognitive, and societal challenges that impact Sanskrit mathematical astronomy in the seventeenth century.

1.1.1 *Medieval Siddhāntic astronomy, ca. 400–1600*

Towards the end of the first millennium, the computational techniques and planetary models in Indian astronomy were influenced by Hellenistic (non-Ptolemaic) astronomy. These influences can be observed in the different subdisciplines of mathematical astronomy, e.g., in the techniques of trigonometry, spherical geometry, algebraic inequalities, and indeterminate equations (see Pingree 1963, 1978a, 1981b). Their subsequent utility brought the systematisation of the different schools of thought called *pakṣa*s, where each *pakṣa* used a slightly different set of astronomical parameters to describe celestial phenomena (see, e.g., Pingree 1970b). A detailed description of the different *pakṣa*s discussing their parametric differences and mathematical innovations is described elsewhere, e.g., in Pingree (1978a) and Plofker (2009). Table 1.2 lists the main texts (*siddhānta*s and *karaṇa*s) and their authors that represent the different *pakṣa*s. These texts often borrowed, amalgamated, or refuted ideas from among each other, and came to define what we collectively call the corpus of Siddhāntic (*saidhhāntika*/*pākṣika*) astronomy. The adherents of these *pakṣa*s created a medieval knowledge economy in their writings where computational rules were often commutable and parametric values commensurable.[12] In many ways, their proclivity for ideational equipoise fostered the environment where Islamicate ideas could also be discussed and debated in the Siddhāntic astronomy of the early modern age (see, e.g., Pingree 1963, 1978c, 1996a).

1.1.2 *Early modern Siddhāntic astronomy, ca. post 1600*

Beginning from the late fourteenth century and culminating in the early eighteenth century, astronomical tables (*zīj*es), observational instruments, and planetary models of Islamicate origin increasingly influenced Sanskrit astronomy (particularly, in North India).[13] While some Sanskrit works were direct translations of Arabic or Persian texts (see, e.g., Misra 2021, Table 1 on p. 40), others were more subtle in discussing Islamicate theories (see, e.g., Pingree 1978c, 1996a, 1999). The exegetical texts and commentaries that ensued is a hallmark of the tumultuous reception of Islamicate theories in Sanskrit. Of particular note are two seventeenth-century immigrant families of Devarāta and Bhāradvāja Brahmins in Kāśī (modern-day Varanasi), led by Munīśvara

Table 1.2 A chronology of some of the major texts in the different schools (*pakṣas*) of medieval and early modern Siddhāntic astronomy, ca. 400–1700

School (pakṣa)	*Texts*
Brāhmapakṣa	*Paitāmahasiddhānta* of Viṣṇudharmottara Purāṇa (ca. fifth century) *Brāhmasphuṭasiddhānta* of Brahmagupta (628) *Mahāsiddhānta* of Āryabhaṭa II (ca. 950–1000) *Siddhāntaśekhara* of Śrīpati (ca. mid eleventh century) *Siddhāntaśiromaṇi* of Bhāskara II (1150) *Karaṇakutūhala* of Bhāskara II (1183) *Laghukaraṇa* of Bhāvasadāśiva (1598) *Karaṇavaiṣṇava* of Śaṅkara (1766)
Āryapakṣa	*Āryabhaṭīya* of Āryabhaṭa I (499) *Mahābhāskarīya* and *Laghubhāskarīya* of Bhāskara I (ca. early seventh century) *Śiṣyadhīvṛddhidatantra* of Lalla (ca. eight/ninth century) *Vaṭeśvarasiddhānta* of Vaṭeśvara (904) *Karaṇaprakāśa* of Brahmadeva (1092) *Tantrasaṅgraha* of Nīlakaṇṭha Somayājī (1501)
Ārdharātrikapakṣa	*Āryabhaṭasiddhānta* of Āryabhaṭa I, *lost* (ca. 500) *Sūryasiddhānta* (a revised version preserved in the Varāhamihira's *Pañcasiddhāntikā*, ca. sixth century) by Lāṭadeva (ca. 505) *Khaṇḍakhādyaka* of Brahmagupta (665) *Bhāsvatī* of Śatānanda (1099)
Saurapakṣa	*Sūryasiddhānata* unknown authorship (ca. 800), most common recension by Raṅganātha of Benaras (1602) *Laghumānasa* of Muñjāla (ca. 932) *Sūryasiddhāntaṭīkā* of Mallikārjuna Sūri (1178) *Somasiddhānta* of Nṛsiṃha (ca. 1400) *Sūryasiddhāntavivaraṇa* of Parameśvara (1432) *Siddhāntasundara* of Jñanarāja (1503) *Rāmavinoda* of Rāmacandra (1590) *Sūryapakṣaśaraṇa* or *Khacarāgama* of Viṣṇu (1608) *Siddhāntasarvabhauma* of Munīśvara Viśvarūpa (1646) *Siddhāntatattvaviveka* of Kamalākara (1658) *Sauravāsanā* on the *Sūryasiddhānata* by Kamalākara (*post* 1658)
Gaṇeśapakṣa	*Grahalāghava* or *Siddhāntarahasya* of Gaṇeśa Daivajña (1520) *Karaṇakaustubha* of Kṛṣṇa Daivajña (1653)

(b. 1603) and Kamalākara (b. 1610) respectively, who argued the validity of Islamicate ideas vigorously. The rivalry between these two scholastic families saw several texts being authored by extended family members assailing each other (see Minkowski 2014, pp. 122–127). For instance, Raṅganātha, the brother of Kamalākara, wrote his *Lohagola-khaṇḍana* 'Critique of the Sphere of Iron' rejecting the orthodox Indian view that the blue sky was, in fact, a sphere of iron (*loha-gola*) and accepting in place the Persian view (*pārasīka-mata*) that it was a crystalline sphere (*sphaṭika-gola*). In response to this, Gadādhara, Munīśvara's cousin, promptly composed his *Lohagolasamarthana* 'Vindication of the Sphere of Iron' as a dissenting rejoinder.

1.1.3 Siddhāntic astronomy in the Mughal world

The *Gurkānī ᶜĀlam* or the Mughal world (1526–1857) was the rule of the Muslim emperors in South Asia led by monarchs of the Timurid dynasty, beginning with Ẓahīr al-Dīn Muḥammad Bābur. The successors of Bābur extended their dominion over the Indian subcontinent, and in doing so, they created a cosmopolitan society where artistic, literary, and scientific traditions from different cultures thrived and flourished. The practice of translating literary and technical texts between Sanskrit and Persian became an administrative activity under the patronage of successive Mughal emperors. In fact, as early as the fourteenth century, Sanskrit literary works were translated into Persian under the sponsorship of the Turkic Sulṭāns of Delhi; for example, the *Tūtīnāma* 'Tales of a Parrot' of Ẓīyāᵓ al-Din Nakhshabī (d. 1350) was a collection of 52 short stories in Persian adapted from the Sanskrit *Śuka-saptati* 'Seventy tales of the Parrot' (ca. twelfth century) under the patronage of Sulṭān Muḥāmmad bin Tughlaq (1325–1351). Modern-day scholars like Alam (2003), Haider (2011), Truschke (2012, 2015, 2016), and Nair (2020)have described the complex sociocultural world of Mughal India where Sanskrit and Persian literary traditions cohabited and contested for power, privilege, and position. These studies also describe the culture in which technical Sanskrit texts were translated into Persian at the behest of Mughal emperors; for example, Bhāskara II's treatise on arithmetic, the *Līlāvatī* (1150), was translated into the Persian as the *Tarjuma-yi Līlāwatī* in 1587 by Abu ᵓl-Fayḍ Fayḍī at the court of the Mughal Emperor Akbar (r. 1556–1605). By the time Nityānanda's patron, Emperor Shāh Jahān (r. 1628–1658), ascended to the throne, his predecessors had already started a custom of decorating Sanskrit court astrologers with titles likes 'royal astronomer' (*Jotik Rāi* or *Vedāṅgarāya*). Even Shāh Jahān conferred such a title on

Mālajit Vedāṅgarāya (fl. 1643), a reputable Hindu Pandit from Kāśī who entered the service of Shāh Jahān's court at the recommendation of his former patron Rāja Giridhara Dāsa, the Rajput King of Ajmer (Minkowski 2014, pp. 121–122).

Under Shāh Jahān's rule, the Hindu Brahmin community continued to participate in various cultural and courtly affairs. For example, there were two Hindi-speaking Brahmin poets at Shāh Jahān's court, Kavīndrācārya Sarasvatī Vidyānidhāna (fl. ca. 1600/1675) and Jagannātha Paṇḍitarāja (fl. ca. 1620/1660), who ingratiated themselves with the emperor and his retinue by composing panegyrics in Brajabhāṣā and singing Hindustānī *dhrupad* songs at his court (Truschke 2016, pp. 50–53). As Pollock (2001) observes, the seventeenth century saw an 'explosion of scholarly production unprecedented for its quantity and quality' (p. 5) by new intellectuals (*navya*, *navīna*, or *arvāc*) in the fields of grammar, hermeneutics, and logic (the classical Sanskrit *trivium*) as well as literary theory and rhetoric. Moreover, the intellectual curiosity and scientific revolution that pervaded the seventeenth century made the business of intellectual exchange a very profitable venture (see, e.g., Huff 2010). The Hindu Pandit Mālajit Vedāṅgarāya who was decorated by Shāh Jahan was also the author of a *Pārasiprakāśa* (1643), a Sanskrit lexicon for translating Persian astronomical terms. A functional multilingualism in Persian, Sanskrit, and vernacular Hindi would have been a competitive and cultural advantage for any aspiring Hindu astrologers (*jyotiṣī*s) vying for Mughal sponsorship.[14]

However, even as the political and linguistic world of seventeenth-century Mughal India promoted cross-cultural exchanges between Sanskrit and Islamicate astronomy (see, e.g., Misra 2021, pp. 33–42), it encountered the resistance of scientific traditionalism. The revelations of the sages (*śrutivāda*) in the Sanskrit *siddhānta*s were meant to preserve ancient Hindu belief systems, and in many instances, they were in opposition to the opinions of the foreigners (*yāvanika-mata*). This created a conceptual battle between opposing adherents. The rivals Munīśvara and Kamalākara both held differing opinions on Islamicate astronomy: Munīśvara, in his *Siddhāntasārvabhauma* (1646), strongly disparaged certain Islamicate ideas (e.g., the Islamicate theory of zodiacal precession implying a tropical zodiac instead of a sidereal one) for being understood subjectively (*svamata*) and defying the revelation of the sages (*ṛṣimata-viparīta*), whereas Kamalākara offers a more tolerant, utilitarian, and pluralistic acceptance of the Islamicate imports in his *Siddhāntatattvaviveka* (1658) (Pingree 1978c, pp. 320–323). As Section 1.2 describes, Nityānanda's efforts in his *Sarvasiddhāntarāja* are more ameliorative as he attempts to reconcile Islamicate and Siddhāntic astronomy.

1.2 The *Sarvasiddhāntarāja*, 'The King of all *siddhāntas*'

According to the *Mulakhkhaṣ-i Shāhjahān Nāma*, the chronicles of Emperor Shāh Jahān, the astronomical tables called the *Zīj-i Shāh Jahānī* of Mullā Farīd al-Dīn Dihlavī (d. ca. 1629/1632) were to be translated into 'the language of Hindustan by Indian astronomers in consultation with Persian astronomers, for the sake of public utility'.[15] Mullā Farid, the 'wonder of the age' (*nādir al-zamān*), was a noted Islamicate astronomer (*munajjim*) who was recruited to compose the *Zīj-i Shāh Jahānī* at the behest of—or as the chronicles declare 'under the expert supervision' of—Āṣaf Khān, the prime minister (*vazīr-i āᶜẓam*) of Shāh Jahān and the 'right hand of the state' (*yamīn al-dawla*) (Begley and Desai 1990, p. 35).[16]

In his writings, Mullā Farid differentiated between astronomical tables (*zīj*es) that were composed by making actual observations (*raṣad*) of the distances, sizes, and positions of celestial objects and those that were simply based on mathematical calculations (*ḥisāb*): the former was called a *zīj-i raṣadi*, while the latter was known as a *zīj-i hiṣābi* (Ghori 2008, pp. 385–386). Mullā Farīd's *Zīj-i Shāh Jahānī* was a *zīj-i hiṣābi* where he modified the set of astronomical tables of Ulugh Beg (*Zīj-i Jadīd-i Sulṭānī*, better known as *Zīj-i Ulugh Beg*) composed by the consortium of ᶜAlī Qūshjī, Jamshīd al-Kāshī, and Qāḍīzāda al-Rūmī at the observatory in Samarqand in 1438/1439 under the supervision of Sulṭān Ulugh Beg himself.[17] The *Zīj-i Shāh Jahānī*, following the *Zīj-i Ulugh Beg*, contains four discourses (*maqāla*s) on four different subjects each containing several chapters (*bāb*s) that are further divided into sections (*faṣl*s).[18]

Mullā Farīd is believed to have completed the *Zīj-i Shāh Jahānī* and presented it to the emperor around October/November of 1629 (month of *Rabīᶜ al-awwal* in the Hijrī year 1039).[19] This work instituted a new calendar, the *Tāʾrīkh-i Ilāhī Shāhishānī*, commencing on 14 February 1628, the day of the coronation of Shāh Jahān (Begley and Desai 1990, p. 35). The *Zīj-i Shāh Jahānī* included the astronomical tables (and canons) of *Zīj-i Ulugh Beg* with little modification; however, it also contained an extensive list of eras (including Akbar's Ilāhī era and the Indian Vikrama Saṃvat) and a comprehensive list of festivities and occasions (Blake 2013, p. 64). The Mughal court chronicler Muḥammad Sāleḥ Kambūh praised the *Zīj-i Shāh Jahānī* as being superior to the *Zīj-i Ulugh Beg* and hence supplanting the latter in regular use. This prompted Shāh Jahān to decree that Mullā Farīd's *Zīj-i Shāh Jahānī* be translated into Sanskrit for use among the Hindu astronomers in his kingdom (Ghori 2008, p. 397).

The responsibility of translating the *Zīj-i Ulugh Beg* into Sanskrit fell upon the Hindu Pandit Nityānanda Miśra (fl. 1630/50), a Gauḍa Brahmin of Mudgala *gotra* (patronymic) from Indrapurī (Old Delhi).[20] Nityānanda was commissioned to this task by Shāh Jahān's Prime Minister Āṣaf Khān. He dedicated himself to the task and soon after, in the early 1630s, he presented his Sanskrit translation the *Siddhāntasindhu*

'The Ocean of *siddhāntas*'.[21] Misra (2021) examines the linguistic affinity between the Sanskrit verses in Nityānanda's *Siddhāntasindhu* and corresponding Persian passages in Mullā Farīd's *Zīj-i Shāh Jahānī* to suggest how technical familiarity, conceptual affordance, and vernacular conviviality all played their parts in this exercise of translating Islamicate astronomy into Sanskrit.

Despite his efforts, Nityānanda's *Siddhāntasindhu*, with its Sanskritised presentation of an Islamicate table-text, failed to make a discernible impact on the Sanskrit astronomers of his time. In the years that followed, Nityānanda made a second more dedicated attempt to adapt Islamicate astronomy to Sanskrit and called it the *Sarvasiddhāntarāja*, 'The King of all *siddhāntas*'.[22] The *Sarvasiddhāntarāja* differed from the *Siddhāntasindhu* in its presentation of Islamicate astronomy, in particular, in its use of metrical Sanskrit verses in composition (as was expected of a traditional Sanskrit *siddhānta*) and citing Islamicate parameters and procedures from allonymous works.

1.2.1 *Nityānanda, the Hindu Brahmin*

There is very little biographical information known about Nityānanda. On the basis of the colophon seen at the end of the manuscripts of the *Sarvasiddhāntarāja*, and cited in secondary sources like S. Dvivedī (1933, pp. 101–102), S. B. Dikshit (1981, pp. 165–166), and CESS A3 (1976a, p. 173b), we learn that Nityānanda was a Gauḍa Brahmin of the Mudgala *gotra*.[23] He was a resident of Indrapurī, an epithet for the city of Old Delhi, a place known to be in close proximity to the historic city of Kurukṣetra (in modern-day Haryana). The colophon also states that Nityānanda was trained in the tradition of Ḍulīnahaṭṭa, which, according to S. Dvivedī (1933), could indicate Nityānanda's place of origin. Dulīna is a village (perhaps, identified with the city of Jhajjar) in the district of Jhajjar/Rohtak in Haryana, around sixty kilometres west of Old Delhi (Indrapurī).[24]

The patrilineage of Nityānanda is stated as follows: Nityānanda, son of Devadatta, son of Nārayaṇa, son of Lakṣmaṇa, son of Icchā. Beyond their names, we know nothing about Nityānanda's ancestors or any texts they may have authored. Nityānanda, however, is believed to have authored the following four texts.

1. The *Siddhāntasindhu* (ca. early 1630s), see CESS A3 (1976a, pp. 173b), CESS A4 (1981a, p. 141ab), and CESS A5 (1994, p. 184a).
2. The *Sarvasiddhāntarāja* (1639), see CESS A3 (1976a, pp. 173b–174a), CESS A4 (1981a, p. 141b), and CESS A5 (1994, p. 184a).
3. The *Śāhajahāṃgaṇita* (unknown date), see CESS A3 (1976a, p. 174a).

4 The *Aṃrtalaharī* (ca. 1649/50), see CESS A1 and A2 (1970a, p. 46a in Volume A1) where this work is (erroneously) catalogued under the name 'Aṃrtalāla'.

1.2.2 Structure and contents of *Sarvasiddhāntarāja*

Nityānanda composed his *Sarvasiddhāntarāja* (in Vikrama Saṃvat 1696, Śaka 1561, 1639 CE) in the style of a traditional Sanskrit *siddhānta* in *jyotiḥśāstra*. It includes discussions on the topics of mathematics, astronomy, and calendrics that are grouped under three chapters (*adhyāyas*), namely, the *gaṇitādhyāya* 'chapter on computations', the *golādhyāya* 'chapter on spheres', and the *yantrādhyāya* 'chapter on instruments'. The *gaṇitādhyāya* is the largest chapter in this work and it includes separate topical sections (*adhikāras*) that discuss the astronomical parameters and mathematical procedures governing planetary motion. The *golādhyāya*, typically following the *gaṇitādhyāya*, discusses the underlying geometry and theory behind the computational methods described in the first chapter. The *yantrādhyāya* (sometimes regarded as being of the *golādhyāya*) is the last chapter on the methods and materials of constructing astronomical instruments. Taken together, these three chapters offer a systematic examination of all the major topics discussed in Sanskrit mathematical astronomy. Table 1.3 lists the topics discussed in the different sections of the three chapters of the *Sarvasiddhāntarāja*.[25]

Each chapter begins with a different benedictory verse (*maṅgalācaraṇa*) that announces an auspicious beginning and also demarcates the chapters (as separate parts of the book). The end of the chapters, and the end of each topical section in the *gaṇitādhyāya*, contains a (near) similar colophon that reads

> *ity etasyām indrapuryāṃ vasan sannityānando devadattasya putraḥ* |
> *sāroddhāre sarvasiddhāntarāje* ____ *prāpayat tatra pūrtim* ||
>
> In this manner, the wise Nityānanda, the son of Devadatta, residing in this city of Indrapurī, caused the ____ to attain completion there in the quintessential *Sarvasiddhāntarāja* [lit. king of all treatises].[26]

The blank space in the verse above is the name of the particular chapter or topical section just completed. The colophons at the end of the individual sections of the *yantrādhyāya* read

> *iti yantraraje* ____*-adhyāyaḥ* |
>
> Thus ends the reading of ____ in the (*Yantrarāja*) [lit. king of all instruments].

Table 1.3 The contents of the *Sarvasiddhāntarāja*

Sarvasiddhāntarāja (1639) of Nityānanda		
Chapter (adhyāya)	*Section (adhikāra)*	
I. *gaṇita* 'computations'	1 *mīmāṃsā*	'philosophical reflections'
	2 *madhyama*	'mean positions of planets'
	3 *spaṣṭa*	'true positions of planets'
	4 *tripraśna*	'three question: direction, place, and time'
	5 *candragrāsa*	'lunar eclipses'
	6 *sūryagrāsa*	'solar eclipses'
	7 *śṛṅgonnati*	'elevation of the lunar cusps'
	8 *saṃkrānti*	'conjunctions'
	9 *spaṣṭakrānti*	'true declination'
	10 *dṛkkheṭa*	'visibility of the planets'
	11 *bhānāṃ jñāna*	'knowledge of the stars'
II. *gola* 'sphere'	*bhūgola*	'sphere of Earth'
(list of topic, see Section 1.3.1)	*bhagola*	'right celestial sphere'
	khagola	'oblique celestial sphere'
	dṛggola	'visible celestial sphere'
III. *yantra* 'instruments'	1 *gaṇita*	'computations'
	2 *ghaṭanā*	'fabrication'
	3 *racanā*	'composition'
	4 *śodhana*	'corrections'
	5 *nirīkṣaṇa*	'examination'

1.2.2.1 *An overview of the Sarvasiddhāntarāja*

1 Nityānanda begins the *gaṇitādhyāya* by discussing the philosophical rationales (*mīmāṃsā*) of Siddhāntic astronomy. There, he begins by making his epistemic standards explicit (more on this in Section 1.2.3), and then goes on to describe his preference for the tropical (*sāyana*) zodiac over the sidereal (*nirayaṇa*).[27] Right at the beginning, these discussions reveal Nityānanda's consonance with foreign opinions (*yāvanika-mata*) which then becomes more evident in the ensuing technical discourse. The remaining sections in the *gaṇitādhyāya* discuss several Islamicate ideas in a complex narrative that tries to syncretise the Greco-Islamicate and Sanskrit Siddhāntic traditions of mathematical astronomy. For example, in the section on true declination (I.9 *spaṣṭakrānti*), Nityānanda proposes three methods to compute the correct declination of a

celestial object. These methods are described in Sanskrit verses that are copied near-verbatim from his *Siddhāntasindhu* (II.6), which, in turn, contains Sanskrit translations of corresponding Persian passages from Mullā Farīd's *Zīj-i Shāh Jahānī* (II.6) (see Misra 2022).

In the various sections of the *gaṇitādhyāya*, Nityānanda also provides several proofs/demonstrations (*upapatti*s) and examples (*udāharaṇa*s) to supplement his technical discussions. For example, verses I.3.19, 42, 50, 153, 200; I.4.42, 45, 53; and I.5.56, 61, and 82 are *upapatti*s while verses I.3.83; I.4.31, 115, 131, 146, 154; and I.5.12 are *udāharaṇa*s.[28] The proofs and examples in these discussions suggest that the *Sarvasiddhāntarāja* was not just a scholarly text, but it was also intend for pedagogical use.

The overarching theme of the *gaṇitādhyāya* is reconciliatory: Nityānanda attempts to bring the astronomical parameters and models of the *Brāhmapakṣa* and the *Saurapakṣa* to harmonise with what he understands as Islamicate equivalents from the Roman *zīj* or *Romakasiddhānta*. For example, in the section on mean positions (I.2 *madhyama*), he provides numerical corrections (*bīja*s) to make the epoch mean longitudes of the planets in the *Romakasiddhānta* equal to those attested in the *Brāhmasphuṭasiddhānta* and *Sūryasiddhānta* (see Pingree 2003b, Table 9.7 on p. 275). Elsewhere, e.g., in the section on true positions (I.3 *spaṣṭa*), Nityānanda introduces Islamicate model of planetary motions with their equants, protective spheres, and lunar crank mechanisms, as well as the Aristotelian description of the elemental spheres that constitute the physics of the universe using novel technical vocabulary and subtle mathematical trickery (see Pingree 1978c, p. 325).

2 The numerical algorithms and mathematical procedures described in the *gaṇitādhyāya* are based on the geometrical properties and trigonometric proportions of the various circles drawn on a sphere. Understanding the science of spheres allows an astronomer to understand the foundations of the astronomical calculations, and, accordingly, be better qualified to practise professionally. In the *golādhyāya*, Nityānanda describes various aspects of the celestial sphere, beginning with a detailed description of the sphere of Earth (*bhūgola*) from an astronomical, geographical, and mythohistorical (Purāṇic) perspective. Following this, the various orientations of the celestial sphere, namely, the right celestial sphere (*bhagola*), the oblique celestial sphere (*khagola*), and the visible celestial sphere (*dṛggola*), are discussed assiduously. Section 1.3.1 provides a more elaborate description of the contents of the *golādhyāya*.

3 The *yantrādhyāya* (sometimes referred to by its own title the *Yantrarāja* 'The King of Instruments') includes topics that discuss the mathematics, fabrication, calibration, precision, and use of various astronomical instruments.

1.2.3 *The episteme of the Sarvasiddhāntarāja*

In the Sanskrit *jyotiṣa* tradition, *siddhānta*s are a class of canonical astronomical treatises with an epoch date corresponding to the beginning of the *kaliyuga* 'age of *kali*', around 3102 BCE.[29] The *Sarvasiddhāntarāja* is one of the earliest Sanskrit *siddhānta*s that discuss Islamicate astronomy. It does so in a language the acknowledges the plurality (*anaikya*) of opinions (*mata*) in the various traditions (*pākṣa*s) of Sanskrit Siddhāntic astronomy. While Nityānanda's *Sarvasiddhāntarāja* is based on his table-text *Siddhāntasindhu*, it is also different from the latter in being a historical exegesis of Siddhāntic astronomy. By presenting multiple opinions in parallel (Greco-Islamicate, being one of them), Nityānanda offers an overview of the Indian and Islamicate astronomical theories prevalent in his times. The comparability of these theories facilitates their syncretism in the *Sarvasiddhāntarāja*, and this, in turn, motivates the epistemological question: what is a valid source of (astronomical) knowledge and how is it tested?

1.2.3.1 *Revelation as divine testimony*

In classical Indian philosophy, the means (*pramāṇa*) to acquire true knowledge (*pramā*) has been a subject of great discussion.[30] In describing testimony (*śabda*, lit. word) as an instrument of true knowledge, Akṣapāda Gautama, in his *Nyāyasūtra* (I.1.7, ca. between sixth century BCE and second century CE), says *āptopadeśaḥ śabdaḥ* '*śabda* is the testimony (*upadeśa*) of an authoritative or credible person (*āpta*)'. Commenting on this statement, Vātsyāyana (fl. ca. 450–500) describes *āpta* as someone who directly apprehends the truth and is driven by a natural desire to communicate it to others without distortion. Thus, trustworthiness (*āptabhāva*) can be extended to all credible sources: seers (*ṛṣi*s), noble men (*ārya*s), and foreigners (*mleccha*s).[31]

In Sanskrit *jyotiḥśāstra*, foreign ideas were (partially) accepted from very early times, and often, the authors of these ideas were designated as ancient seers to justify their testimony as divine revelations. For example, Varāhamihira, in his *Bṛhat Saṃhitā* (ca. sixth century), accords the status of a seer (*ṛṣi*) to those foreigners (*yavanas*, *pārasīkas*, or *mleccha*s) who are well versed in the study of sciences (*śāstra*s).[32] According to Varāhamihira, an education in the *śastra*s is what makes a *yavana* a sagacious scholar—despite their lowly (*mleccha*) origins—and comparable to a Brahmin Pandit. While some Siddhāntic scholars agreed with this secular meritocratic view, there were many who maintained the orthodox (*smārtika*) position that the words of the *yavana*s were blasphemous and worthy of disavowal.

One of Nityānanda's contemporary, Balabhadra criticises this position in his *Hayanaratna* (1649).[33] In his critique, he relies on value of traditional testimony, predictive utility, and mythological narratives to make the teachings of the *yavana*s authoritative, efficacious, and revelatory. For example, by including Yavanācārya (lit. the preceptor called Yavana), the author of a *tājika* text in the lineage of semi-divine sages, Balabhadra makes Yavanācārya's statements authoritative, prophetic, and divine, even if they use Persian (*pārasīka*) words (see, e.g., Pingree 1997, pp. 86–87).[34] In a sense, Balabhadra's position on accepting foreign ideas in Sanskrit *jyotiḥśastra* is not a denial of traditionalism but a nuanced way of justifying the *śāstric* tradition itself.[35] When applied to *jyotiḥśāstra*, practical (*vyāvahārika*) and utilitarian (*aupayaugika*) concessions allow for verifiable improvements to mundane sciences (*laukika-śāstra*s). In matters of spiritual (*ādhyātmika*), moral (*dhārmika*), and ritual (*vaidhika*) concerns, the canonical authority of the *smṛti* texts is infallible and immutable.

Like Balabhadra's Yavanācārya, Nityānanda also situates Romaka among a lineage of gods (*sura*s) and semi-divine sages (*ṛṣi*s). In his *Sarvasiddhāntarāja* (I.1.2), he claims that

> *śrīsūryasomaparameṣṭhivasiṭhagargācāryātriromakapulastyaparāśarādyaiḥ* |
> *tantrāṇi yāni gaditāni jayantāni sphūrjaddvidhāgaṇitagolaparisphuṭāni* ||
>
> Those treatises that were proclaimed by the excellent [gods] Sūrya, Soma, Parameṣṭhi [and the *ṛṣis*] Vasiṣṭha Garga, Atri, Romaka, Pulastya, and Parāśara, all of them are victorious, bursting into the two parts prominently seen as computations and spheres.

Here, Romaka is elevated to join the ranks of the ancient Hindu sages Vasiṣṭha, Garga, Atri, Pulastya, and Parāśara, and therefore, his *Romaka-siddhānta* is to be regarded as a revelatory text.

To extol Romaka's divine antiquity, Balabhadra, in his *Hayanaratna*, invokes the myth of the Hindu Sun god Sūrya who, having learnt astronomy from Brahmā and being afflicted by a curse, is born among the *yavana*s (or *mleccha*s) in a place called Romaka (or Roma). There, he teaches what he learnt from Brahmā to a *yavana* and thus it comes to pass that the revelation of Brahmā finds in way into the writing of Romaka (variously called *Romakazīj*, *Romakatājika*, or *Romakasiddhānta*).[36] Nityānanda repeats this myth in his *Sarvasiddhāntarāja* (I.1.16–18) saying

> *atha kaḥ kila romako bhavan munidevādiṣu gaṇyate tu yaḥ* |
> *kathayāmi tadīyam uttaraṃ śṛṇu sūryāruṇapūrvasammatam* ||

> *itihāsakathāprasaṅgato vidito bhāskara eva romakaḥ |*
> *puruhūtavirañciśāpato yavano romakapattanaṃ bhavat ||*
>
> *punar eva tayor anugrahāt cyutaśāpaḥ savitā svayaṃ purā |*
> *kṛtavān iha tantram uttamaṃ śrutirūpaṃ kila romakachalāt ||*
>
> Now, who indeed is [this] Romaka who is counted among the sages, the gods, etc.? I shall speak of him; listen [then to what has been] previously agreed by Sūrya and Aruṇa,...
>
> ...on account of their fondness for legendary stories. Indeed Bhāskara [i.e., Sūrya] was known as Romaka due to a curse of Puruhūta [i.e., Indra] and Virañci [i.e., Brahmā]. Yavana [i.e., Romaka] being in the city of Romaka,...
>
> ...[and] then, becoming free of the curse due to the favour of those two gods, Savitṛ [i.e., Sūrya] himself composed in ancient times the foremost treatise here, which has the form of the [sacred] *śruti* texts even if [it is] in the guise of [a text composed by] Romaka.

Nityānanda attributes this story to a previous work, the *Jñānabhāskara* or *Sūryāruṇasamvāda*, a dialogue between the Sūrya and his charioteer Aruṇa, wherein we are told that Sūrya, being cursed by Brahmā to be born among the *yavana*s, revealed the *Romakasiddhānta* to one Romaka, who then revised it while living in a city called Romakanagara (Pingree 1978c, p. 324). For Balabhadra, the *Romakasiddhānta* is, perhaps, an astrological text (that is part of a *Śrīṣavayaṇasaṃhitā*); however, for Nityānanda, the *Romakasiddhānta* appears to an Islamicate text on astronomy (see, e.g., Misra 2022). As Pingree (1978c, p. 324) speculates, Nityānanda's Romaka could well be Qāḍīzāda al-Rūmī, one of Ulugh Beg's astronomer at Samarqand, who, along with ᶜAlī Qūshjī and Jamshīd al-Kāshī, composed the *Zīj-i Ulugh Beg*. The influence of the astronomy of Ulugh Beg (perhaps, mediated through its recension in Mullā Farīd's *Zīj-Shāh Jahānī*) in Nityānanda's works is increasingly well documented (see, e.g., Montelle, Ramasubramanian and Dhammaloka 2016, Montelle and Ramasubramanian 2018, Misra 2021, 2022).[37]

The historicity of Romaka notwithstanding, Nityānanda's and Balabhadra's repetitions of mythopoeic tales involving him indicate the importance of such narratives. If revelation (*śruti*), understood as divine testimony (*śabda*), is to become an epistemic mean (*pramāṇa*) that is infallible (*amogha*), then the revelator (*śrutavid*, lit. knower of revelation) themself must be an authoritative or credible person (*āpta*). By being counted among the sacred *ṛṣi*s, Romaka acquires the moral authority (*dharmādhikāra*) to be considered a revered preceptor (*ācārya*). A bit further in his discussion on philosophical rationales (*mīmāṃsā*) in his *Sarvasiddhāntarāja* (I.1.55–56), Nityānanda defines his position on divine testimony by saying

> *daivatair munibhir abhyudīritaṃ yat tadeva sakalaṃ samaṃ bhavet |*
> *mānavā 'vyabhicaranty atattvataḥ satyam atra gaṇitaṃ tu yuktimat ||*
>
> *astu vā munivacaḥ pramāṇatā yuktibhiḥ saha tathopalabdhiḥ |*
> *anyathā na katham ādṛtā budhaiḥ sūryaśītakiraṇoparāgatā ||*
>
> Indeed whatever is proclaimed by gods and by the sages, all of that is equal. Men deviate on account of untruth but [the method of] computations is, in this case, the truth with rationales.
>
> May the words of the sages have probative validity, be accompanied by rationales, [and] thus, [lead to] perception. Otherwise, how will the state of the eclipsing of the Sun and the Moon be respected by wise men?

He reiterates his certitude in the words of the gods and the *ṛṣi*s by adding (in I.1.94)

> *yad yad uktam ṛṣibhiḥ devais tat tad atra sakalaṃ saphalaṃ hi |*
> *pūruṣair aviditāgamatattvaiḥ kṣiptam ūharahitaṃ tad asatyam ||*
>
> Whatever is proclaimed by the sages and the gods, all of that, in this case, is indeed fruitful. What is added by men on account of not knowing the truths of the traditional doctrines [and] what is deprived of examination, that is false.

It is, therefore, the rectitude (*samatā*), propitiousness (*saphalatva*), and probative potential (*pramāṇatā*) of divine revelation that makes it true (*satya*). In contrast, any additions by men who are ignorant of the true principles of traditional doctrines (*āgamatattva*s), as well as those additions that are devoid of critical deliberation (*ūharahita*), all of these are veritably untrue (*asatya*).

Interestingly, Nīlakaṇṭha Somayājī, in his *Jyotirmīmāṃsā* (ca. 1504, I.2), offers what is, perhaps, the most rational view on revelation: *granthakaraṇe devatāprasādaḥ mativaimalyahetuḥ na sākṣād upadeśaḥ* 'in composing [one's] text, divine grace is the reason for clarity of thought; not [for providing] manifest instructions'. According to Nīlakaṇṭha, receiving the favour of the gods (*devatā-prasāda*) is revelatory: it offers perspicuity of thought and not a divine dictation, and, accordingly, enables the author's own work to be considered as credible testimony (*āptavacana*) for successive generations of learners. Nīlakaṇṭha makes this argument in critique of the view that Brahmā revealed astronomy to Āryabhaṭa I (founder of the *Āryapakṣa*) having been pleased by Āryabhaṭa's religious austerity (*tapas*) (see, e.g., K. V. Sarma 1977, pp. 2–3).

1.2.3.2 *Predictive and observational prowess*

The power to predict celestial motions accurately was also an important criterion in determining the validity of a source in Sanskrit *jyoiḥśāstra*. Many Sanskrit authors recognised Islamicate astronomy for its power to predict celestial phenomenon precisely. For example, Gaṇeśa Daivajña (fl. ca. 1550–1600), in his *Tājīkabhuṣaṇa* (I.4), encouraged the use of the Tataric science (*tārtīyikaṃ śāstraṃ*) in Sanskrit astrology despite its origins among the Brahmin-hating Turks (*brahmadveṣi-turuṣka*), and justified it by the analogy of appreciating a lotus even if grows in dirt (Gansten 2020, pp. 84–85). Nīlakaṇṭha Somayājī goes further by emphasising the importance of observational concordance, in his *Jyotirmīmāṃsā* (I.4):

> *pañcasiddhāntās tāvat kvacitkale pramāṇam eva ity avagantyam | api ca yaḥ siddhānto darśanāvisaṃvādī bhavati so 'nveṣaṇīya | darśanasaṃvādaś ca tadānīn taiḥ parīkṣakair grahaṇādau vijñātavyaḥ | ye punar anyathā prāktana siddhāntasya bhede sati yantraiḥ parīkṣya grahāṇāṃ bhagaṇādisaṃkhyāṃ jñātvā abhinavasiddhāntaḥ praṇeya ityarthāt | tat ta iha loke 'hasanīyāḥ paraloke 'daṇḍanīyāś ca iti |*
>
> The five *siddhānta*s [i.e., the ones described in Varāhamihira's *Pañcasiddhāntikā*] were indeed valid means of knowledge up until a certain point in time, this should be known. Moreover, the *siddhānta* that agrees with observation is [the one] to be desired. And the agreement with observation should be inspected by the examiners at the time of beginning of eclipses. Again, what is different [between observation and theory], evident in the schism in an ancient *siddhānta*, having examined by instruments [and] after having known the number of revolutions of the planets, a modern *siddhānta* should be composed on account of that. Nobody will be ridiculed in this world nor punished in the next [for doing that].

Among the more philosophical astronomers, Bhāskara II's commentator Nṛsiṃha Daivajña, in his commentary *Vāsanāvārttika* on the *Siddhāntaśiromaṇi* (I.1.*bhagaṇādhyāya*.1–6), discusses the value of direct perception (*pratyakṣa-upalabdhi*) in accepting a doctrine that teaches planetary motions (*graha-gati-pratipādaka-śātra*) as probative (*prāmāṇya*). For Nṛsiṃha, the cause (*kāraṇa*) of this perception may be unreal (*atattvika*) as they arise from human reasoning (*puruṣa-buddhi*); however, the effect (*phala*) of this perception is faultless due to the very non-existence of faults (*doṣa-abhāva*) in any effect. Faults lie in the human reasoning that leads to false notions (*vikalpas*) about the cause, just as seeing the individual soul (*jīva*) separate from God (*iśvara*) is false (*mithyā*) in Advaita Vedānta.[38]

1.2.3.3 *Appeal to rationales*

For some Sanskrit astronomers, particularly those following Mādhava of Saṅgamagrāma (ca. 1340–1425) in the Kerala (Niḷā) school of mathematics, the continuity of tradition became closely associated with its contemporaneity. Nīlakaṇṭha Somayājī, in his *Jyotirmīmāṃsā* (I.3), quotes from the *Tantravārttika* (I.3.2) of Kumārila Bhaṭṭa, a seventh-century exegesis of Jaimini's *Mīmāṃsāsūtra*, to emphasise the role of tradition (*saṃpradāya*) and inference (*anumāna*). According to Nīlakaṇṭha

> *'jyotiśāstre 'pi yugaparivṛttiparimāṇadvāreṇa candrādityādigativibhagena tithinakṣatrajñānamavicchinnasaṃpradāyagaṇitānumānamūlam' iti vārttikakāro 'pi grahagatijñānam anumanenāha* | (*kumārilabhaṭṭaḥ, tantravārttikam*, 1.3.2) |
>
> Even in the science of astronomy, the computation of *tithi*s and *nakṣatra*s by means of the motion of the Sun and the Moon, which is determined through the number of their revolutions in a *yuga*, is based on the continuity of tradition and deduction.
>
> (Kumārila Bhaṭṭa, *Tantravārttika*, 1.3.2)
> (Subbarayappa and Sarma 1985, pp. 57–58)[39]

Other Siddhāntic authors have also spoken about provability or demonstrability (*upapattitva*) as a standard for validating astronomical computations; see, e.g., Bhāskara II's statement in his auto-commentary *Vāsanābhāṣya* on his *Siddhāntaśiromaṇi* (I.2.1–6) in endnote 11 on page 48.

1.2.3.4 *Epistemic pluralism*

Nityānanda professes a pluralistic epistemology where authoritative speech (*āptavāc*), appeal to antiquity (*prācīna-matāvalambana*), concordance with observations (*dṛktultyatva*), and rational explanations (*yuktārtha-vyākhyā*s) are all regarded as valid epistemic means (*pramāṇa*s). In his view, different epistemic criteria should be applied appositely to assess the validity of a particular position (or proposition). Accordingly, in his *Sarvasiddhāntarāja* (I.14–15), he offers three distinct and individual justifications for accepting the three *siddhānta*s (i.e., the *Romakasiddhānta*, the *Sūryasiddhānta*, and the *Brāhmasphuṭasiddhānta*):

> *dṛṣṭvā romakasiddhāntaṃ sauraṃ ca brāhmaguptakaṃ* |
> *pṛthak spaṣṭān grahāñ jñātvā siddhāntaṃ nirmame sphuṭam* ||
>
> *romakoditakhacāracāturī dṛktulāṃ vrajati sarvathā sadā* |
> *sauratantram iha vedavid vidur jiṣṇujoktam api yuktayuktiyuk* ||
>
> Having consulted the *Romakasiddhānta*, the *Sūryasiddhānta* and [the *siddhānta* of] Brāhmagupta individually, [and] having known the true [positions of the] planets, I composed the true *siddhānta*.

> The motion of the heavens described by Romaka always agrees with observation in every respect. Men versed in the Vedas recognised [their doctrine] in this treatise of the Sun [i.e., the *Sūryasiddhānta*], [and] the very words of Jiṣṇuja [i.e., the *Brāhmasphuṭasiddhānta*] are endowed with appropriate rationales.

In many ways, the *Sarvasiddhāntarāja* is a treatise that breaks away from doctrinal partisanship (*saiddhāntika-pākṣatā*) in favour of syncretism (*samanvayavāda*). By discussing Islamicate astronomical parameters and computations in the *Sarvasiddhāntarāja*, and occasionally doing so with a completely novel vocabulary, Nityānanda brings the science of the *yavanas* to meet the traditions of Sanskrit *jyotiḥśāstra*. Neat the beginning of the text, in I.9–10, he recognises the novelty (*nutanatva*) of his composition, and then goes on to state what he thinks a *siddhānta* is meant to be:

> *ihānekaiś cakre svakṛtir akṛtārtheva kṛtibhīḥ*
> *purā pāramparyād anavaracanaiś cāruvacanaiḥ* |
> *mamāpi grantho 'yaṃ navanavasuyuktyuktisahito*
> *vilokyo dhīmadbhiḥ kutukam iva siddhāntakuśalaiḥ* ||
>
> *siddhāntaityanugatārthapadaprayogād*
> *yad vastu sūkṣmataram asti tadeva siddham* |
> *nānyac ca golagaṇitadvayayuktihīnaṃ*
> *kiṃvopalabdhirahitaṃ sudhiyeti cintyam* ||
>
> In this regard, up until now, the works authored by many experts are just unaccomplished with no novel compositions [and merely] beautiful words following tradition. Although this book of mine is composed with completely new rationales and words, [it] will be seen by learned men competent in the doctrines with a little curiosity.
>
> Because the use of the word *siddhānta*, the [etymological] meaning of which is pertinent, only that topic which is most precise is to be considered as established by a learned man, not anything else which is devoid of the rationales of both spheres and computations or devoid of perception.

According to his statements, a genuine astronomical *siddhānta* should include only those methods that yield the most precise (*sūkṣmatara*) results, following the very etymology of the word *siddhānta*, the 'established end'. If traditional methods are lacking in the rationales (*yukti-hīna*) of spheres (*gola*) and computations (*gaṇita*) or are perceptively deficient (*upalabdhi-rahita*), a mere repetition of these methods in one's work does not merit the title *siddhānta*. Such irrational works should be decried by learned men even if they pander to tradition with beautiful words.

1.3 The *golādhyāya* in Nityānanda's *Sarvasiddhāntarāja*

The *golādhyāya* 'chapter on spheres' in Nityānanda's *Sarvasiddhāntarāja* describes the terrestrial sphere (*bhūgola*) and three different orientations of the celestial sphere (*bhagola*, *khagola*, and *dṛggola*). The configuration of the great circles on these spheres forms the geometrical basis for the computational methods described in the *gaṇitādhyāya* 'chapter on computations'. Most Siddhāntic authors have discussed spheres (*gola*) as separate sections or chapters in their own works, and in particular, authors from the Kerala (Niḷā) school of mathematics have discussed spheres in separate ancillary texts (*tantras*) to their technical treatises; see, Table 1.4.[40]

The list of topics in the *golādhyāyas* of most *siddhāntas* vary, albeit only slightly. The enumerated list below indicates the types of topics discussed in Siddhāntic *golādhyāyas*. (This list is drawn from Lalla's

Table 1.4 List of Siddhāntic and Tāntric works discussing spheres (*gola*)

Title	*Work*
golapāda	*Āryabhaṭīya* of Āryabhaṭa I (499)
gola	*Brāhmasphuṭasiddhānta* of Brahmagupta (628)
bhūgolādhyāya	*Sūryasiddhānata* unknown authorship (800)
golābandhādhikāra	*Śiṣyadhīvṛddhidatantra* of Lalla (ca. eight/ninth century)
golādhyāya	*Vaṭeśvarasiddhānta* of Vaṭeśvara (904)
golādhyāya	*Mahāsiddhānta* of Āryabhaṭa II (ca. 950–1000)
golavāsanā and *golavarṇana*	*Siddhāntaśekhara* of Śrīpati (ca. mid eleventh century)
golādhyāya	*Siddhāntaśiromaṇi* of Bhāskara II (1150)
bhāṣya	*Siddhāntadīpikā* of Parameśvara (1432)[a]
bhāṣya	*Goladīpikā* I of Parameśvara (1443)[b]
bhāṣya	*Goladīpikā* II of Parameśvara (ca. 1450)
sāragrantha	*Golasāra* of Nīlakaṇṭha Somayājī (ca. 1500)
golādhyāya	*Siddhāntasundara* of Jñanarāja (1503)
golādhyāya	*Sarvasiddhāntarāja* of Nityānanda (1639)
avanigolaja	*Siddhāntasarvabhauma* of Munīśvara Viśvarūpa (1646)
gola	*Siddhāntasamrāṭ* of Jagannātha (1732)
gola	*Paramasiddhānta* of Premavallabha (1882)
golādhikāra	*Siddhāntadarpaṇa* of Sāmanta Candraśekhara (1899)

a a super-commentary on Govindasvāmin's *Mahābhāskarīyabhāṣya* of Bhaskara I.

b includes the author's self-commentary (*vivṛti*).

Śiṣyadhīvṛddhidatantra, Chapters XV–XXI, and Bhāskara's *Siddhānta-śiromaṇi*, Book II, on spheres.)

1 Advantages and importance of studying the spheres (*gola-praśaṃsā*).

2 Questions about the form of the sphere (*gola-svarūpa-praśna*).

3 Motion of the celestial sphere (*graha-bhrama*) with discussions of the seven winds (*sapta-vayu*), the fixity of the Earth (*bhuva-vinyāsa*), and the motions of the Sun and the Moon (*ravi-candra-cāra*).

4 Conception (*bandha*) of the armillary sphere (*khagola*), the sphere of asterisms (*bhagola*), the sphere of the planets (*graha-gola*), and the general sphere (*samānya-gola*).

5 Rationales (*vāsanā*) of mean motion (*madhya-gati*), including questions about sidereal day-lengths (*sāvana-dina-māna*), ascensional difference due to latitudes (*cara*), longitudinal difference of day-lengths (*deśāntara*), visibility corrections (*dṛkkarma*) and their corrections (*dṛkkarma-saṃskāra*), eclipses (*grahaṇa*), phases of the Moon (*candra-sitāsita*), elevation of the lunar horns (*śṛṅgonnati*), visibility of the body of the planets (*graha-bimba*), etc.

6 Rationales on questions of direction, place, and time (*tripraśna*).

7 Form of the Earth-sphere (*bhūgola-svarūpa*), along with discussions on terrestrial geography and cosmography (*bhuvanakośa*).

8 Graphical methods of representing planets (*chedyaka-vidhi*).

9 Miscellaneous discussion on dispelling false (astronomical) notions (*mithyā-jñāna-nirākaraṇa*), describing the seasons (*ṛtu-varṇana*), sundry questions (*praśnādhyāya*), worldly advice (*lokopadeśa*), etc.

10 Use of astronomical instruments (*yantrādhikāra*).

1.3.1 *Structure and contents of Nityānanda's golādhyāya*

Table 1.5 provides a list of the contents of the *golādhyāya* in Nityānanda's *Sarvasiddhāntarāja* alongside the respective verse numbers. Nityānanda groups the 135 verses in the *golādhyāya* under four broad sections on the sphere of the Earth (*bhūgola*), the sphere of asterisms (*bhagola*), the oblique celestial sphere (*khagola*), and the visible celestial sphere (*dṛggola*). In essence, the first part of the *golādhyāya* includes a description of the Earth-sphere, including a dedicated section on Purāṇic cosmography.[41] The second part of the *golādhyāya* describes three orientations of the celestial sphere based one three different frames of references (more on

Table 1.5 A list of the contents of the *golādhyāya* in Nityānanda's *Sarvasiddhāntarāja* alongside the respective verse numbers

Verse	*Content*
incipit	Introducing the *golādhyāya*
1	Extolling the divine for benediction
2–3	Justifying the study of the science of spheres
4	On the training and authority of a mathematician
5–6	On the contents of the four kinds of spheres
	The Earth-sphere (*bhūgola*), Part I
incipit	Introducing the four kinds of spheres, beginning with the Earth-sphere
7	On the nature and composition of the Earth-sphere
8–12	Geodetic methods to determine the height and the distance of a mountain
13–15	Relative positions of the four equatorial cities and the two polar regions
16	On the perspective of equatorial observers
17	On the orthogonal separation of the cardinal localities
18–20	On the perspective of polar observers
21	Locating the terrestrial horizon and the antipodes
22–23	Locating the equatorial cities and polar localities
24	Locating the terrestrial equator
25	Dimensions of the oecumene
26–27	Describing the inhabitable arctic regions
28	List of the cities along the prime meridian
29–30	On the change in the elevation and depression of the pole star with changing terrestrial latitudes
31–32	Describing antipodal terrestrial locations in egocentric coordinates
33–34	Describing the seven climes on Earth
	Purāṇic cosmography
incipit	Introducing Purāṇic cosmography
35	Justifying a planar Earth
36–37	On the Lokāloka mountain as the visible horizon
38	On the Udayācala mountain as the eastern horizon
39	On the Sun revolving around Mt Meru
40	Locating the central Jambūdvīpa continent and the first saline ocean on Earth
41–42	Describing the seven oceans on Earth
43	Locating and describing the submarine Vaḍavānala
44–45	List of the seven island-continents on Earth
46–47	Locating the northern mountain ranges of Jambūdvīpa

Table 1.5 (Continued)

Verse	*Content*
48–49	Geography of the *varṣa*s and central mountain ranges of Jambūdvīpa
50–51	Geography of the central *varṣa* of Jambudvīpa
52–53	Describing the central Ilāvṛta *varṣa* and Mt Meru
54	List of the buttress mountains to Mt Meru with their signature trees
55	Describing the Jambūnada gold found in Jambūdvīpa
56–57	List of the forests and lakes in the buttress mountains
58	Locating the abode of the gods on Mt Meru
59–60	Extolling the river Gaṅgā and her four distributaries
61	Locating the nine divisions of Bhārata *varṣa*
62	On the segregation of men in Bhārata *varṣa*
63	List of the seven principle mountain-ranges in Bhārata *varṣa*
64–65	List of the seven celestial *loka*s in the universe
	The Earth-sphere (*bhūgola*), Part II
66	On the time of day at equatorial cities relative to one another
67	Determining the north and south directions at a terrestrial location
68–69	Justifying the spherical shape of the Earth
70	Calculating the surface area of the climes on the surface of the Earth
71	Describing parallels of latitude at a non-equatorial location
colophon	Ending the section on the Earth-sphere
	The sphere of asterism (*bhagola*)
incipit	Beginning the section on the sphere of asterisms
72	On the nature of the sphere of asterisms
73.1 and 73.2	Describing the circle of asterisms
74	Describing the ecliptic or zodiacal circle
75	On the configuration of the ecliptic and its poles
76	On the inclination of the ecliptic and the solstitial points
77	On the latitudinal circle being the orbit of the planet
78	Describing the apogeal and perigeal points of a planet's orbit
79	Describing the ecliptic latitude and declination of a planet
80–81	On the constellations and their constituent stars
82	Describing the circle of ecliptic latitude or diurnal motion of a planet
83	On the illusion of great circles on a sphere
84	On the arrangement of the orbits of the planets up to the outermost zodiacal circle

(*Continued*)

Table 1.5 (Continued)

Verse	*Content*
	The oblique celestial sphere (*khagola*)
incipit	Beginning the section on the oblique celestial sphere
85	Describing the oblique celestial sphere and the astronomical horizon
86	Describing the prime vertical and the meridian
87	Describing the vertical circle of altitude and the intercardinal circles
88	Describing the secondary circle to the ecliptic and the equinoctial colure (six o'clock hour circle)
89	Describing the circle of day-radius and the secondary to the prime vertical
90	Describing the central meridian of a celestial body and the foremost parallel of meridian
91	Describing the parallel of altitude (circle of zenith-distance)
92	Describing the twelve astrological houses
93–94	Describing the Ascendant, Descendant, Medium Coeli, Imum Coeli, and the intermediate astrological houses
95–97	On the three systems of domification
98–102	Describing the oblique ascensions and sidereal day lengths at the Arctic circle
103	On the visibility of a planet in the oblique celestial sphere
104–106	On the latitudinal variations in day-lengths in the Arctic
107	On the variation in the degrees of solar elevation with respect to solar declination
108	On the nychthemeron for the gods and demons constituting a solar year
109	On the midnight day-reckoning system for polar day and night
110	On the nychthemeron for the lunar manes constituting a synodic month
	The visible celestial sphere (*dṛggola*)
111–112	Ending the section on the oblique celestial sphere (section colophon) and beginning the section on the visible celestial sphere (section incipit)
113	On the equinoctial and solstitial points of the ecliptic
114	On the greatest solar declination and an arc of declination
115	On the true declination of a planet
116	On the other (second) declination and the latitude of a planet
117	On the curve of true declination of a planet
118	On the position of the celestial poles and the celestial equator at a terrestrial latitude
119	On the inclination of the celestial equator and the depression of the meridian ecliptic point
120	On the maximum elevation and ortive amplitude of a planet

Table 1.5 (Continued)

Verse	*Content*
121	On the azimuth and the altitude of a planet
122	On the zenith distance and the ascensional difference of a planet
123	Describing the arcs of day and night on the diurnal circle
124	On the variation in the lengths of day and night
125	On the measure of day-length
126	On the equality of day and night at the terrestrial equator
127	On the difference in the rising times of an arc of the ecliptic
128	On the difference in the angles of ascension of the zodiacal signs
129	On the measure of elevation with respect to the nonagesimal point
130	On the measure of elevation with respect to the ecliptic pole
131	On the intercardinal and prime vertical altitudes
132	On the sine of the altitude and zenith distance of the Sun at its equinoctial colure transit
133	On an astronomer's ability to demonstrate using the rule of three
134	On the material for constructing an armillary sphere
135	Ending the chapter on spheres (chapter colophon)

this in Section 1.3.4). Each section begins with a short sectional incipit (in prose), and, in one instance, ends with a corresponding sectional colophon (also in prose). All other statements are composed in versified poetry in an assortment of Sanskrit metres (described in Section 1.3.7).

Typically, earlier Siddhāntic authors divided the contents of their *golādhyāyas* into topical sections (*adhikāras*); for example, Bhāskara II, in the *golādhyāya* of his *Siddhāntaśiromaṇi*, groups his discussions on graphical methods to represent planetary motions and descriptions of great circles on a sphere under the headings *chedyakādhikāra* 'on the topic of drawings' and *golabandha* 'construction of the armillary sphere' respectively. In contrast, Nityānanda's sectioning of the contents based on the orientations of the celestial sphere is different from what is seen in the *golādhyāyas* of earlier Siddhāntic works. However, beyond these minor differences in taxonomy, the nature of the topics in Nityānanda's *golādhyāya* follows what is commonly and conventionally discussed in most Siddhāntic texts.

There are a few instances where Nityānanda's discussions are completely novel. For example, in his statements on Purāṇic cosmography, he adds a few remarks taken from Islamicate sources—most notably, the discussions on the seven climes on Earth (*sapta-khanḍa* in Sanskrit or *haft iqlīm* in Arabic; see notes to verses 33 and 34 on page 216) and on calculating the surface area of a clime; see notes to verse 70 on page 237. There are also instances where Nityānanda's discussions appear a little misplaced: for example, verses 8–12 discuss geodetic methods to compute the distance and height of a mountain maieutically in contrast to the otherwise factive assertions found in the rest of the chapter; see notes to verses 8–12 on page 196. Such didactic discussions are more typically found in a chapter on computations rather than in a chapter on celestial and terrestrial spheres.

1.3.2 *Relation to Purāṇic discourse in Nityānanda's golādhyāya*

In his *golādhyāya*, Nityānanda discusses Purāṇic cosmography fairly elaborately (in thirty-one verses). A large part of this is a repetition of Purāṇic geography commonly found in most Sanskrit *siddhāntas*. However, for those topics that have astronomical significance (see, e.g., verses 35 and 39), he argues against their Purāṇic positions.[42] Earlier Siddhāntic authors have also similarly criticised Purāṇic views on astronomical phenomena. For example, Lalla, in his *Śiṣyadhīvṛddhidatantra* (XX.22), condemns the Purāṇic view that eclipses are caused by a demon (*asura*):

> *asuro yadi māyayā yuto niyato 'tigrasatīti te matam |*
> *gaṇitena kathaṃ sa labhyate grahakṛtparva vinā kathañcana ||*
>
> If you are of the opinion that an artful demon is always the cause of eclipses by swallowing (the Sun or Moon), then how is it that an eclipse can be determined by means of calculation? Moreover, why is there not an eclipse on a day other than the day of the new or full moon?
>
> (Chatterjee 1981, p. 272)

The use of rhetorical questions in arguing against Purāṇic opinions is a common Siddhāntic trait; it allowed Siddhāntic authors to oppose traditional Purāṇic views without being irreverent to the Purāṇas. In his writings, Nityānanda refers to those who held Purāṇic views as wise men (*budha, subudha, vibudha, budhajana, vicakṣaṇa*) and scholars of the Purāṇas (*purāṇapaṇḍita, paurāṇika*). Their opinions in matters of terrestrial and celestial phenomena varied, and it was this variation that was the cause of contradiction (*virodha*) and doubt (*bhrānti*) in the minds of men.[43] By questioning certain Purāṇic views in his *golādhyāya*, Nityānanda attempts to get a clear and concise understanding (*saṃjñā*) of the topic free of incongruous statements (*vacana-virodha-mukta*). Here, he positions himself as a seeker of truth (*pāramārthika*) and not an adversary (*prativādin*) of traditional beliefs. His discourse follows in the Siddhāntic tradition of reconciliation (*virodha-parihāra*, lit. removal of contradictions)—see, e.g., Minkowski (2004) and T. Knudsen (2021)—and his attempts at 'reconciling different opinions' (*pratipattibheda-samarthana*) in his *golādhyāya* go beyond a simple rejection of Purāṇic views to also accept certain Islamicate ideas.

1.3.3 *The cardinal localities on the Earth-sphere in Nityānanda's golādhyāya*

In his discussions on the Earth-sphere (*bhūgola*), Nityānanda specifies six cardinal localities (four equatorial cities and two polar regions) relative to one another. The positions of these cardinal regions help orient various geometrical perspectives on the Earth-sphere. See Figure 5.5 for the relative arrangement of the cardinal localities on the Earth-sphere. Most earlier Siddhāntic texts also describe a similar configuration of these cities on the Earth-sphere.

1.3.3.1 *Laṅkā, the cupola of the Earth*

The central city of Laṅkā was considered a place on the Earth's equator corresponding to zero degrees of longitude (i.e., 0° N, 0° E). In Indian astronomy, Laṅkā is regarded as a mythical equatorial city and, as such, has no relation to the country of Ceylon (Siṃhaladvīpa, Tāmradvīpa, or Śrīlaṅkā) or the golden island of the demon king Rākṣasarāja Rāvaṇa (Suvarṇalaṅkā). Its astronomical significance is in its location at the origin of a terrestrial coordinate system.

Islamicate astronomers often described the central place of the inhabited world (oecumene), with equal longitudinal extension to its east and west along the equator, as the *qubbat al-ᶜarḍ* 'cupola of the Earth', comparable with the Greek ὀμφαλὸς θαλάσσης and Latin *umbilicus Terræ* (Sachau [1910] 2013, p. 306). This term referred to a central point—like the top of a cupola or a tent—from which all places of the oecumene were equidistant.[44]

Muslim astronomers placed the city Bārah on a mythical island located at the cupola of the Earth.[45] In the Indian tradition, the city of Laṅkā was designated as the Earth's cupola. In fact, the (indirect) identification of Bārah and Laṅkā was pointed by ᶜAlī ibn Sulaymān al-Hāshimī (fl. 890) in his *Kitāb fīᶜilal al-zījāt* 'Book of the Reasons behind Astronomical Tables' (see Kennedy and Ḥaddād 1981, p. 91 (f. 93v: 8–9)).

1.3.3.2 *Yamakoṭi or Yavakoṭi, east of Laṅkā*

The equatorial city of Yamakoṭi (or Yavakoṭi) was considered to be east of Laṅkā (at a longitude of 90° E). Al-Bīrūnī identified the place Yamakoṭi—etymologically understanding its name to mean the castle (*koṭi*) of the angel of death (*yama*)—with Kangdizh 'the fortress (*dizh* in Persian) of Kang', a mythical paradise-like fortress of Iranian folklore; see, e.g., Minorsky and Bosworth (1982, p. 189, with the references therein), Kennedy and Kennedy (1987, pp. 15 and 17), and Sachau ([1910] 2013, pp. 303–304).[46] In fact, in his book *Kitāb fī taḥqīq mā liᶜl-Hind* on India, al-Bīrūnī also mentions Yamakoṭi (identified as Kangdizh) as the prime meridian of Abū Maᶜshar al-Balkhī, the Persian astronomer from the Abbasid court (Sachau [1910] 2013, p. 304).[47] (Compare with the description of the city of Gaṅgadujda in Section 1.3.9.3.) The name Yavakoṭi, when etymologically understood as the tip (*koṭi*) of a barley-corn (*yava*), has been suggested to mean Yavadvīpa (the Indonesian island of Java) (Yule and Burnell 2013, pp. 282–283).

1.3.3.3 *Romaka, west of Laṅkā*

The equatorial city of Romaka was considered to be west of Laṅkā (at a longitude of 90° W). The location of the city of Romaka is speculative. Even as the exonymous word *romaka* suggest the city of Rome, Nityānanda's Romaka (or Romakapattana), following earlier Siddhāntic authors, is perhaps better identified with the city of Alexandria (*al-iskandariyya*, Egypt). In some Islamicate sources, the equatorial location ninety degrees west of the Earth's copula was identified as the island of Būlā (Pula/Thule?) in the Green Sea (Atlantic Ocean?)[48]

1.3.3.4 *Siddhapurī, antipodal to Laṅkā*

The equatorial city of Siddhapurī (or Siddhapura, lit. the city of the blessed) was considered a (mythical) antipodal city to Laṅkā (at a longitude of 180° E). In his *Kitāb fī taḥqīq mā liᶜl-Hind*, al-Bīrūnī acknowledges this Indian city of Siddhpura that was believed to be diametrically opposed to Laṅkā but expresses his disbelief at the thought of such a city located in a place beyond the 'inhabited half-circle' (i.e., the oecumene) and surrounded by 'unnavigable seas' (Sachau [1910] 2013, p. 304).

1.3.3.5 *Mt Meru, the terrestrial North Pole*

According to Purāṇic descriptions, Mt Meru (or Mt Sumeru) was regarded as the sacred mountain-abode of the gods and the *axis mundi*, the cosmic central axis around which all the planets and luminaries revolved. However, in most Siddhāntic texts, it was associated with the northern polar regions, or, more specifically, the terrestrial North Pole. The words *sumeru* and *kumeru* were sometimes used to refer to the north and south poles respectively.

1.3.3.6 *Vaḍavānala, the terrestrial South Pole*

By Purāṇic accounts, Vaḍavānala (allophonic to Vaḍabānala) was a location (submerged in the austral sea) where a submarine cavern (called *vaḍavāmukha*, a 'mare's mouth') contained the apocalyptic fire of destruction (*vaḍavāagni*, the 'mare's fire') within it. In most Siddhāntic works, Vaḍavānala was simply taken to be the southern polar regions, or, more specifically, the terrestrial South Pole.

1.3.4 *Three orientations of the celestial sphere in Nityānanda's golādhyāya*

In talking about spheres, Nityānanda arranges various topics related to the celestial sphere under three broad sections: the *bhagola* 'sphere of asterisms' or 'right celestial sphere', the *khagola* 'oblique celestial sphere', and the *dṛggola* 'visible celestial sphere'. The distinction between these three types of orientations of the celestial sphere is related to the choice of one's reference frame.

1 The *bhagola* is the right celestial sphere with (i) its centre coincident with the centre of the Earth (*bhūmadhya*); (ii) its north and south celestial poles extending in the direction of Earth's north and south poles respectively; and (iii) its equator, called the celestial equator, being an extension of the Earth's equator. In essence, the *bhagola* presents the perspective of a hypothetical geocentric observer oriented along the direction of the north-south pole of the Earth; see Figure 1.1(A).

2 The *khagola* or the oblique celestial sphere is also a geocentric celestial sphere but with two main differences, namely, (i) its north and south poles are oriented along the direction of the observer's zenith and nadir; and (ii) its equator, called the celestial or astronomical horizon, is a great circle containing the centre of the earth in its plane but inclined to the celestial equator by an amount equal to the co-latitude of the observer. In the *khagola* perspective, the hypothetical geocentric observer is oriented along a direction that is inclined to the north-south pole of the Earth; see Figure 1.1(B). If the inclination is zero, the *khagola* and *bhagola* perspectives are identical.

3 The *dṛggola* or the visible celestial sphere is a hemispherical topographic projection of the *khagola* with (i) its centre being coincident with the point on the surface of the Earth (*bhūpṛṣṭha*) where an observer stands; (ii) its visible pole being the observer's zenith (the other pole is at the observer's nadir and hence not visible); and (iii) its equator being the geographic horizon along the observer's line of sight (ignoring any visible obstructions); see Figure 1.1(C).

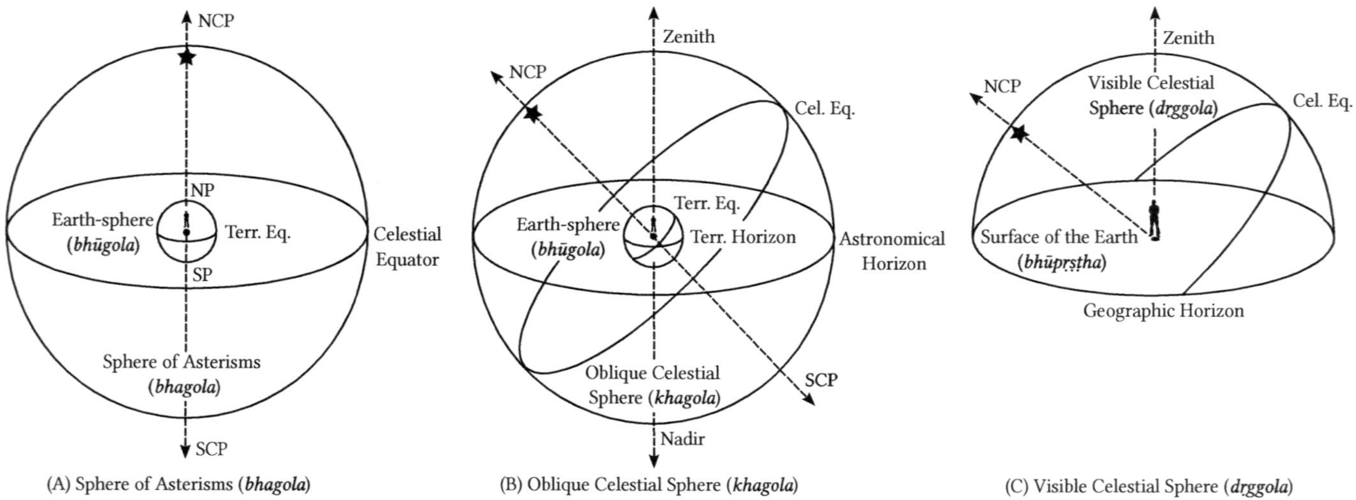

Figure 1.1 Three orientations of the celestial sphere based on three different frames of reference: (A) the right celestial sphere (*bhagola*) with the north celestial pole (NCP), the south celestial pole (SCP), and the celestial equator; (B) the geocentric oblique celestial sphere (*khagola*) with the zenith, the nadir, and the astronomical horizon; and (C) the topocentric visible celestial sphere (*dṛggola*) with the zenith and the geographic horizon of a terrestrial observer

1.3.5 *Grammar of Nityānanda's golādhyāya*

1.3.5.1 *On the polysemy of the word gola*

Nityānanda uses the word *gola* to mean variously a circle, a sphere, the celestial sphere, the visible dome of the heavens, the vast expansiveness of space, and even the study of spheres. In each instance, the implied meaning of this word is to be understood contextually. The use of polysemic words in Sanskrit technical literature is a noted feature, especially in texts that are composed in metrical poetry.[49]

1.3.5.2 *On nominating technical terms*

Nityānanda often uses the affixial word *abhidhā* (in compounds) to indicate the conventional denotation (*saṃketāgrahaṇa*) of technical terms; for example, *bhūjābhidha* 'the horizon' (in verse 6), *romakābhidha-pura* 'the city called Romaka' (in verse 23), or *madhyābhidha-sūtra* 'the secondary circle to the meridian' (in verse 90).[50] In other places, he uses (near-)synonymous affixial words like *ākhyā* (e.g., *vaḍavānalākhya* 'the place called Vaḍavānala' in verse 19), *nāma* (e.g., *ramyakanāma-khaṇḍaka* 'the region called Ramyaka' in verse 49), *nāmadheya* (e.g., *dṛṅmaṇḍala-nāmadheya* 'the object called the vertical circle of altitude' in verse 87), or *saṃjñā* (e.g., *madhyāhnasaṃjña-samaya* 'the time called midday' in verse 66) to designate (or 'call') technical terms. In fact, in verse 112, he specifically uses the term *lakṣaṇanāmaka* 'characteristic name' to signify the technical names of different arcs of celestial circle (in the *dṛggola*) that he then subsequently describes (in the following verses). In all of these instances, the relation between the formal expressions (signifier or *vācaka*) and their technical meaning (signified or *vācya*) is established by convention (and therefrom considered established).

1.3.5.3 *On apophonic variations of technical words*

Nityānanda refers to the celestial equator as *viṣuvavṛtta*, *viṣuvatvṛtta*, or *vaiṣavavṛtta*, in essence, a great circle (*vṛtta*) that possesses equality on both sides, i.e., the equatorial perimeter. The indeclinable *viṣu* is compounded (*upapada-tatpuruṣa*) to form *viṣuva*, or, alternatively, derived into the secondary nominal possessive form (*matvarthīya-tatdhita-vṛtti*) *viṣuvat* (see, e.g., Bhaṭṭācārya 1962, p. 4934a). In two particular instances (verses 20 and 29), the stem *viṣu* is changed to the apophonic form *vaiṣava* following *viṣu* → *vaiṣu* (*vṛddhi* of *vi-* to *vai-*) + *a* (*matvarthīya-pratyaya* or possessive affix) → *vaiṣava* (with terminal vowel *sandhi*). This change is perhaps *metri causa*: *viṣuva* has a metrical signature ⏑ ⏑ ⏑ whereas *vaiṣava* is – ⏑ ⏑. However, the meaning of the words *viṣuva* and *vaiṣava* are the same.

1.3.5.4 On the choice of declarative verbs

In the verses of the *golādhyāya*, Nityānanda uses an assortment of declarative verbs (with various conjugations) to state a position or proposition: for instance, √*vad* (cl. 1), √*vac* (cl. 2, with directional prefixes like *pra* or *niś*), *pra*-√*cakṣ* (cl. 2), *udā*-√*hṛ* (cl. 1), *ud*-√*īr* (cl. 2), and *ā*-√*mnā* (cl. 1) 'to declare, describe, mention, proclaim, relate, speak, state, or tell'.

On some occasions, he uses the verb √*gai* (cl. 1) 'to sing, speak, or recite' in its perfect active plural third-person form *jaguḥ*, e.g., in verses 37, 86, and 109. When interpreted as 'described or declared', it conjures the need of an agent who has described or declared the given statement. Nityānanda silently attributes this agency to anonymous wise men of antiquity (*budhajana*s) or mythohistorical sages (*ṛṣi*s or *muni*s)—in both instances, their authority on matters of canonical knowledge being infallible. By transferring the agency from human intellect to (semi-)divine testimony, he skilfully transforms foreign knowledge into traditional wisdom.[51] Indeed, the probative force of the verb *jaguḥ* (understood as 'sung, recited, or proclaimed') validates an aphorism or statement as revelatory. See Section 1.2.3.1.

1.3.5.5 On the use of figurative language

The use of figurative language is an important aspect of knowledge communication in oral societies, and Nityānanda's language is visually evoking in several places in the *golādhyāya*. For instance, in the second *pāda* of verse 85, Nityānanda describes the oblique celestial sphere (*khagola*) as *yatkhaṃ khalu sarvato vijayate* '[the] space that extends in every direction' using language that reveals a sense of pathetic fallacy. The verb *vijayate* (present middle singular third-person form of the reflexive verb √*viji*, cl. 1, 'to conquer, defeat, or overpower') takes the inanimate *kha* 'sky' as its subject, and, in doing so, personifies the sky as possessing the ability to conquer itself, i.e., extends outwards. Or in verse 87 where, while describing the vertical circle of altitude (*dṛṅmaṇḍala*) of a planet, Nityānanda uses words like *iṣṭa-diś* 'the desired direction' and *tad-viparītakāṣṭhā* 'the opposite region from that'. Here, the 'desired direction' is a hypallage: the desired direction is, in fact, the direction of the desired object, i.e., the planet being studied.

More commonly, Nityānanda invokes many similes to compare geometrical objects (on the celestial sphere) with everyday (and hence, familiar) objects. For instance, in verse 85, he compares the astronomical horizon (i.e., the equatorial region) of the oblique celestial sphere to a girdle or waist-band (*parikara*); in verse 77, he considers the diurnal spirals of a planet (north and south of the celestial equator) as convolutions of a conch-shell (*śaṅkhāvarta*); and in verse 83, he likens all great circle on the celestial sphere, being imaginary themselves, to the whirling circles of firebrand (*ālāta-cakra-bhrami*) made by playful children.

1.3.6 *Technical lexicon in Nityānanda's golādhyāya*

1.3.6.1 *On the use of word-numerals*

Following convention, Nityānanda uses the *bhuṭasaṃkyā* system of word-numerals to represent numbers in the metrical verses of his *golādhyāya*.[52] For instance, *kha-guṇa* for '30' in verse 8; *arka* for '12' in verse 75; *gaja-veda* for '48' in verse 80; or *khāṅka* for '90' in verse 105. Occasionally, he constructs the *bhuṭasaṃkyā* numerals more elaborately: for example, *viśarāṃśakhāga* implying *vi-śara-aṃśa* 'less by the fifth of a degree' and *kha-aga* '70', i.e., 70° – ⅕° or 69° 48′ in verse 104; or *sārdhāṣṭaśaila* implying *sārdha* 'with half [of a degree]' and *aṣṭa-śaila* '78', i.e., 78°+ ½° or 78° 30′ in verse 105.

1.3.6.2 *On arithmetic operations and mathematical objects*

In the *golādhyāya*, Nityānanda uses typical expressions to describe different arithmetic operations. For example, in verses 9–12, he uses *āḍhya* 'augmented or increased by'; *vihīna* (or *hīna*) 'deprived of or subtracted from'; *ūnā* 'less than or diminished by'; *guṇita, ghna,* or *nighna* 'multiplied by'; *āpta* or *bhakta* 'divided by'; *varga* 'the square of [a quantity]'; and *mūla* 'square-root of [a quantity]'.[53] In these same verses, he also describes different mathematical objects using commonly attested Sanskrit words; for example, *hāra* 'divisor'; *labdha* 'obtained quotient of division'; *śeṣa* 'the remainder of division'; *māna* or *unmiti* 'measure'; *adhvan* 'distance'; *āṃśa, aṃśaka, bhāga,* or *lava* 'parts' or 'degree'; *vivarāṃśa* 'the difference in degrees'; and *vyāsārdha* (lit. half the diameter) 'radius'. The *sinus totus* (sine of 90°), or, effectively, the radius of the base circle is synonymously referred to as *tribhajyā* (lit. sine of three zodiacal signs), or *trimaurvī*/*trijyakā* (lit. three sines).

1.3.6.3 *On the measure of arc*

The measure of arc (*cāpa* or *dhanu*, lit. bow) in degrees corresponds to the chord-length (*jyā* or *jīva*, lit. bowstring) in a circle of a specified radius. While the measure can be considered equivalent to a central angle subtended by the arc, Indian geometers considered the measure of arc in association with its chord-lengths, in particular, the Sine (*jyāardha* 'half chord-length'), the Cosine (*koṭijyā* 'perpendicular of the chord-length'), and Versine (*utkramajyā* 'excess [to the perpendicular of the] chord-length'). See Brummelen (2009, pp. 94–98) for a fuller discussion on the development of trigonometry in India.[54] Occasionally, Nityānanda refers to the measure of arc (in degrees) by synonyms of *cāpa*: for example, in verse 115, he refers to the *cāpa* 'bow' as *śarāsana*, lit. the resting place (*āsana*) of an arrow (*śara*).

In verse 117, he talks about the *sphuṭatara-apama-aṅka* or the 'curve of true declination'. Ordinarily, the word *aṅka* refers to a 'number', 'measure', or 'sign'; however, in this case, Nityānanda uses the word *aṅka* to designate a 'curved line' or an 'arc' of a great circle. See notes to verse 117 on page 294 for a similar interpretation of *aṅka* in Nityānanda's *Siddhāntasindhu* (II.6), and also in the *Sarvasiddhāntarāja* (I.9.*spaṣṭakrānti*).

1.3.6.4 *On identifying the directions*

Nityānanda uses a variety of Sanskrit synonyms to indicate egocentric, temporal, and cardinal directions in his *golādhyāya*.[55] For example,

1 *prācī* (from *prañc*) 'forwards' or 'eastwards', *puratas* 'before' or 'eastwards', *pūrva* 'prior' or 'eastern';

2 *pratīcī* (from *pratyañc*) 'backwards' or 'westwards', *paścima/apara* 'later' or 'western';

3 *udak* (from *udañc*) 'upwards' or 'northwards', *uttara* 'to the left' or 'northern';

4 *avācī* (from *avañc*) 'downwards' or 'southwards', *dakṣiṇa* 'to the right' or 'southern'.

In other instances, he uses more culture-specific terms for directions: for example, *saumya* (lit. belonging to Soma, the regent of the north) for 'northern' or *yāmya* (lit. belonging to Yama, the regent of the south) for 'southern'. The deification of the cardinal and intercardinal directions by regents or guardians (*lokapāla*s or *dikpāla*s), and, correspondingly, the identification of these direction by the names of the reigning deity are a common practice in most Indian *śāstra*s.[56] In verse 58, Nityānanda lists the abodes of these directional regents on Mt Meru (according to Purāṇic cosmography), and elsewhere, he refers to the cardinal directions by their names: for example, *kuberadiś* (lit. direction of Kubera, the regent of the north) and *śaśidiś* (lit. direction of Śaśi or the Moon god, synonymous with the regent of the north) for 'northern' and *yāmyadiś* (lit. direction of Yama, the regent of the south) for 'southern'.

1.3.6.5 *On the synonyms of terrestrial objects*

In discussing Purāṇic geography, Nityānanda uses a variety of synonyms to describe different geographical features. For example, oceans are called *adbhi*, *ambudhi*, *arṇava*, *sindhu*, or *samudra*, while mountains (or mountain ranges) are called *giri* or *parvata*. The choice of the synonym appears to be *metri causa*.

Moreover, the annular island-continents that are encircled by the oceans are called the *dvīpas*; while the physiographic regions (in the central island-continent) that are naturally separated from one other by mountain ranges are called *varṣas*. See notes to verses 44–51, beginning on page 224. These words, along with other words like *khaṇḍa* 'clime/region'; *pradeśa* 'country'; *kṣetra* (in its geographic context) 'region'; *nagara* 'town or city'; *grāma* 'village'; *loka* 'settlement, realms, or world'; *śakala* 'region or area', represent the enthogeographic division of the world understood in the historical, mythological, and sociocultural context of India. Their modern translations in English are often approximate owing to a difference in the cognitive geographies of people separated by language, space, and time.

1.3.6.6 *On the synonyms of celestial objects*

In his *golādhyāya*, Nityānanda uses several Sanskrit synonyms to refer to celestial objects (planets, stars, constellations, etc.). For example, planets are called *khacara, khaga, khecara*, or *kheṭa* 'sky-movers', and, in a few instances, *graha* 'seizing or grasping' and *dyuṣad* 'situated in the heavens'.[57] The stars or luminaries are called *bha* (related to *bhā* 'light' or 'splendour'); *tāraka* '[that which is] liberating'; or *uḍu* (perhaps, related to *uḍ*-$\sqrt{}$*ḍī* (cl. 1) 'to fly'). Correspondingly, the constellations are called *rāśi* 'zodiacal sign'; *bhamūrti* 'configuration of stars [fixed in the celestial sphere]' (comparable with the Arabic *al-kawākib al-thābita* 'fixed stars'); *nakṣatra/ṛkṣa* 'asterism'; and *bhavana* 'lunar mansion or astrological house'. The Earth is personified as a nurturing (*poṣin*) and life-supporting (*prāṇabhṛt*) feminine entity, and is synonymously called *bhū, bhūmi, dharā, dharaṇi, dhātrī, ilā, kṣamā, kṣiti, ku, pṛthivī, vasumatī*, and *vasuṃdharā*.[58]

Nityānanda also refers to the Sun and the Moon with different names. For instance, the Sun, as the one possessing luminosity, is called *arka, bhāsvat, bhānu*, or *bhānumat*; as a yellow object, *hari*; as the day-maker, *divākara*; as the jewel of the sky, *dyumaṇi*; as a benevolent saviour, *taraṇi*; as a lord or king, *ina*; and as a deified being, *ravi* (synonymous with *sūrya* or *āditya*). Similarly, the Moon is called *indu*, the bearer of (the drink) Soma, and occasionally *vidhu* on account of being a solitary celestial object.

1.3.6.7 *On the use of technical (astronomical) terms*

Bodies of asterisms Nityānanda uses the word *bhagola* to refer to the sphere of asterisms (i.e., the right celestial sphere); however, he uses the words *bhacakra, bhamaṇḍala, bhavṛtta*, and *bhavalaya* interchangeable to refer to the ecliptic (i.e., the circle of the zodiacal signs or *rāśis*) as well as the wheel of asterisms (understood as the aggregation of star, planets, etc. within the *bhagola* that move eastwards with their own daily motions). See, e.g., notes to verses 73.1 and 73.2, beginning on page 242. More generally, he uses polysemous words like *gola, valaya, maṇḍala, vṛtta, cakra,* and *kakṣa* variously to mean 'a circle, hemisphere, orb, orbit, ring, sphere, wheel, zone etc.'

Direction circles The direction circles are reference secondary circles on a sphere along which particular measurements are made. In other words, the direction circles are perpendicular to a specified great circle and pass through the poles of that great circle. Nityānanda uses the word *sūtra* or *sūtraka,* lit. line, cord, or string, to identify these directions circles; for example, *bāṇasūtra* (in verse 75), *śilīmukhasūtraka* (in verse 79), *viśikhasūtra* (in verse 116) for the direction circle of ecliptic latitude (*bāṇa, śilīmukha,* or *viśikha*); *samasūtra* (in verse 89) for the secondary circle to the prime vertical (*samavṛtta*); *madhyasūtra* (in verse 90) for the secondary circle to the meridian (*madhyāhnavṛtta*); or *apamasūtra* for direction circle of declination (*apama*). See notes to verses 75, 79, 89, 90, 114, and 116 on pages 244, 249, 259, 289, and 291 respectively.

Orbit of planets Nityānanda uses the word *vikṣepa* 'ecliptic latitude' in the compound *vikṣepavṛtta* to refer to the 'circle of ecliptic latitude'—in other words, the actual orbit of the planet inclined to the ecliptic. The projection of these planetary orbits onto the outermost sphere of asterisms (*bhagola*) is called the *bāṇavṛtta* or 'latitudinal circle'. See notes to verse 84 on page 253.

Diurnal circles and circles of day-radius Nityānanda differentiates between the moving spirals of diurnal motion of a planet, identified as *dyuvṛtta* or *dyumaṇḍala* (see notes to verses 73.1/2, 77, and 82 on pages 242, 244, 246, and 252 respectively), and the fixed small circles of day-radius of a planet identified as *dyujyāvṛtta* or *dyujyāmaṇḍala* (see notes to verses 82 and 89 on pages 252 and 259 respectively).

Zenith and nadir Nityānanda indicates the direction of the zenith (*samt al-raʾs* 'direction of the head' in Arabic) by words like *khasvastika* 'the auspicious mark in the sky'; *naramastaka* (lit. head of man) 'directly overhead'; or *śirasthita* 'situated overhead'. Similarly, he uses words like *adaḥ-svastika* 'the auspicious mark below' or *aṅghritala* (lit. sole of the feet) 'directly below' to indicate the direction of the nadir (*samt al-naẓīr* 'opposite to the direction [of the head]' in Arabic). Occasionally, he calls the zenith *nabhatala* 'base of the sky' in relation to tenth astrological house (*bhāva*) known as the Medium Coeli (mid-heaven); see, e.g., notes to verse 29 on page 214.

1.3.7 Prosody of Nityānanda's golādhyāya

In composing the 135 verses in the *golādhyāya* of his *Sarvasiddhāntarāja*, Nityānanda has used an assortment of Sanskrit metres of different lengths. Table 1.6 lists the name of these different metres alongside their corresponding verse numbers.[59] The sequence of five verses 113–117 is the longest sequence of homometric verses in the *golādhyāya*. All five verses are composed in the twelve-syllable *jagatī* metre called *drutavilambita* (lit. fast and slow), and they all begin with the same initial word *viṣuva*. The initial-word repetition in successive verses creates an anaphoric effect which then gets amplified by the iambic cadence (*tāla*) of the *drutavilambita* metre. It is perhaps not surprising to find verse 116 introducing, for the very first time, the arc of second declination (an Islamicate astronomical concept) inserted between these homometric verses.

A literary analysis of Sanskrit texts in the exact sciences is not very commonplace; however, in recent times, select studies have attempted to explore the poetic dimensions of Sanskrit astronomical *siddhānta*s: for example, Misra, Montelle, and Ramasubramanian (2016)discuss the tropology (*alaṅkāra*) and aesthetics (*rasa*) in the *golādhyāya* of Nityānanda's *Sarvasiddhāntarāja* in an effort to highlight his compositional prowess (*kāvya vaiśiṣṭya*).

Table 1.6 The list of Sanskrit metres used in Nityānanda's *golādhyāya* alongside their corresponding verse numbers

Metrical Class	*Subclass (syllables)*	*Meter*	*Verse*
varṇavṛtta-samacatuṣpadī	*anuṣṭubh* (8)	*anuṣṭubh*/*śloka*	37, 41, 45, 48, 81, 82
		pramāṇikā	109
	triṣṭubh (11)	*indravajrā*	14, 15, 18, 21, 25, 26, 64, 69, 70, 102
		upendravajrā	94, 99
		rathoddhatā	23, 31, 32, 51, 131
		śālinī	10, 11, 12, 28, 36, 42, 54, 135
	jagati (12)	*drutavilambita*	38, 50, 57, 79, 113, 114, 115, 116, 117
		vaṃśastha	9, 29, 30, 39, 44, 49, 52, 107, 108
		vasantatilakā	40, 47, 53, 58, 63, 66, 67, 95, 96, 132, 134
	atiśakvarī (15)	*mālinī*	78

(*Continued*)

Table 1.6 (Continued)

Metrical Class	*Subclass (syllables)*	*Meter*	*Verse*
	atyaṣṭi (17)	*pṛthvī*	71
		mandākrāntā	72, 73.1, 73.2, 74
		śikhariṇī	7
	atidhṛti (19)	*śārdūlavikrīḍita*	1, 2, 3, 4, 20, 24, 55, 59, 60, 61, 77, 85, 86, 133
	prakṛti (21)	*sragdharā*	5, 8, 110
varṇavṛtta-ardhasama-catuṣpadī	(11-11)	*ākhyānakī* (*bhadrā*)	43, 46, 56, 80, 121
		viparītākhyānakī (*haṃsī*)	13, 22, 87, 123, 127
	(10-11)	*viyoginī* (*sundarī*)	35
varṇavṛtta-viṣamadvipadī	*gāthā-upajāti* (11)	*dvipadī*	126
varṇavṛtta-viṣama-catuṣpadī	*upajāti* (11)	*ārdrā*	98
		ṛddhi	104, 128
		kīrti	17, 19, 33
		jāyā	27, 105
		premā	65, 100
		bālā	6, 16, 76, 118, 129
		buddhi	124
		mālā	34, 88, 93
		rāmā	101, 125
		śālā	68, 83, 130
		vāṇī	84, 119, 120, 122
	upajāti (12)	*vāsantikā*	103
		śaṅkhacūḍā	106
	upajāti (17)	*mandakrānta-citralekhā-miśrita*	75
mātrājāti-tripadī	*miśrajātīya*	*āryā* in 1 *pāda*, *udgiti* in 2-3 *pādas*	90

Table 1.6 (Continued)

Metrical Class	*Subclass (syllables)*	*Meter*	*Verse*
mātrājāti-catuṣpadī	*āryā*	*āryā* (12-18-12-15)	97
		udgīti (12-15-12-18)	111, 112
		upagīti (12-15-12-15)	62, 92
		gīti (12-18-12-18)	89, 91

1.3.8 *Śāstric appropriations in Nityānanda's golādhyāya*

In some of the verses of his *golādhyāya*, Nityānanda appears to have appropriated grammatical constructions from earlier Śāstric literature, in particular, notable literary, religious, and scientific texts. By virtue of imitation, this appropriation reveals his familiarity (*paricaya*), acceptance (*anumodana*), and justification (*samarthana*) of those particular disciplinary opinions (*śāstrika-mata*). The following three examples (from the *golādhyāya*) illustrate this repetition of tradition (*āmnāya*).

1 In verse 44ab, Nityānanda uses the expression *dvayor dvayor antaragaṃ samudrayor udāhṛtaṃ dvīpam udārabuddhibhiḥ* 'between each pair of oceans, an island-continent is declared by those with great intelligence'. This is a near-identical restatement from Bhāskara II's *Siddhāntaśiromaṇi* (II.3.25cd) which states *dvayor dvayor antaram ekam ekaṃ samudrayor dvīpam udāharanti* '[they] declare each island-continent lying between each pair of oceans'.

2 Verse 46 opens with the declaration *himācalo nāma nagādhirājaḥ* 'the king of mountains called Himācala'. This is also near-identical to the opening line *himālayo nāma nagādhirājaḥ* 'the king of mountains called Himālaya' from *Kumārasambhava* (V.1), a Sanskrit *śravya mahākāvya* (an epic poem worthy of recitation) composed by the famous classical Sanskrit poet and dramatist Kālidāsa in the fifth century.

3 In verse 60, Nityānanda extols the river Gaṅgā with several epithets, many of which resemble those composed by the Advaita Vedāntic scholar Śaṅkarācārya in his *Gaṅgāṣṭakam* (an eight-stanza hymn to Gaṅgā) in the eight century.

   For instance, Śaṅkarācārya refers to the river Gaṅgā with epithets like *svargasopānagaṅge* (verse 1b) 'Gaṅgā, the stairway to heaven'; *bhavalīlāmaulimāle* (verse 2a) '[she who is] the elegant head-garland of Śiva'; *kanakagiriguhāgaṇḍaśailāt skhalantī* (verse 3a) '[she who] stumbles through the mountains, caves, and large rocks [on her way] from Mt Meru'; *bhagavataḥ pādodakaṃ pāvanam* (verse 5a) '[she who is] the purifying waters washing the feet of Viṣṇu'; and *śambhujaṭāvibhūṣaṇamaṇīḥ* (verse 5b) '[she who is] the ornamental jewel adorning the matted locks of Śiva'.

Near-identically, in verses 59 and 60, Nityānanda calls the river Gaṅgā *yā sopānaparampareva viditā svargonmukhānāṃ nṛṇām* '[she] who is known as the staircase for men who desire heaven'; *śambhormastakabhūṣaṇāya vimalā muktāvalīvatsthitā* 'the stainless one [who is] established like the pearl necklace of the head-ornament of Śiva'; *gacchantī girigahvarādiṣu* '[she, who descended from Mt Meru] going through the mountains, caves, etc.'; and *viṣṇupadato jātā* '[she who is] born from the foot of Viṣṇu'.

1.3.9 Islamicate imports in Nityānanda's *golādhyāya*

Nityānanda's *Sarvasiddhāntarāja* incorporates several Islamicate ideas in its canonical (*saiddhāntika*) presentation of mathematical astronomy. As briefly described in Section 1.2, the *Sarvasiddhāntarāja* attempts to syncretise Greco-Islamicate ideas (via the *Romakasiddhānta*) with the traditional views (*pākṣika-mata*) of the *Brāhmasphuṭasiddhānta* and *Sūryasiddhānta*. In the chapter on computations (*gaṇitādhyāya*), Nityānanda discusses planetary models and different mean-motion parameters in a manner that harmonises them (albeit, awkwardly) to evince an underlining unity of the different methods despite their apparent plurality (see, e.g., Pingree 1978c, p. 325 and 2003b). When a comparative explanation cannot be offered, he simply states the foreign idea without an apology or justification. In the *golādhyāya*, Nityānanda's references to Islamicate ideas follow this same form of impassive declarations.

1.3.9.1 *Sinus totus* as 60

Nityānanda's *Siddhāntasindhu* and *Sarvasiddhāntarāja* are among the earliest Sanskrit astronomical texts to use the 60 (an Islamicate parameter) as the *sinus totus* (*tribhajyā*, lit. sine of three signs, i.e., sine of 90°). The *Siddhāntasindhu*, following Mullā Farīd's *Zīj-i Shāh Jahānī*, tabulates the Sine values (along with their successive differences) for every minute of arc between 0° and 360°, with a maximum value (corresponding to 90°) of 60.[60]

Nityānanda also uses 60 as the *sinus totus* in his *Sarvasiddhāntarāja*. In fact, in his discussions on the geometrical methods to compute sine values (*jyākathanam*) in the *Sarvasiddhāntarāja* (I.3.19–85), he uses sixty as the radius of the base circle (equal to the *sinus totus*) (Montelle, Ramasubramanian, and Dhammaloka 2016, verse 20 on p. 13).[61]

Interestingly, in the *Hayatagrantha*, an anonymous mid-sixteenth century Sanskrit translation of ᶜAlī Qūshjī's Persian *Risāla dar Ilm al-Hayʾa* (1458), we find the following statements justifying the use of sixty as the radius of the base circle.

> *yathā paridhau 360 saptaguṇe 2520 dvāviṃśati hṛte labdho vyāsaḥ 114 sūryasiddhāntamatena vyāsārdho 'yaṃ 57 tasya sugamagaṇitārthaṃ ṣaṣṭim iti vyāsārdhaḥ kṛtaḥ svalpāntaratvāt | taddvigoṇo vyāso jātaḥ 120 | vyāse triguṇe paridhiḥ |*
>
> Bhaṭṭācārya and Upadhyaya (1967, p. 16–17)
>
> Just as a circumference [of] 360 multiplied by 7, [equalling] 2520, [and] divided by 22 results in a diameter [of] 114 [as the] quotient according to the opinion of the *Sūryasiddhānta*, 57 [as the] half of that diameter [can be taken as] 60 to simplify calculations. In this manner, the radius is fixed [as 60] on account of the smallness of difference [between 57 and 60]. Double that, the diameter becomes 120. In tripling the diameter, the circumference [becomes 360].

1.3.9.2 *Circumference of the Earth as 6000*

In the Indian tradition, Siddhāntic authors have stated different measures for the Earth's circumference in their calculations; for instance,

1 Āryabhaṭa I, in his *Āryabhaṭiya* (I.7), uses 3299 or 3300 *yojanas*;

2 Varāhamihira, in *Pañcasiddhāntikā* (XIII.18), uses 3200 *yojanas*;

3 Brahmagupta, in his *Brāhmasphuṭasiddhānta* (I.37) and Khaṇḍakhādyaka (*uttarabhāga*, I.6), uses 5000 *yojanas*, but in his Khaṇḍakhādyaka (*pūrvabhāga*, I.15), he uses 4800 *yojanas*;

4 Bhāskara I, in his *Laghubhāskarīya* (I.24), uses 3299 *yojanas*, but in his *Mahābhāskarīya* (II.3–4), he uses 3299-$8/25$ *yojanas*;

5 Lalla, in his *Śiṣyadhīvṛddhidatantra* (I.43), uses 3300 *yojanas*;

6 Śrīpati, in his *Siddhāntaśekhara* (II.94), uses 5000 *yojanas*;

7 Bhāskara II, in his *Siddhāntaśiromaṇi* (II.2.52), uses 4967 *yojanas*; and

8 the *Sūryasiddhānta* (I.59) uses $\sqrt{10} \times 2 \times 800$, or (approximately) 5060 *yojanas*.

In talking about the geodetic methods to compute the height and distance of a mountain in verses 8–12, Nityānanda uses 6000 as the circumference (*pariṇāha*) of the Earth (measured in *yojanas*). This value appears to be unique, and, possibly, of foreign origin. However, at the time of writing, I have been unable to locate this value in any Islamicate or Indian source. Nityānanda's verses on the geodetic methods are an adaptation of al-Bīrūnī's method to determine the radius of the Earth (see Appendix A), and hence, the value of the Earth's circumference (as 6000) may simply be a number selected for computational convenience.

1.3.9.3 Equatorial antipodal cities of Khāladāta and Gaṅgadujda

Nityānanda describes the oecumene (the inhabited world) in verse 25 as being the northern quadrant of the Earth-sphere, bound to the north by Mt Meru (the north pole) and extending east-west along the equator bound by the antipodal equatorial cities of Khāladāta and Gaṅgadujda (see Figure 5.10). A spherical quadrant-like division (of the Earth-sphere) as an inhabited continent is an Islamicate import; see notes to verse 25 on page 210. Nityānanda's *Sarvasiddhāntarāja* is among the earliest (if not the only) Sanskrit Siddhāntic text to mention these two cities as the equatorial shores of the oecumene. In verse 23, he describes Khāladāta lying 22° west of Romaka and Gaṅgadujda being antipodal, i.e., 180° from Khāladāta.

On the location of Khāladāta According to Islamicate geographers, the Fortunate Islands (*Jazāʾir al-Saʿādāt*) and the Eternal Islands (*al-Jazāʾal-Khālidāt*) were thought to lie off the coast of al-Maghrib at a distance of 200 *farsakhs* (based on al-Bīrūnī's observation) (Jwaideh [1959] 1987, p. 60). The Fortunate Islands were located at the extreme end of the oecumene (*al-maʿmūrah*). The prime meridian (in the Greco-Islamicate tradition) was thought to pass through these islands, e.g., see Pliny the Elder's *Naturalis Historia* (ca. 77/79), Ptolemy's *Geōgraphikḕ Hyphḗgēsis* (ca. 150), al-Bīrūnī's *al-Qānūn al-Masʿūdī* (ca. 1036), or Jamshīd al-Kāshī's *Khāqāni Zīj* (ca. 1413/1414). Nityānanda's Khāladāta appears to be a Sanskrit calque of the Arabic word *al-Jazāʾal-Khālidāt*, the Eternal or Immortal Islands.

In the modern day, these islands can be identified as part of the Macaronesian archipelagos (from the Greek name μακάρων νῆσοι 'isles of the blessed') in the North Atlantic Ocean, off the coasts of Africa and Europe. More specifically, the westernmost island of El Hierro (*Isla del Meridiano*, with a longitude of 18° 1′ 27″ W) in the modern-day Canary Islands is known to have been regarded as the end of the inhabited world (see, e.g., Meliá 2001). At the time of writing, I have been unable to locate any Indian or Islamicate text that specify the longitude of Khāladāta. Incidentally, Kamalākara, in his *Siddhāntatattvaviveka* (II.172), also describes Khāladāta as lying 22° W of Romaka, and, hence, with a longitude of 112° W of Laṅkā (since Romaka is 90° W of Laṅkā) (K. C. Dvivedī 1993–98, p. 56 in Volume 1).

On the location of Gaṅgadujda According to Nityānanda, the equatorial city of Gaṅgadujda is antipodal to Khāladāta; in other words, 68° E of Laṅkā (since Khāladāta is 112° W of Laṅkā). This places the city of Gaṅgadujda twenty-two degrees west of the city of Yamakoṭi. Although al-Bīrūnī identifies Yamakoṭi with the mythical fortress-city of Kangdizh (see Section 1.3.3.2), I suspect Gaṅgadujda to be a Sanskrit calque for the city of Gangdizh, an erroneous rendering of the fortress-city of Kangdizh.[62] The terminal affix *-da* (in the Sanskrit loanword *gaṅgaduj-da* for *gangdizh*) could be a grammatical addition to accommodate Sanskrit nominal inflections.

However, notwithstanding the conjectural identification of Yamakoṭi or Gaṅgadujda with Kangdizh, what is clear from Nityānanda's statements is that the cities of Yamakoṭi and Gaṅgadujda are geographically distinct. Yamakoṭi is the cardinal city located 90° E of Laṅkā, whereas Gaṅgadujda, at 68° E of Laṅkā, forms the easternmost perimeter of the oecumene. The former is a location on the Earth-sphere derived from Siddhāntic sources while the latter appears to be an Islamicate import.

1.3.9.4 *Geographical and mathematical descriptions of the seven climes on Earth*

In verses 33 and 34, Nityānanda describes the seven *climes* as distinct latitudinal regions on the Earth-sphere (north of the equator) with differing hours of daylight on the day of the summer solstice. The first clime (closest to the equator) receives 31;52,30 *nāḍikās* (i.e., 12^h45^m) of daylight on the summer solstitial day, and each successive clime therefrom (going northwards towards the north pole) receives an additional 1;15 *nāḍikās* (i.e., 30^m) of daylight. To my knowledge, Nityānanda's *Sarvasiddhāntarāja* is the first Sanskrit text to discuss the Greco-Islamicate concept of climes; see notes to verses 33 and 34, beginning on page 216. Notably, Nityānanda renders the word clime (*κλίμα* in Greek or *ʾiqlīm* in Arabic, meaning 'latitude', 'region', or 'territory') as *khaṇḍa* 'fragment', 'region', or 'continent' in Sanskrit. In verse 70, he describes a mathematical expression to compute the surface area of a clime on the Earth's surface. Nityānanda's expression is a restatement of Archimedes' hatbox theorem—a theorem discussed at length by several prominent Islamicate scholars, including, *inter alia*, al-Qūhī, al-Bīrūnī, and al-Ṭūsī. The *Sarvasiddhāntarāja* is the first (and perhaps, only) Sanskrit *siddhānta* to expressly state the formula to compute the surface area of a clime; see notes to verse 70 on page 237.

1.3.9.5 *Physics of the celestial sphere*

In verse 72, Nityānanda describes the interior (*antar*) of the sphere of asterisms (*bhagola*) as transparent (*svaccha*) like a crystalline pitcher (*sphaṭika-ghaṭa*), like water (*nira*), and with [expansive] space (*sāvakāśa*). This material description of the celestial sphere is Greco-Islamicate in its origin. Other Sanskrit authors have also discussed whether the sphere of heaven was in fact a crystalline sphere near-contemporaneously. However, Nityānanda's description is among the earliest Sanskrit writings on the materiality of the celestial sphere.[63] As Section 1.1.2 briefly describes, Nityānanda's successors at Kāśī (relatives of the rival families of Munīśvara and Kamalākara) variously argued for and against the Greco-Islamicate view of a crystalline celestial sphere, choosing the alternative Indian view that the sphere of the heavens was, in fact, a sphere of iron (*lohagola*).

In the *gaṇitādhyāya* of his *Sarvasiddhāntarāja* (I.3.180–200), Nityānanda devotes twenty-one verses to describe the formation of the universe (*brahmāṇḍa-nirmāṇa*). In these verses, he describes the concentric spherical shells of the orbs of the planet in which the bodies of the respective planets are embedded, with the outermost (ninth) sphere being the right celestial sphere that impels the seven inner spherical shells of the planets westwards as the *Primum Mobile*. This description is clearly Greco-Islamicate in its origin; it appears to be derived from earlier sources including ʿAlī Qūshjī's Persian *Risāla dar Ilm al-Hayʾa* (1458) and its anonymous Sanskrit translation *Hayatagrantha* (ca. mid-sixteenth century).[64] Nityānanda's statements in these verses, like his discussion in verse 72 of the *golādhyāya*, are based on the Aristotelian idea that solid spherical crystalline shells held the planets in their respective orbits. The cosmology of these spherical shells is completely different to any Purāṇic or Siddhāntic accounts found elsewhere.[65]

1.3.9.6 *Obliquity of the ecliptic*

In verse 76, Nityānanda suggests the obliquity of the ecliptic to be 23° 30′. Most Siddhāntic authors use a value of 24° as the ecliptic obliquity. The value 23° 30′ is a (near-similar) value of Islamicate origin; e.g., al-Bīrūnī uses 23° 35′, whereas Ulugh Beg, Mullā Farīd (in his *Zīj-i Shāh Jahānī*), and Nityānanda (in his *Siddhāntasindhu*) all use a value of 23° 30′ 17″ in their table-texts. See notes on verses 33 and 34 on page 216.

1.3.9.7 *Numbers and configurations of constellations*

In verses 80 and 81, Nityānanda describes a collection of 48 constellations containing 1022 stars, arranged in a manner that 21 constellations are north of the zodiac, 12 constellations are zodiacal, and the remaining 15 are south of the ecliptic. These numbers are of Greco-Islamicate (Ptolemaic) origin, and, to my knowledge, not attested in any earlier Siddhāntic text. See notes to verses 80 and 81 on page 250.

1.3.9.8 *Arc of second and true declination*

In verses 116 and 117, Nityānanda defines the arc of second declination (*dvitīya-krānti*, also known as *para-apama*, lit. other declination) and the curve of true declination (*sphuṭatara-apama-aṅka*) respectively. These two concepts play an important role in calculating the true declination of a celestial object in Islamicate astronomy. Nityānanda states the geometrical definitions of these Islamicate quantities in the *golādhyāya*. These quantities and the algorithms to compute the true declination using them are described in Nityānanda's *Siddhāntasindhu* (II.6) and *Sarvasiddhāntarāja* (I.9 *spaṣṭakrānti*) (see Misra 2021, 2022).

Notes

1 See Neugebauer ([1951] 1969), Pannekoek ([1961] 1989), Pedersen ([1974] 1993), and Evans (1998)for a review of the history of astral sciences—in particular, the history of mathematical astronomy—and Rochberg (1999), Wright (2000), Rochberg (2002), Krupp ([1983] 2003), and Dreyer and Stahl ([1906] 2019)for contributions from philosophy, folklore, and divination in shaping its early history.

2 Throughout this book, all Gregorian dates are from the Common Era (CE) unless specified otherwise, in which case they are marked by the standard abbreviation for Before the Common Era (BCE). All other dates using other calendar systems (like Vikrama Saṃvat, Śāka, Hijrī, etc.) are indicated appropriately.

3 In the Sanskritic tradition, the words, *gaṇaka, piṇḍila, nākṣatra, nakṣatradarśa, jyotiṣa, jyotirjña, hiṇḍika, jñānin, kālajña, kartāntika, bhojaka, śāstratattvajña, varājīvin, bhārgava* and the variations thereof, are used to refer to technically trained professionals in mathematics, astronomy, and astrology. These epithets were accorded to them from early times on account of their ability to predict auspicious lunar mansions (via their calculations) under which the sacrificial fires were to be established; see, e.g., *Śatapatha Brāhmaṇa*, II.1.2.xix. (The *Vājasaneyi Saṃhitā* of the Śukla Yajurveda uses the word *nakṣatradarśa* (XXX.10) and *gaṇaka* (XXX.20) to identify those individuals who understand the stars.) Over time, and following the social federalism of traditional Indian society, these men constituted the hereditary (*vaṃśa*) or ethnic (*jāti*) social class of priests, teachers, and learned men (variously called *ācārya, bhaṭṭa, paṇḍita, vaidya*, etc.).

4 The word Islamicate (instead of Islamic) refers to the artistic, literary, and scientific works produced in Muslim societies in the Arabic and Persian. Being secular, these works are not directly related to Islam or a particular geographic region; see discussions on Islamicate Secularities in Dressler, Salvatore, and Wohlrab-Sahr (2019).

5 The Sanskrit word *śāstra* refers to a collective body of knowledge. Its use in a compound (e.g., in *jyotiḥśāstra*) is comparable to the Greek suffix -λογία denoting the study of a topic or a branch of knowledge of a discipline. Colloquially, *śāstra* is considered synonymous with *vidyā* (a word variously meaning 'science', 'learning', or 'scholarship') that has four main types: *trayī* or the study of the Vedas and their auxiliary texts, *ānvīkṣikī* or the study of logic and metaphysics, *daṇḍanīti* or the science of government and jurisprudence, and *vārttā* or the practical arts of agriculture, medicine, economics, etc. The study of astral sciences (*jyotiḥśāstra* or *jyotirvidyā*) is considered one of the auxiliary disciplines (*vedāṅga*) of Vedic studies.

6 The word *yavana* referred to the Ionians or Greeks, but in the later periods of history it came to be associated with the Muslims (Arabs and Persians), or, more generally, any foreigner (Vasunia 2013, footnote 6 on p. 92).

7 Some of the more comprehensive studies on the history of Sanskrit *jyotiḥśāstra* include S. B. Dikshit (1969), Pingree (1978a, 1978b), S. B. Dikshit

(1981), Pingree (1981b), Subbarayappa and Sarma (1985), Kak (2000), Sen and Shukla ([1985] 2000), Minkowski (2002) and Delire (2012).

8 For example, see Neugebauer (1975), Dicks (1985), Hunger and Pingree (1999) and Ossendrijver (2012) for studies on Babylonian astronomy; Dicks (1985), Gutas (1998) and Bowen and Rochberg (2020) for studies on Hellenistic astronomy; D. King (1986), Saliba (1995) and Berggren (2016) for studies on Islamicate astronomy; and Huxley (1964), Pingree (1973a, 2001) and Mercier (2004) for studies on the interactions between these cultures of science.

9 For example, see Plofker (2005, 2008) for discussions on the sociocultural and linguistic complex in which Sanskrit mathematical astronomy functioned, and Narasimha (2007, 2013) for studies on the ethico-onto-epistemic basis of Indian scientific thinking.

10 The allegiance of authors to their chosen *pakṣa*s was occasionally mutable and multiple, e.g., Āryabhaṭa I composed both the *Āryabhaṭīya* (499) and the *Āryabhaṭasiddhānta* (ca. early sixth century) that became the basis of the *Āryapakṣa* and the *Ārdharātrikapakṣa* respectively. In fact, some topics like the differences in the calculation of a planet's longitude due to double-epicyclic motion not only caused friction between astronomers from rival *pakṣa*s but also among those adhering to the same *pakṣa*.

11 In Sanskrit *jyotiḥśāstra*, proof or demonstration (*upapatti*) was often thought to be one of the more rigorous means (*pramāṇa*) of validating true knowledge (*pramā*). Bhāskara II, in his twelfth-century auto-commentary *Vāsanābhāṣya* on his *Siddhāntaśiromaṇi* (I.2.1–6), states *yady evam ucyate gaṇitaskandhe upapattimān evāgamaḥ pramāṇam* 'for all that is mentioned in the chapter on computation, only the traditional text (or doctrine) that is supported by *upapatti* is considered *pramāṇa*'. See Srinivas (2008) for discussions on *upapatti* in Sanskrit mathematical astronomy, and Keller (2012) for the role of verification (*pratyaya-karaṇa*) in the process of building conviction (as discussed by Bhāskara I in his seventh-century commentary on Āryabhaṭa I's *Āryabhaṭīya*).

12 Agathe Keller (in Keller and Volkov 2014, § 2. Mathematics Education in India on pp. 70–79) reviews the history of mathematical education in ancient, medieval, and premodern India. Therein, her discussions on the interconnected pedagogy of mathematics and mathematical astronomy in India suggest that successive generations of Sanskrit astronomers were indeed familiar with several mathematical and astronomical works authored by earlier Siddhāntic authors.

13 It is speculated that as early as the tenth century, traces of foreign influence could be seen in the *Laghumānasa* of Muñjāla; however, the major bulk of translations of Persian astronomical treatises into Sanskrit started around the sixteenth and seventeenth centuries (Pingree 1981b, p. 30).

14 Describing the vernacular literary culture of early modern North India, Orsini (2012) remarks that 'it is better to understand the literary culture in fifteenth-century north-India as a multilingual and multilocation literary culture—with a trend towards Persian-Hindavi bilinguality in the domains of politics and literature of the various regional Sultans and in the Sufi religious and literary practices' (pp. 238–239).

15 The *Mulakhkhaṣ-i Shāhjahān Nāma* 'Summary of the Chronicles of the King of the World' is an abridged history (*al-mulakkhaṣ*) of Emperor Shāh Jahān written by his seventeenth-century court chronicler Muḥammad Ṭāhīr Khān, alias ᶜInāyat Khān. The larger work, *Pādshāhnāma* 'Chronicles of the Emperor', was written by the seventeenth-century Mughal chronicler ᶜAbd

al-Hamīd Lāhūrī and Muḥammad Wāris, and presented to the ᶜInāyat Khān who was then the superintendent of the royal library (*dārogha-i kutub-khāna*). See Ansari (2015) for Muḥammad Ṭāhīr Khān's Persian text (Appendix II. A4 on p. 597) and its English translation (p. 584a).

16 Mullā Farīd was a student of the Indo-Persian polymath ʾAmīr Fatḥallāh Shīrāzī (d. 1589) at the court of the Mughal Emperor Akbar. From what is known about him, he served under Mirzā ᶜAbdʾl-Raḥīm Khān-i Khānān, a prominent Mughal nobility during the reigns of the Mughal emperors Akbar and Jahāngīr (Ghori 2008, pp. 396–397).

17 The abbreviated name *Zīj-i Shāh Jahānī* derives from *Kārnamāh-i Ṣaḥib Qirān-i Thānī, Zīj-i Shāh Jahānī* 'Grand Accomplishment of the Second Lord of the Conjunction, the *Zīj* of Shāh Jahān' where Mullā Farīd uses the royal epithet of Shāh Jahān, the 'Second Lord of the Conjunction' born on the auspicious conjunction (*qirān*) of Jupiter and Venus at his natal hour on 5 January 1592. This title traces Shāh Jahān's direct descent from the First Lord of Auspicious Conjunction Sulṭān Amīr Tīmūr (Tamerlane, in Latin), the founder of the Timūrid dynasty (Chann 2009, pp. 1105–1106).

18 See Ghori (2008, pp. 396–398), Rosenfeld and İhsanoğlu (2003, pp. 357–358), and Ansari (2015, § 3.2 on pp. 583–585) for a survey of the text and manuscripts of the *Zīj-i Shāh Jahānī*.

19 It has been suggested that Mullā Farīd completed the *Zīj-i Shāh Jahānī* with the assistance of his brother Mullā Ṭayyib and other Hindu and Muslim astronomers, overlooked by Āṣaf Khān. Also, Muḥammad Ṣādiq Dihlavī, in his *Ṭabaqāt-i Shāhjahānī* 'Compendium of biographies of Mughal notables', mentions that Mullā Farīd did not live to see the end of the *Zīj-i Shāh Jahānī* on account of his ailing health and advancement of age (Ghori 2008, p. 397).

20 The Sanskrit term *gotra* refers to a clan-like kinship group where agnatic descendants derive from an eponymous apical ancestor, most commonly considered to an ancient sage (*ṛṣi*).

21 The *Siddhāntasindhu* is an enormous table-text extant in four complete manuscripts, each containing around 440 folia measuring approximately 45 × 33 cm; see CESS A3 (1976a, p. 173b), CESS A4 (1981a, p. 141ab), CESS A5 (1994, p. 184a), and MMSM (2003a, pp. 138–143).

22 According to Pingree (2003b, p. 270),

> [t]he failure of the *Siddhāntasindhu* to find favour among the Hindu adherents of the traditional *siddhānta*s inspired Nityānanda to write an elaborate apology for using Muslim astronomy, the *Sarvasiddhāntarāja* that he completed in 1639.

23 The Gauḍa Brahmins are a caste-subdivision (*jāti*) of Hindu Brahmins who originate from the Gauḍa region, a region extending from Bengal (*vaṅga*) to Kashmir (*kaśmīra*) in India and passing through the modern-day regions of Haryana (*hariyāṇā*). The Mudgala *gotra* (patrynomic) follows the lineage of the sage-king Rājarṣi Mudgala, a mythohistorical ruler of the Pañcāla region (modern-day Uttar Pradesh).

24 Interestingly, in verse 28, Nityānanda lists various places traditionally believed to lie on the Indian prime meridian (passing through Ujjain) and in that list of localities, he includes the city of Rohtak (Rohita) in Haryana and sacred water reservoir of Sannihit Sarovar (Sāṃnihitya) in town of Thanesar in the Kurukshetra district of Haryana. These two places are not included in earlier Siddhāntic lists of prime meridional cities; see notes to verse 28

on page 212. Nityānanda also again mentions Kurujāṅgala, an area covering the modern-day cities of Rohtak, Hansi, and Hisar in Haryana, as a prime-meridional location; see notes to verse 68 on page 235.

25 For the most part, the arrangement of topics in Table 1.3 follows David Pingree's observation in Pingree (1981b, Table 4 on p. 29). However, some additional topics are included under the chapter on computations (*gaṇita*) upon examining the manuscripts of the *Sarvasiddhāntarāja*. See Chapter 2 for a description of the manuscripts of Nityānanda's *Sarvasiddhāntarāja* consulted in this study.

26 From here on forwards, all English translations of Sanskrit passages in this book are my own, unless otherwise noted.

27 The system of tropical zodiac (*sāyana*, lit. with motion) considers the vernal equinoctial point as the beginning of the zodiacal signs (i.e., the first point of Aries, 0° ♈), whereas the system of sidereal zodiac (*nirayaṇa*, lit. without motion) regards a fixed ecliptic point (*meśādi*) as the beginning of the zodiacal signs. In the Indian astronomy, the fixed ecliptic point is taken to be exactly 180° from the star α Virginis (Spica or *citrā*). The precession of the equinoxes implies that the vernal equinoctial point continually drifts westwards at a rate of 50 arcseconds per year, and hence, the sidereal and tropical zodiacs are offset from one another. Indian astronomers typically used the sidereal zodiac as it helped them calculate the mean motions of planets as integer number of rotations in a *kalpa* or *yuga* (large intervals of sidereal time). This was different to the practice of Greco-Islamicate astronomers who preferred the tropical zodiac as their frame of reference.

28 With regard to verses from the *Sarvasiddhāntarāja* cited here and elsewhere through this chapter, the chapter/section numbers follow the order in Table 1.3 while the verses numbers are from MS Np: National Archives Nepal, NAK 5.7255 (NGMCP Microfilm Reel № B 354/15), see, Section 2.2.4.

29 As Monier-Williams, Leumann, and Cappeller ([1899] 1960, p. 1216a) indicate, the lexical meaning of the word *siddhānta* is 'a fixed or established or canonical text-book or received scientific treatise on any subject'. Its etymology derives from its constituent words *siddha–* 'established' and *–anta* 'end' which makes *siddhānta* a fundamental tenet, or in other words, the truth established by those means (*pramāṇa*) of knowledge that are considered epistemologically valid.

30 In Indian philosophy, there are seven types of *pramāṇa*s that are considered as valid means of acquiring true knowledge, namely, perception (*pratyakṣa*), inference (*anumāna*), comparison or analogy (*upamāna*), circumstantial implication or postulation (*arthāpatti*), non-perception or non-existence (*anupalabdhi* or *abhāva*), scriptures (*āgama*s), and testimony (*śabda*). However, different philosophical schools (*darśana*s) argued on the epistemic validity of these means. For example, the theist schools (*āstika*s) considered testimony (*śabda*) as one of the valid means of knowledge, whereas some atheistic schools (*nāstika*s) disregard this instrument (see, e.g., Bilimoria 2017).

31 Vātsyāyana's commentary to *Nyāyasūtra*, I.1.7, reads

> *aptaḥ khalu sākṣātkṛtadharmā yathādṛṣṭasyārthasya cikhyāpayiṣayā prayukta upadeṣṭā | sākṣātkaraṇamarthasyāptistayā pravartata ityāptaḥ | ṛṣyāryamlecchānāṃ samānaṃ lakṣaṇam | tathā ca sarveṣāṃ vyavahārāḥ pravartanta iti |*
>
> That person is called '*āpta*', 'reliable', who possess the direct and right knowledge of things, who is moved by a desire to make known (to others) the things as he knows it, and who is fully capable of

speaking of it. The word *'āpta'* is explained as denoting one who acts or proceeds, through *'āpti'*, i.e., through the direct knowledge of things. This definition applies to sages, as well as to *Ārya*s and *Mleccha*s; the activities of all these people are carried on through such 'Words'.

(Jhā 1939, p. 30)

32 Varāhamihira's *Bṛhat Saṃhitā* (II.14) quotes Garga's statement:

mlecchā hi yavanās teṣu samyak śāstram idaṃ sthitam |
ṛṣivat te 'pi pūjayante kiṃ punar daivavid dvijaḥ ||

For the Yavanas are foreigners; [yet] this science is well established among them, and they are venerated like sages. How much more, [then], a twice-born astrologer!

(Gansten 2020, pp. 82–83, quoted in the *Hayanaratna*)

33 Before laying out his critique, Balabhadra describes why the orthodox proponents reject the impure science of the foreigners (*yavanaśāstra*), even if it is translated into Sanskrit. According to him, they make their case based on the following two statements: *na vaded yāvanīṃ bhāṣāṃ prāṇaiḥ kaṇṭhagatair api* 'one should not speak the Yavana language even when the vital [breaths] are in one's throat [i.e., even when one is about to die]' (*Bhaviṣyapurāṇa*, III.28.53) and *mūlāśuddhyā sarvam aśuddham* 'if the foundation is impure, all is impure' (a *nyāya* or maxim) (Gansten 2020, pp. 80–81).

34 Among Indian genethlialogical texts, the *tājika* texts (written from about the thirteenth century) are Indian adaptations of Arabic/Persian astrology that are different from traditional *jātaka* texts in several technical aspects. The word *tājika/tājaka* derives from the Persian word *tāzīg*, a Pahlavī term used to refer to Arabs (by calling them with their tribal epithet Ṭayyiʾ). Since the eight century, the Arabs and Persian who occupied the west coast of India were called *tājika*s; and at other times, the Greeks (*yavana*s) from the northwest India, Turkish (*turukṣa*), or even the Tatar Muslims (*tārtīyaka*s) were collectively called *tājika*s (see, e.g., Gansten 2020, p. 3 and Pingree 1981b, p. 97).

35 In his *Hayanaratna*, Balabhadra says *na vaded yāvanīṃ bhāṣām etad vacanaṃ tu yāvanīyakāvyālaṃkārādiviṣayakam iti siddhāntaḥ* 'But the statement "One should not speak the Yavana language" applies [only] to Yavana poetry, rhetoric and so forth: this is the conclusion'. (Gansten 2020, pp. 84–85). This then allows Balabhadra to argue that the science (*śāstra*) of the *yavana*s could be accepted even if, perhaps, their arts (*kalā*s) should not be spoken about.

36 Balabhadra quotes this story from Hillaja's *Hillājadīpikā*, I.6, in his *Hayanaratna*; see Gansten (2020, pp 84–85).

37 Qāḍīzāda al-Rūmī also wrote a commentary, the *Sharḥ al-Mulakhkhaṣ*, on Jaghmīnī's *Mulakhkhaṣ fī al-hayʾa al-basīṭa* (ca. 1205/1206) in 1412 and dedicated it to Ulugh Beg. Pingree suspected that al-Rūmī's commentary could have been Nityānanda's *Romakasiddhānta* (Pingree 1978c, p. 324); however, this claim remains unverified presently.

38 In Nṛsiṃha Daivajña's commentary to *Siddhāntaśiromaṇi* (I.1.*bhagaṇādhyāya*.1–6), we find his statement:

nanu vastunirvikalpāsaṃbhavāt kathaṃ parasparaviruddhasya gatikāraṇapratipādakaśāstrasya prāmāṇyam ucyate | grahagatipratipādakaṃ śāstraṃ tāvatpramāṇaṃ gateḥ pratyakṣopalabdhatvāt | tatkāraṇam puruṣabuddhi prabhavatvād atāttvikaṃ bhavatu tathāpi na ko'pi doṣaḥ phalato doṣābhāvāt | tattvajñānopayoginyāḥ jīveśvarādivibhāgakalpanāyā api puruṣabuddhiprabhavatvena vikalpasya dṛṣṭatvāt |

(Chaturvedi 1981, p. 48)

But, [merely] because of an impossibility of not admitting alternative things, how can the authoritativeness of a mutually contrary doctrine that teaches the cause of motion be declared [valid]? A doctrine that teaches the motion of the planets is authoritative insofar as motion is understood by direct perception. The cause of this may be regarded as unreal on account of arising from the intellect of man, yet there are no faults in the effect on account of the [very] non-existence of faults. [This is] because a false notion is seen arising from the intellect of man on account of applying the knowledge of the principles and on account of conceiving the divisions like the individual soul, God, etc.

39 Ajitā, a later commentator on the *Tantravārttika*, explains the idea of 'continuity of tradition' more elaborately. Ajitā says

tatrāvicchinnasampradayapadamapyevaṃ vyācaṣṭe–'gaṇitonnitasya candrādeḥ deṣaviśeṣānvayasya pratyakṣeṇa saṃvādaḥ tato niścitānvayasya parasya gaṇitaliṅgopadeśaḥ, tatstasyāptopadeśavaśāvagatānvayasya anumānam, saṃvādaḥ, parasmai copadeśaḥ iti sampradāyāvicchedāt prāmāṇyam' iti | (ajitā vyākhyā) |

Here, the expression 'the continuity of tradition' is explained thus: 'The correlation of the Moon etc.as observed in a place and as computed (using extant texts in the science) is done first; this information is then discussed with astronomical rationale with one rooted in the science; there is then deductions by that person rooted in the science; then follows discussions and again instruction to another (on the basis of such deductions). Thus the authority of the science (along with the new deductions and revisions) is due to this continuity of tradition'. (Com. *Ajitā*) (Subbarayappa and Sarma 1985, p. 58, cited by Nīlakaṇṭha Somayājī in his *Jyotirmīmāṃsā*, ca. 1504).

40 The works listed in Table 1.4 contain sections or chapters on spheres that are explicit and separate. There are, however, other Siddhāntic texts that implicitly discuss spheres, e.g., Varāhamihira's *Pañcasiddhāntikā* (in Chapters XIII and XIV) or Kamalākara's *Siddhāntatattvaviveka* (in Chapters V, VI, and XV). For simplicity, Table 1.4 only lists works that discuss spheres as a separate section or chapter, or as a separate self-titled work.

41 In Indian literature, the Purāṇas are regarded as mytho-historical accounts of antiquity that include discussions on various subjects including, *inter alia*, the five indicatory topics (*pañcalakṣaṇa*): cosmogony or the creation of the universe (*sarga*); cataclysm or the destruction of the universe (*pratisarga*); heroic and prophetic genealogies (*vaṃśas*); the reigns of the descendants of Manu (*manvantara*); and the history of the kings of the solar and lunar dynasties (*vaṃśyānucarita*); see Rocher (1986, pp. 24–30). Appendix B includes a fuller discussion on Purāṇic cosmography.

42 The history of Siddhāntic astronomers challenging Purāṇic views, especially those that counter astronomical (mathematical) calculations, is as old as the tradition of composing astronomical *siddhānta*s themselves. Lalla, in his

Śiṣyadhīvṛddhidatantra (ca. eight/ninth century), includes a separate chapter on removing erroneous notions (*mithyājñānanirākaraṇa*). At the beginning of this chapter (XX.3–7), he lists several erroneous Purāṇic views on astronomical phenomena, which he then argues against (and dismisses) in the course of the chapter (see Chatterjee 1981, footnote on verses 1–7 on pp. 268–269 includes Lalla's list of erroneous Purāṇic beliefs).

43 For example, Lalla, in his *Śiṣyadhīvṛddhidatantra* (XX.1–2), says

> *saṃsthānam aśeṣam īritaṃ brahmāṇḍodaravartinām idam |*
> *prativādivacāṃsi śṛṇvato manasi bhrāntir ivāvatiṣṭhate ||*
>
> *yata evam ataḥ kuhetumadvacanāni prathamaṃ bravīmyaham |*
> *upapattimadāgamaṃ tato manasaḥ sthairyakaraṃ parisphuṭam ||*
>
> Different situations have been ascribed (by different people) on matters relating to the universe. When one listens to the different statements made by opposing schools, doubts, as it were, will assail one's mind.
>
> For this reason, I shall first enumerate the irrational views; and then I shall put forward the truths together with proofs. That will make things clear and thus steady the mind.
>
> (Chatterjee 1981, p. 268)

44 al-Khwārizmī, in his encyclopedia *Mafātīḥ al-ᶜUlūm* 'Keys to the Sciences' (ca. tenth century), describes the cupola as

> القبة وسط الأرض أعنى مابين نقطة المشرق المفروضة وبين نقطة المغرب المفروضة وذلك مائة ومثانون درجة وبين نقطة نهاية ناحية الجنوب وبين نقطة نهاية ناحية الشمال وذلك أيضا مائة ومثانون درجة
> (van Vloten [1895] 1968, p. 218)
>
> The cupola is the middle of the Earth, that is, the point between the supposed eastern point and the supposed western point that are one hundred and eighty degrees [apart], and between the extreme point in the southern quarter and the extreme point in the northern quarter that are also one hundred and eighty degrees [apart].

45 For example, al-Khwārizmī, in his *Mafātīḥ al-ᶜUlūm*, describes Bārah as

> باره اسم مدينة في جزيرة البحر الاعظم قريبة من القبة
> (van Vloten [1895] 1968, p. 218)
>
> Bārah is the name of the city on an island in the greatest ocean close to the cupola.

46 Also, in his *Mafātīḥ al-ᶜUlūm*, al-Khwārizmī describes Kangdizh as

> كنكدز هى أقصى مدينة فى المشرق وهي فى أقاصى بلاد الصين والواقوا
> (van Vloten [1895] 1968, p. 217)
>
> Kangdizh is the farthest city in the east, and in the farthest country of China and [the island] Vāqvāq.

47 Also, Abū al-Fidā' (fl. late thirteenth century), in his *Kitāb taqwīm al-buldān* 'Book of the Sketch of the Countries', identifies Yamakoṭi as الفرس جماكود (*al-furs jamākūd*) 'Jamākūd [of] the Persians' (Reinaud and de Slane) 1840, p. 367).

48 ʿAlī ibn Sulaymān al-Hāshimī, in his *Kitāb fīʿilal al-zījāt*, states

والطرف الاخر البحر الاخضر في جزيرة يقال لها بولا

> The other end [of the Earth's copula, at a longitude of 90° W] is in the Green Sea at an island called Būlā (Thule?).
>
> (Kennedy and Ḥaddād 1981, p. 91 (f. 93v: 10–11))

49 Aussant (2014) provides a succinct review of the theories of homonymy and polysemy in Sanskrit language theories; in particular, the philosophy of plurivocality in Sanskrit.

50 In Sanskrit grammar, the three modes (*vṛtti*s) of understanding the relationship between words (*śabda*) and their meanings (*artha*) are called primary or literal (*abhidhā*), indicative or figurative (*lakṣaṇa*), and suggestive or implied (*vyañjana*). The primary or matter-of-fact meaning of words (variously called *abhidhā, śāstrokti, sanketārtha, vācyārtha*, etc.) was thought to be established by convention. For example, Mammaṭa, in his *Kāvyaprakāśa* (II.7cd, ca. eleventh century), says *sākṣāt saṃketitaṁ yo 'rtham abhidhatte sa vācakaḥ* 'a literal designator is a word that directly denotes a conventional meaning'. The conventions that were thought to establish (or teach) the meaning of words included its colloquial usage by elders (*vṛddhavyavahāra*), direct statements by a trustworthy authority (*āptavākya*), grammar (*vyākaraṇa*), analogy (*upamāna*), lexicon (*kośa*), context (*vākyaśesa*), explanation (*vivṛti*), and syntactic connection (*siddhapadasāṃniddhya*). See Raja ([1963] 1977) for a comprehensive review of Indian theories of words and their meanings.

51 A similar intention of transferred agency is also implied in Nityānanda's use of reflexive verbs $\sqrt{}$*kḷp* (cl. 1, with directional prefixes like *pra*), $\sqrt{}$*man* (cl. 4, with directional prefixes like *pra*), *vi-*$\sqrt{}$*cint* (cl. 10), and $\sqrt{}$*vid* (cl. 1) 'to believe, consider, suppose, or think'.

52 The *bhuṭasaṃkyā* system is a Sanskrit numeration system where words (referring to objects, ideas, or collections of things) that have physical or mythohistorical significance are used to denote numbers (see, e.g., Petrocchi 2016). A collated reference list of *bhuṭasaṃkyā* numerals can be found at http://www.hamsi.org.nz/p/concrete-number-system.html.

53 See Shukla (1976, Appendix III on pp. 340–344) for a comprehensive list of Sanskrit expressions of mathematical operations (taken from Bhāskara I's works, the *Mahābhāskarīya* and the *Laghubhāskarīya*, as well as his commentary on Āryabhaṭa's *Āryabhaṭiya*).

54 Throughout this book, capitalised trigonometric functions indicate a non-unitary radius, i.e., Sin = $\mathscr{R}$ sin and Cos = $\mathscr{R}$ cos, where the radius $\mathscr{R}$ is the *sinus totus* (sine of 90°). In Mullā Farīd's *Zīj-i Shāh Jahānī* and Nityānanda's *Siddhāntasindhu* and *Sarvasiddhāntarāja*, the *sinus totus* is 60. More on Nityānanda's choice of *sinus totus* in Section 1.3.9.1.

55 Throughout the *golādhyāya*, the words *diś, dṛś, āśā, kāṣṭhā, pada, aṅghri*, etc.variously indicate 'direction, region, or space'. The intercardinal or angular directions are called *koṇa* 'corner or angle', e.g., in verse 87.

56 The deification of the directions is also seen in the Greek concept of Anemoi (Ἄνεμοι) where various Aeolian deities (wind gods) presided over the directions individually and controlled their respective winds that influenced weather patterns and seasonal conditions.

57 The power of the planets (*khecara* 'sky-movers') to seize (*graha*) the destiny of men made them the objects of astrolatry in India. For example, see Sivapriyananda (1990, p. 62) for a discussion on the socioreligious and iconographic aspects of the nine planets (*navagrahas*). Among the nine planets, seven planets are visible (*dṛśya*): namely, the Sun (*arka*), the Moon (*candra*), Mars (*maṅgala*), Mercury (*budha*), Jupiter (*bṛhaspati*), Venus (*śukra*), and Saturn (*śani*), whereas the other two planets, namely, the ascending lunar node (*rāhu*) and the descending lunar node (*ketu*), are invisible (*adṛśya*). The aspect of these planets (and their influence on human destiny) is signified with several Sanskrit synonyms, e.g., *daśā, ulkā, dṛṣṭi*, etc. In verse 84, Nityānanda calls planets *dyuṣad*, a word more commonly used to refer to the heaven-dwellers or gods, in reference to their divine (and deified) potency.

58 The earth-element (i.e., soil or clay) called *mṛd* or *ku* is understood differently to the Earth-sphere; see notes to verse 7 on page 195.

59 A technical description of Sanskrit prosody (*chanda*) can be found in Apte ([1890] 1965, Appendix I on pp. 1035–1042) and Velankar (1949, pp. 114–160).

60 See, e.g., Mullā Farīd's *Zīj-i Shāh Jahānī*, MS Or. 372 from the British Library London, ff. 21r–29v, *jadval al-jayb* 'table of sines'; and Nityānanda's *Siddhāntasindhu*, MS 4962 from the Khasmohor Collection at the City Palace Library of Jaipur, ff. 29r–37v, *pratikalājyākoṣṭhakā* 'table of Sines for every minute of arc'. In MS 4962 (f. 37v), the tabulated Sine value (corresponding to 89°59′) is '59;59,59,59,27,6' with a forward increment of '0;0,0,0,32,54', which suggests that the maximum Sine value (*sinus totus* corresponding to 90°) is '1;0,0,0,0,0', or simply, '60'.

61 See Gupta (1977) for discussions on the different values of the *sinus totus* in Indian astronomy, and more generally, see Brummelen (2009) for the history of trigonometric functions (including the *sinus totus*) across scientific traditions. The table of Indian *sinus totus* values, seen in Gupta (1977, p. 136) and Subbarayappa and Sarma (1985, p. 71), attribute the use of sixty as the *sinus totus* to Kamalākara (*Siddhāntatattvaviveka*, 1658) and Jagannātha (*Siddhāntasamrāṭ*, a Sanskrit translation of Naṣīr al-Dīn's Arabic version of Ptolemy's *Almagest* from 1732). These lists miss Nityānanda's use (in both his texts) from an earlier decade (ca. 1630s).

62 For example, the fortress-city of Kangdizh was erroneously reported as Gangdizh in the *Shāhnāma* 'The Book of Kings', an epic poem written in the tenth century by the Persian poet Abuʾl-Qāsem Ferdowsī. See the description of Kangdizh in *Enclyopædia Iranica* at https://iranicaonline.org/articles/kangdez-fortress-of-kang.

63 Nṛsiṃha Daivajña, in his commentary *Vāsanāvārttika*, I.1.*bhagaṇādhyāya*. 1–6 (composed at Kāśī in 1621) on Bhāskara II's *Siddhāntaśiromaṇi*, argued against the opinion of the foreigners (*yāvanika-mata*) that the sphere of asterism was transparent (*svaccha*) like crystal (*kācamaṇi*), supported the westward motion of the ecliptic, and yet remained fixed.

> *yavanamate tūrdhvagasya sajīvākāśasya mūrtimato 'ntarikṣasthititvaṃ mūrtimadādhāraṃ vinānupapadyamānaṃ mūrtimadādhāraṃ kalpayati | tasyāpy ādhārakalpanāyām anavasthābhayād ūrdhvagasyākāśasyāntarikṣasthititvaṃ śaktyaiveti mantavyam | tasya niyataparāśābhimukhagamanaśaktikalpaneti śaktidvayam | akḷptakalpanā ca bhacakrādhārabhūtasyākāśasya ca kācamaṇyādivat svacchatvakalpanāpi gauravam āvahatītyādyā doṣāḥ vartanta iti parihāryo yāvanaḥ panthāḥ |*
>
> (Chaturvedi 1981, p. 46)

> But, in the opinion of the foreigners, the fixity of space of a sentient sky situated above [and] having a material form is without a support that has a material form; [this opinion] conceives a support of a material form as being unjustified. And even in conceiving its support, on account of fearing instability, it should [still] be supposed that the fixity of space of the sky situated above is only with energy. The two-fold energy is conceived as the energy to move in the western direction and [also be] fixed. And the imperfect conception of [it] being the support of the ecliptic and of the sky, and the conception of [it being] perfectly transparent like [being made of] crystal etc.leads to cumbrousness—with beliefs such as these, errors are circulated. Accordingly, the doctrine of the foreigner should be avoided.

64 I am currently preparing a critical edition (with English translations and technical annotations) of the twenty-one verses in Nityānanda's *Sarvasiddhāntarāja* (I.3.180–200) on the formation of the universe (*brahmāṇḍanirmāṇa*) along with Christopher Minkowski.

65 See Saliba (2007, pp. 73–100) for an excellent review of Islamicate encounters with Greek astronomy.

2 Manuscript Sources and Stemma

In preparing a critical edition of the *golādhyāya* in Nityānanda's *Sarvasiddhāntarāja*, I have consulted eight manuscript witnesses. The present chapter includes:

1 a description of the source manuscripts and their metadata (i.e., the palaeographical, codicological, and historical aspects of the manuscripts, along with the information about their provenance, access, and location);

2 a description of the structure and content of the individual manuscripts;

3 a stemma of the manuscript witnesses;

4 a description of the editorial conventions in preparing the critical edition; and

5 a description of the structure and format of the *apparatus criticus*.

2.1 Catalogues, holding institutions, and manuscript sigla

The manuscripts of Nityānanda's *Sarvasiddhāntarāja* are described in the following six manuscript catalogues listed below with their names and abbreviated keys.

1 CESS: Census of the Exact Sciences in Sanskrit: Series A. Volumes 1–5 by David Pingree (CESS A1 and A2 1970a, CESS A3 1976a, CESS A4 1981a, and CESS A5 1994).

2 NGMCP: Nepalese-German Manuscript Cataloguing Project: Catalogue of Indic manuscripts (NGMCP 2002–16).

3 SATE: Sanskrit Astronomical Tables in England by David Pingree (SATE 1973c).

DOI: 10.4324/9780429506680-2

4 SATIUS: Sanskrit Astronomical Tables in the United States by David Pingree (SATIUS 1968).

5 SOI: A descriptive catalogue of manuscripts in the Scindia Oriental Institute, Vikram University, Ujjain (SOI 1983–85).

6 SSS: The Shanti Saroop Saith Catalogue of Sanskrit Manuscripts in the Punjab University Library, University of Punjab, Lahore (SSS 1941).

For the purpose of this study, seven institutions provided copies of eight extant manuscripts containing the *golādhyāya* of Nityānanda's *Sarvasiddhāntarāja*; namely,

1 the Saraswati Bhawan Library of Sampurnanand Sanskrit Vishwavidyalaya in Varanasi (Benaras),

2 the Bhandarkar Oriental Research Institute in Pune,

3 the National Archives of Nepal in Kathmandu in conjunction with the Nepalese-German Manuscript Cataloguing Project maintained by Asia-Africa Institute of University of Hamburg,

4 the John Hay Library at Brown University,

5 the Fergusson College Library in Pune,

6 the Rajasthan Oriental Research Institute in Jaipur, and

7 the Scindia Oriental Institute at Vikram University in Ujjain.

Moreover, the Maharaja Sawai Man Singh II Museum Library at the City Palace in Jaipur provided a copy of Nityānanda's table-text the *Siddhāntasindhu* from their royal (*khāsmohor*) collection. Table 2.1 lists the sigla I have used to identify the eight manuscripts of Nityānanda's *Sarvasiddhāntarāja* throughout this book (including in the critical edition).

All of these manuscripts are opisthographs written *scriptura continua* in the Devanāgarī script (with Nāgarī variations) on hand-made paper.[1] I inspected four out of the eight manuscripts in their original before acquiring their digital copies, namely, Bn.I, Bn.II, Br, and Sc. The other four manuscripts, i.e., Np, Pb, Pm, and Rr, were remotely sent to me as digital or microfilm copies by their holding institutions. The digital images of their folia, in conjunction with their descriptions in the manuscript catalogues, confirm that these four manuscripts were also written on hand-made paper.

Section 2.2 provides a detailed description of the metadata, structure, and content of the eight manuscripts individually. The main peculiarities of these manuscripts are also noted and commented upon. In the descriptions that follow,

Table 2.1 List of manuscript sigla

Bn.I	Benares (1963) 35741 from Saraswati Bhawan Library, Varanasi.
Bn.II	Benares (1963) 37079 from Saraswati Bhawan Library, Varanasi.
Br	BORI 206 of A 1883/84 from Bhandarkar Oriental Research Institute, Pune.
Np	NAK 5.7255 from National Archives of Nepal, Kathmandu identical to NGMCP Microfilm Reel № B 354/15 from Nepalese-German Manuscript Cataloguing Project, Asia-Africa Institute of University of Hamburg.
Pb	PUL (II) 3870 from Punjab University Library, Lahore. Acquired via the David E. Pingree Collection (from Box № C7) held at John Hay Library, Brown University.
Pm	Poona Mandlik Jyotisha 15/BL 368 from Fergusson College Library, Pune.
Rr	RORI (Alwar) 2619 from Rajasthan Oriental Research Institute, Alwar.
Sc	SOI 9409 from Scindia Oriental Institute, Ujjain.

- any foliation number (or page number) enclosed in angular brackets or chevrons < > indicates a correction or appropriation in place of a missing or unnumbered folio (or page) in the original manuscript.
- Multiple folia (or pages) with the same number are subscripted to distinguish between them, e.g., when two folia are marked as 23, the notations 23_a and 23_b are used to distinguish the antecedent from the subsequent.
- The manuscripts that are separated into parts, with each part differently foliated or paginated, are subscripted with Greek letters in the standard alphabet order α, β, γ, etc. In contrast, manuscripts where the foliation (or pagination) is continuous and chapter-breaks are made explicit by reading concluding-verses (colophon)[2] (and subsequent resetting of the verse numbers back to 1 in the following chapter) are labelled with subscripted uppercase letters A, B, C, etc.
- The four quarters (*pāda*s) of a verse (*padya*) are labelled with affixes a, b, c, and d; e.g., 'verse 12b' indicates the second *pāda* of verse 12.
- And finally, the transliterated Devanāgarī text is simply italicised without any enclosing marks.

2.2 Metadata, structure, and content of the manuscripts

2.2.1 *Bn.I: Saraswati Bhawan Library, Benares (1963) 35741*

Catalogue description in the CESS:

> Benares (1963) 35741. 84 ff. Copied in Saṃ. 1804 = A.D. 1747. (CESS A3 1976a, p. 174)

The cover page of the manuscript identifies the work as *Sarvasiddhānta-rāja* authored by Nityānanda Miśra, son of Devadatta. Table 2.2 describes the metadata of the manuscript (as attested on the cover page). The cataloguer's remark *granthādhikārī agastigotrotpannaḥ raṅganāthaḥ ā◦ 1860* on the cover page (in handwritten Devanāgarī) identifies one Raṅganātha belonging to the Agasti *gotra* as the owner of this manuscript in the Āṣāḍha month of Saṃ. 1860 (around June–July 1804 CE).

The manuscript appears to have been completed by a second hand: there is a difference in penmanship on certain folia.

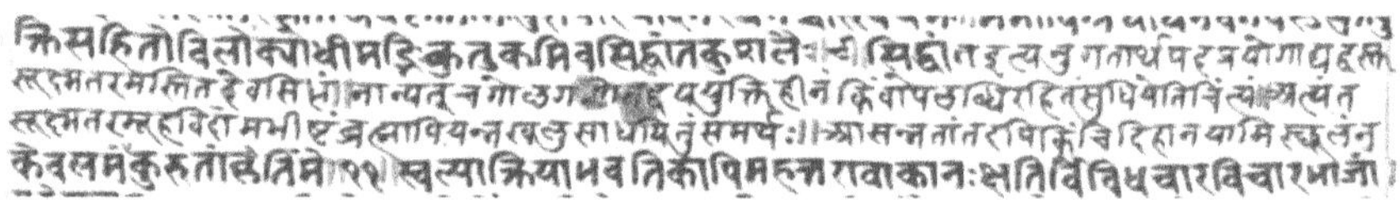

For example, the sentence beginning with *siddhāṃ* of verse 9 on f. 2r (excerpted digital trace shown above) includes the words

Table 2.2 Metadata of Bn.I

Library	Saraswati Bhawan Library of Sanskrit University Varanasi
Title	*sarvasiddhāntarājaḥ*
Author	*nityānandamiśraḥ* (*devadattaputraḥ*)
Accession Number	44110
Subject	*jyotiṣam*
Catalogue Number	35741
Number of folia	84
Lines per page	12 (approximate)
Letters per line	39 (approximate)
Folia Size (in inches)	10.5 × 4.7
Script	Devanāgarī
Material	Paper
Completeness	Complete
Date of composition	Saṃ. 1696 (1639 CE)
Date of copying	Saṃ. 1804 (1747 CE)

taitya∘...∘*sthūlaṃna* (verses 9a–11b, f. 2r) in a different handwriting, and subsequently end with the words *kevalamaṃ*∘... (verse 11b–, f. 2r) written in the original hand. There are aligned blank spaces on certain folia (e.g., f. 17r) with the handwritten text wrapped around it. Overall, the manuscript is continuous over 84 pages without any chapter breaks in the handwritten text. However, the presence of benedictory verses (*maṅgalācaraṇa*s), inceptive and conclusive statements (*ārambha*s and *śeṣa*s)for each section, colophons (*puṣpikā*s), and changes in verse numbering suggest a tripartite content-based division of Bn.I, namely, parts Bn.I$_{A}$, Bn.I$_{B}$, and Bn.I$_{C}$. These are described below.

2.2.1.1 *Content-based divisions of Bn.I*

1 Bn.I$_{A}$ (68 pages, ff. 1v–67v) includes the chapter on computations (*gaṇitādhyāya*). The verso folia have *si*∘*rā* along with the page number (written below), both placed at the top-left corner of each folio. The word *rāma*∘ appears along with the page number (written below) at the bottom-right corner of each folio verso. The page numbers on the first six verso folia (written in the top-left and bottom-right corners) are enclosed by the double *daṇḍa*s on both sides, e.g., f. 1v. has its page number written as || 1 ||.

There are occasional irregularities to this structure, namely, (i) the word *sirā* appears in the top-right corner of ff. 11v, 12v, 14v, and 43v; (ii) f. 10v has the bottom-right page number enclosed by the double *daṇḍa*s on both sides; (iii) the text and page number (at the bottom-right corner) are missing on f. .18v; and (iv) the text at the bottom-right corners of ff. 2v and 5v are enclosed with the double *daṇḍa*s, e.g., the bottom-right corner of f. 2v. has || *rāma*∘ ||.

On some folia, the text appears positioned alongside reserved white spaces, presumably to draw geometrical diagrams; however, no drawings of any kind are found in this manuscript. Prominent blank spaces with text wrapped around them are found on ff. 15rv, 17r, 32v, 36r, 47r, and 49v. Ff. 16r, 19v, 20r, 25v, 30r are completely blank and without any text. There are scattered marginal corrections to the text (in the form of vocalic marks, consonant ligatures, words and sentence corrections) on various folia.

There are underlined and semi-bracketed portion (with *bhūtasaṁkhyā* word-numerals and associated number) all throughout f. 19v; an excerpted digital trace of the folio is shown to the right.

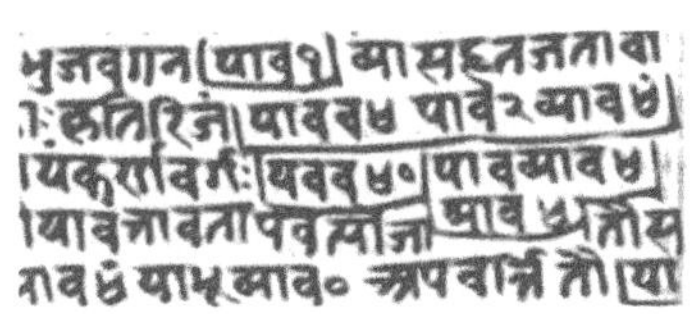

2 Bn.I$_B$ (8 pages, ff. 67v–75v) contains the chapter on spheres (*golādhyāya*). The verso of each folio has *si∘rā* along with the page number (written below), both placed at the top-left corner of the folio. The word *rāma∘* appears along with the page number (written below) at the bottom-right corner of each folio verso. There are no blank spaces within the text block on these pages.

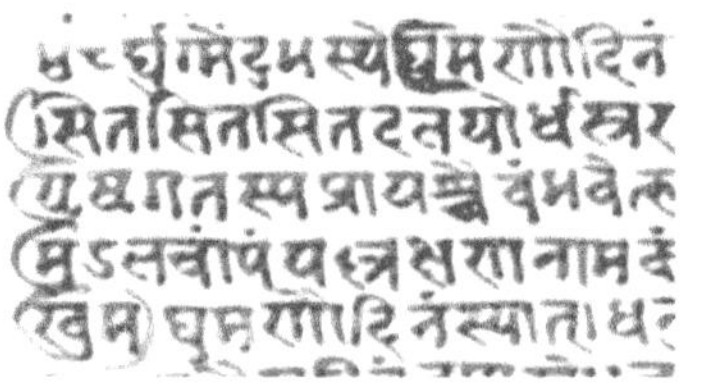

f. 73v has bracketed content indicating a scribal copying error through verse 103, where parts of other verses (verses 109b–112a) have been erroneously copied into verse 103. An excerpted digital trace of the folio is shown to the left.

3 Bn.I$_C$ (9 pages, ff. 75v–84r) contains the chapter on instruments (*yantrādhyāya*). The verso folia have *si∘rā* along with the page number (written below), both placed at the top-left corner of each folio. The word *rāma∘* appears along with the page number (written below) at the bottom-right corner of each folio verso. f. 76v has only four lines of text followed by an empty blank space. There are no scribal corrections, just an ink smear on f. 83: line 5.

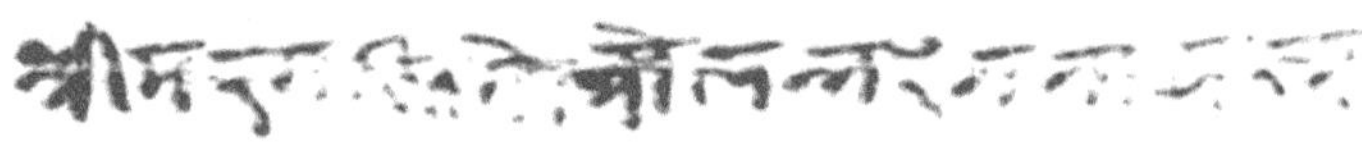

At the end of the colophon on f. 84r, to the right, there appears an entry by a second hand (shown in the trace above). This reads *śrīmadagastigotrotpannaraṃganāthasya* 'Belonging to Raṅganātha born in the venerable Agasti *gotra*'.

2.2.1.2 General comments on Bn.I

The three chapters on computations (in Bn.I$_A$), on spheres (in Bn.I$_B$), and on instruments (in Bn.I$_C$) are arranged as a continuous text following the traditional format of a Sanskrit *siddhānta*. The presence of blank spaces (with justified text wrapped around them) hints at a similar structure in an intermediary hyparchetypal manuscript from which this manuscript was presumably copied. Although no diagrams or drawings are found in this manuscript, the aligned blank spaces suggest the presence of geometrical diagrams in an antecedent manuscript. As noted before, there appears to be a difference in penmanship in this manuscript: incomplete verses with lacunae in the original writing are completed by a different scribal hand. The excerpted digital trace of ff. 39r–39v below shows two different scribal hands with scattered corrections and marginal emendations, mostly, by a second hand.

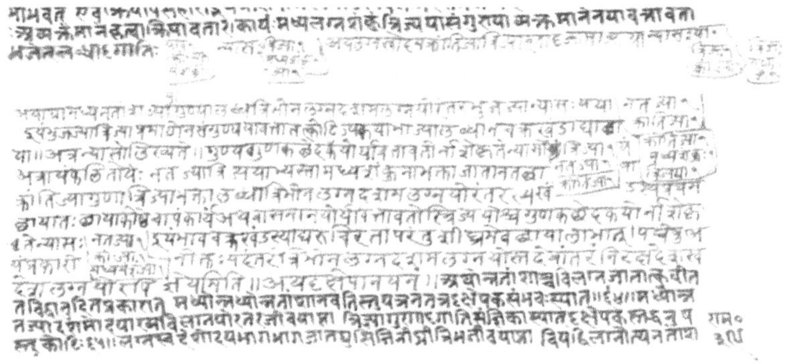

The handwriting of the manuscript is relatively clear; however, some ligatures are occasionally difficult to disambiguate. There are extended ink smears across words to indicate deletion, e.g., on f. 12r: lines 1–3, f. 22v: line 6, f. 41v: line 12, and f. 58v: line 11. Sporadically, there are some lacunae seen between the text, e.g., in  on f. 17v: line 7.

The verses of the three chapters are numbered individually, and moreover, the different sections of the chapter on computations (in Bn.I_A) also have different verse numbers. The half-verse and change-of-verse breaks use the punctuation *daṇḍa* or double-*daṇḍa* inconsistently; sometimes, the change in verse number at the end of a verse is simply appended at the end of last word of the line (*pāda*) without any punctuation mark.

On f. 84r of part Bn.I_C, we find a colophon that reads *iti śrīmiśranityānandakṛtaṃ sarvasiddhāntarājaṃ samāptaṃ* 'Thus is the complete *Sarvasiddhāntarāja* composed by Nityānanda Miśra'. This author-work identification agrees with the cataloguer's remark on the cover page of Bn.I.

2.2.1.3 *On Raṅganātha [Dīkṣita?]*

According to the colophon on f. 84r of Bn.I, as well as the cataloguer's remark on the cover page, Raṅganātha of the Agasti *gotra* is believed to be the owner of this manuscript. I suspect him to be also the primary scribe of this manuscript.

In Pingree's CESS, a Raṅganātha Dīkṣita of the Āgastya *gotra* is identified as the copyist of a manuscript of Viśvanātha's *Pātasādhanavivṛti* (ca. 1631), namely, 'Kathmandu (1965) 147 (2980)' copied on Monday 14 *kṛṣṇapakṣa* of Māgha in Saṃ. 1787 (Śāka 1652), or equivalently, 25 January 1731 CE (CESS A5 1994, p. 675a). The colophon of Bn.I states Saṃ. 1804 (1747 CE) as the date of copying, and the cataloguer's remark (on the cover page) places this manuscript in Raṅganātha's possession by Saṃ. 1860 (1804 CE). These events indicate that Raṅganātha Dīkṣita was a scribe who was active sometime between 1731 and 1804.

2.2.1.4 *Salient features of Bn.I$_B$ (ff. 67v–75v)*

In preparing a critical edition of the *golādhyāya*, I observed the following palaeographic, structural, and orthographic features of Bn.I$_B$ on ff. 67v–75v.

1 Bn.I$_B$ consists of 134 verses (including the *maṅgalācaraṇa*, sectional *ārambhas* and *śeṣas*, and the *puṣpikā* verse) written over ff. 67v–75v. The distribution of the irregularly numbered verses on the different folia is as follows:

   a) f. 67v: verses 1–4b;
   b) f. 68r: verses 4b–10d and f. 68v: verses 10d–20b;
   c) f. 69r: verses 20b–28d and f. 69v: verses 28d–38b;
   d) f. 70r: verses 38b–48c and f. 70v: verses 48c–57c;
   e) f. 71r: verses 57c–64d and f. 71v: verses 64d–72c;
   f) f. 72r: verses 72c–80d and f. 72v: verses 80d–89a;
   g) f. 73r: verses 89a–99a and f. 73v: verses 99a–105c;
   h) f. 74r: verses 105c–115b and f. 74v: verses 115b–124b;
   i) f. 75r: verses 124b–133c and f. 75v: verse 133c and the colophon (unnumbered).

2 Table C.1 (in Appendix C) lists the order of the verses in Bn.I$_B$ vis-à-vis the numbering seen in the critical edition (in Chapter 3).

3 In Bn.I$_B$, verse 103 (corresponding to verse 104 in the critical edition) conflates the words *vanidalamataḥ*◦...◦*yormukhama* from verses 109 to 112 (corresponding to verses 110c–113c in the critical edition) erroneously. This error is marked by enclosing the extra lines in parenthesis, starting at the middle of the sixth line and continuing through to the first two letters of the tenth line on f. 73v. After this contamination, the words *dyumaṇau*◦...◦*dhanurmṛgasthe* (from verse 103, second and third *pāda*s)are repeated for a second time.

   Similarly, verse 120 (corresponding to verse 121 in the critical edition) includes the words *paścima*◦...◦*nnataṃśā adha* (from verse 120, second, third, and fourth *pāda*s)twice. This appears to be a dittography following the near-similar terminal words *digaṃśkāḥ* (second *pāda*) and *adharaṃśkā* (fourth *pāda*).

4 Some of the more common orthographic features of Bn.I$_B$ include the following:

- *anusvāra*s are used as the standard notation for all allophonic nasalisation occurring at morpheme boundaries;
- *visarga*s are shown for allophonic pausa as well as appropriate case-endings of words;
- *avagraha*s are not (always) explicitly indicated, e.g., *vaḍavānalovāk* for *vaḍavānalo' vāk*;
- the morpheme boundaries are generally ligated (with the corresponding consonantal change based on euphonic rules), e.g., *udagvivardhane*;
- tri-conjuncts are written explicitly in a ligated (or sometimes, non-ligated) form with all three participating consonants, e.g., *tatstham*; the (optional) doubling of the consonants of the same class in tri-conjuncts is maintained throughout, e.g., *syāddyujyākhya*;
- the doubled consonants *tta* and *kka* have distinct ligatures;
- the conjuncts *dya* and *hna* are often written in an inverted manner as *yda* and *nha* respectively;
- *cha* is substituted for *ccha*, *ddha* for *dha*, *yya* for *ya* (when succeeding the approximant consonant *r*), and *scha* for *stha*; and
- the letters *ma*, *pa*, and *ya* appear identical in some instances.

5 The digital traces of select graphemes from Bn.I_B (ff. 67v–75v) alongside their equivalent Devanāgarī glyphs are noted below. The equivalence is based on my editorial reading of the text. The verse numbers, however, follow what is attested on the manuscript folia.

 as ह्न (verse 1 on f. 67v)

 as ड्व (verse 17 on f. 68v)

 as ष्ट्या (verse 26 on f. 69r)

 as ब्धि (verse 41 on f. 70r)

 as ट्य (verse 49 on f. 70v)

 as ष (verses 51, 56, and 59 on ff. 70v-71r)

 as ह्व (verse 59 on f. 71r)

 as ष्ण (verse 60 on f. 71r)

 as त्क (verse 67 on f. 71v)

 as र्द्ध (verse 78 on f. 72r)

 as ख (verse 92 on f. 73r)

 as खा (verse 93 on f. 73r)

 as द्व (verse 94 on f. 73r)

 as ङ्मं (verse 120 on f. 74v)

 as क्क्षे (verse 128 on f. 75r)

2.2.2 *Bn.II: Saraswati Bhawan Library, Benares* (*1963*) *37079*

Catalogue description in the CESS:

> Benares (1963) 37079 = Benares (1878) 68 = Benares (1870-1880) 9. 85 ff. Copied in Saṃ. 1895 = A.D. 1838 (in Benares (1878) said to have been copied in Saṃ. 1936 = A.D. 1879). (CESS A3 1976a, p. 174)

The cover page of the manuscript identifies the work as *Siddhāntarāja* authored by Nityānanda. Table 2.3 describes the metadata of the manuscript (as attested on the cover page). The cover page suggests two copying dates of this manuscript: Saṃ. 1895 and Saṃ. 1936. The colophon (on f. 85r) mentions *saṃvat 1895 mitī phāguṇa ṣudī 12 saṃvat 1936* '[the year] Saṃ. 1895, the 12th [*tithi*] of the bright [fortnight] of Phālguna Saṃ. 1936' (i.e., Monday, 25 February 1839 CE).

The meaning of the second date Saṃ. 1936 (1879 CE) is unclear. It is possible that the scribe copied the exact date of Saṃ. 1895 (1839 CE) from an antigraph, in which case the second date Saṃ. 1936 (1879 CE) may refer to the date of copying on this apograph. However, I take the date of copying to be Saṃ. 1895 (1839 CE) on the basis of the attested colophon.

The manuscript is written by a single scribal hand and appears complete. There are, however, few lacunae found in between the text, e.g.,  seen on f. 18r: line 9. Also, blank spaces surrounded by text are seen on various folia of this manuscript. Similar to Bn.I, the present manuscript is also continuous over 85 pages without

Table 2.3 Metadata of Bn.II

Title	~~*sarva*~~ *siddhāntarājaḥ*
Author	*nityānandaḥ*
Accession Number	2919; a second 68
Subject	*jyotiṣam*
Catalogue Number	37079 (with ~~33917~~)
Number of folia	85
Lines per page	13 (approximate)
Letters per line	39 (approximate)
Folia Size (in cm)	10.3 × 6.8
Script	Devanāgarī
Material	Paper
Completeness	Complete
Date of composition	Saṃ. 1696 (1639 CE)
Date(s) of copying	Saṃ. 1895 (1838 CE); Saṃ. 1936 (1879 CE)

any chapter breaks in the handwritten text. The presence of benedictory verses (*maṅgalācaraṇas*), inceptive and conclusive statements (*ārambhas* and *śeṣas*)for each section, colophons (*puṣpikās*), and changes in verse numbering suggest a tripartite content-based division of Bn.II, namely, parts Bn.II$_{A}$, Bn.II$_{B}$, and Bn.II$_{C}$. These are described below.

2.2.2.1 *Content-based divisions of Bn.II*

1. Bn.II$_{A}$ (68 pages, ff. 1v–68v) includes the chapter on computations (*gaṇitādhyāya*). Page numbers are written at the top-left corners of verso folia. The text appears positioned alongside reserved white spaces on some folia, presumably to accommodate geometrical diagrams; however, there are no drawings in this manuscript. The following folia contain blank spaces with text wrapped around them: ff. 14r–16r, 18v–20r, 21rv, 25v–26r. 27v–30r, 35v, 36v, 37v–39v, 40v, 47v, 49v, 53r, 54rv, 55v, 58r, 65v, 67r, and 68r. f. 26v contains just one line of text and the rest of the folio is blank; f. 29v is completely blank and devoid of any text. There are frequent marginal corrections to the text (in the form of vocalic marks and consonant ligatures) on multiple folia.
2. Bn.II$_{B}$ (8 pages, ff. 68v–76r) includes the chapter on spheres (*golādhyāya*). Page numbers are written at the top-left corners of verso folia. There are no blank spaces in the text blocks on any these pages.
3. Bn.II$_{C}$ (9 pages, ff. 76r–85r) includes the chapter on instruments (*yantrādhyāya*). Page numbers are written at the top-left corners of verso folia. Like part Bn.II$_{A}$, certain folia, namely, ff. 77r, 78r, and 84v, contain blank spaces with text wrapped around them. There are sporadic scribal corrections in the margins on multiple folia.

2.2.2.2 *General comments on Bn.II*

Like Bn.I, the three chapters on computations (in Bn.II$_{A}$), on spheres (in Bn.II$_{B}$), and on instruments (in Bn.II$_{C}$) appear as a continuous text with indented white spaces on several folia, presumably reserved for drawing diagrams. The presence of blank spaces (with justified-text wrapped around them) suggests that this manuscript was perhaps copied from an intermediary hyparchetypal manuscript. The scatter of orthographic marginal emendations indicates high acuity in revisions by the scribe. Overall, the manuscript is fairly legible; however, the presence of homoglyphs (identical looking graphemes) makes parsing the text fairly demanding even as there are no ink smears or smudges in this manuscript. All three chapters have separate verse numbering, and the different sections in the chapter on computations (in Bn.II$_{A}$) have their own verse numbers. The half-verse and change-of-verse breaks do not use the punctuation *daṇḍa*: the change in verse number at the end of a verse is simply appended at the end of last word of the line (*pāda*) without any punctuation mark.

2.2.2.3 *Salient features of Bn.II$_B$ (ff. 68v–76r)*

In preparing a critical edition of the *golādhyāya*, I observed the following palaeographic, structural, and orthographic features of Bn.II$_B$ on ff. 68v–76r.

1 Bnii$_B$ consists of 132 verses (including the *maṅgalācaraṇa*, sectional *ārambhas* and *śeṣas*, and the *puṣpikā* verse) written over ff. 68v–76r. The distribution of the irregularly numbered verses on the different folia is as follows:

   a) f. 68v: verse 1c;
   b) f. 69r: verses 1c–5b and f. 69v: verses 5b–13b;
   c) f. 70r: verses 13b–22b and f. 70v: verses 22b–31d;
   d) f. 71r: verses 32a–41c and f. 71v: verses 41c–52b;
   e) f. 72r: verses 52b–59b and f. 72v: verses 59c–68d;
   f) f. 73r: verses 68d–75c and f. 73v: verses 75c-84c;
   g) f. 74r: verses 84c–31c and f. 74v: verses 31c–105c;
   h) f. 75r: verses 105c–14c and f. 75v: verses 14c–124d;
   i) f. 76r: verses 125a–132 (colophon).

2 Table C.1 (in Appendix C) lists the order of the verses in Bn.II$_B$ vis-à-vis the numbering seen in the critical edition (in Chapter 3).

3 Some of the more common orthographic features of Bn.II$_B$ include the following:

   - *anusvāras* are used as the standard notation for all allophonic nasalisation occurring at morpheme boundaries;
   - *visargas* are shown for allophonic pausa as well as appropriate case-endings of words;
   - *virāma* for consonant-breaks in non-ligated conjuncts is occasionally omitted, e.g., *tasmina dvitīya* for *tasmin dvīya*;
   - *avagrahas* are explicitly indicated, e.g., *'mantakale'rkataḥ*;
   - tri-conjuncts are written explicitly in a ligated (or sometimes, non-ligated) form with all three participating consonants, e.g., *tatstham*; the (optional) doubling of the consonants of the same class in tri-conjuncts is omitted, e.g., *syādyujyākhya* for *syāddyujyākhya*;

- Nāgarī letters are occasionally written without their characteristic horizontal overhead line (*śirorekhā*);
- the gap-filler (*pāda-saṃdhi-cihna*) (resembling the *daṇḍa* in the Siddhamātṛkā script) is placed at the end of certain lines to indicate continuity between successive lines of the text, e.g., at the end of line 11 on f. 75v *svadina-* before *māṇaṃ* at the beginning of line 12 on f. 75v;
- the corrective mark is (generally) indicated above the letter(s) with double vertical bars, e.g., *dyu* in verse 39b on f. 71r to indicate deletion;[3]
- the vocalic mark ि for the short vowel *i* is sometimes written without its vertical stroke;
- the vocalic mark ो for the diphthong *o* replaces the mark ौ for the diphthong *au* through most of the text (e.g., *sumero* for *sumerau*);
- the doubled consonants *tta* and *kka* have indistinct ligatures compared to their singular forms *ta* and *ka* respectively;
- *tta* is replaced by *ta*; *cha* is substituted for *ccha*, *ddha* for *dha* (when succeeding the approximant consonant *r*), and *ṣṭa* for *ṣṭha*;
- the ligatures *dva* and *ddha* are written identically;
- the letters *ma*, *pa*, and *ya* appear identical in most instances; and
- the letter *ya* sometimes appears with the diacritic *nuqtā* (underdot) य़ resembling the Bengali semivowel য় (*ya* in Eastern Nāgarī script (*bāṅglā lipi*), an adaption of the Siddhamātṛkā script).

4 The digital traces of select graphemes from Bn.II$_B$ (ff. 68v–76r) alongside their equivalent Devanāgarī glyphs are noted below. The equivalence is based on my editorial reading of the text. The verse numbers, however, follow what is attested on the manuscript folia.

as श्रो (verse 2 on f. 69r)

as घ्रा (verse 11 on f. 69v)

as द्यो (verse 12 on f. 69v)

as ड्डुः (verse 18 on f. 70r)

as या (as a correction, verse 22 on f. 70r)

as ह्लो (verse 34 on f. 71r)

as त्न (verse 50 on f. 71v)

as पि (verse 18 on f. 75v)

2.2.3 Br: Bhandarakar Oriental Research Institute 206 of A, 1883–1884

Catalogue description in the CESS:

> BORI 206 of A 1883/84 47 ff. (CESS A3 1976a, p. 174)

The cover page of the manuscript identifies the work as *Siddhāntarāja* authored by Nityānanda. Table 2.4 describes the metadata of the manuscript (as attested on the cover page). The cover page suggests this manuscript was copied in Saṃ. 1941 (1884 CE). The words *siddhānta rāja prārambhaḥ nityānanda iti* 'thus the beginning of Nityānanda's *Siddhāntarāja*' are written in Devanāgarī on f. 1r, presumably, by the cataloguer. The colophon (on f. 47v) mentions *mitī baiśāṣa sudī 10 saṃvat 1941* '[the year] Saṃ. 1941, the 10th [*tithi*] of the bright [fortnight] of Vaiśākha' (i.e., Monday, 5 May 1884 CE).

The manuscript contains Nāgarī text written on hand-made paper; however, the dimensions of the folia are not clear from the digital photocopy provided by the library. All folia have the same marginal padding and each folio has fourteen lines of text on its recto and verso. There are no aligned blank spaces in this manuscript, and also, no diagrams or drawings are found on any of the folia. The handwritten text appears continuous over 47 pages without any chapter breaks. However, the presence of benedictory verses (*maṅgalācaraṇa*s), inceptive and conclusive statements (*ārambha*s and *śeṣa*s)for each section, colophons (*puṣpikā*s), and changes in verse numbering suggest a tripartite content-based division of Br, namely, parts Br_A, Br_B, and Br_C. These are described below.

2.2.3.1 Content-based divisions of Br

1 Br_A (38 pages, ff. 1v–37v) includes the chapter on computations (*gaṇitādhyāya*). Page numbers are written at the top-left and bottom-right corners of verso folia. Two marginal additions are seen, one on f. 31r (in the top margin) and the other on f. 34v (in the left margin), both indicating a missing syllable of a word.

Table 2.4 Metadata of Br

Title	*siddhāntarāja*
Author	*nityānanda*
Accession Number	206 of A·1883-84
New Number	3
Section Number	3
Number of folia	47
Date of composition	Saṃ. 1696 (1639 CE)
Date of copying	Saṃ. 1941 (1884 CE)

2 Br_B (6 pages, ff. 37v–42v) includes the chapter on spheres (*golādhyāya*). Page numbers are written at the top-left and bottom-right corners of verso folia. There are no marginal additions or (prominent) inline deletions.

3 Br_C (6 pages, ff. 42v–47v) includes the chapter on instruments (*yantrādhyāya*). Page numbers are written at the top-left and bottom-right corners of verso folia. Only one marginal addition (missing syllable in a word) is seen on f. 45r. The text ends in the middle of f. 47v with the colophon (described below). On the same folio, at the bottom-right corner, the words *ślo*[*ka*] *2232* 'verses 2232' are faintly written (with the same handwriting seen on f. 1r).

2.2.3.2 *General comments on Br*

The three chapters on computations (in Br_A), on spheres (in Br_B), and on instruments (in Br_C) appear as a continuous text with no blank spaces or word breaks. The handwriting is clear and distinct with very few corrections or emendations. The graphemes are mostly regular with a few palaeographic peculiarities. There is complete lack of any ink smears or smudges throughout this manuscript. The verses of the three chapters are numbered individually, and moreover, the different sections of the chapter on computations (in Br_A) have different verse numbers. The half-verse and change-of-verse breaks lack punctuation marks (like the *daṇḍa*); the change in verse number at the end of a verse is simply appended at the end of last word of the line (*pāda*) without any punctuation mark.

2.2.3.3 *Colophon on F. 47v*

Br_C includes an extended colophon on f. 47v that reads

> *anyoktavṛttaṃ*
>
> *ye nāma kecid iha naḥ prathayanty avajñāṃ*
> *jānanti te kim api tān prati naiṣa yatna* |
> *utpatsyate ' sti mama ko ' pi samānadharmā-*
> *kālo hy ayaṃ niravadhir vipulā ca pṛthvī* || 1 ||
>
> *mitī baiśāṣasūdī 10 saṃvat 1941 kā*[4]
>
> A metre [i.e., verse] declared by another:
>
> Whosoever by name derides me in this case, whatever they understand: this work [of mine] is not [directed] towards them. There will either be born, or [already] exists, someone whose nature is the same as mine, [for] time is indeed endless and the world vast. 1

The 10th [*tithi*] of the bright [fortnight] of Vaiśākha in Saṃ. 1941.

The consolatory verse above (in the *vasantatilakā* meter) was originally composed by the Sanskrit dramatist and poet Bhavabhūti in the prologue of his *Mālatīmādhava* in the eight century (Bhandarkar 1905, pp. 14–15). Since then, it has appeared in various anthologies on Sanskrit poetry and technical texts on Sanskrit rhetoric; for example, the *Subhāṣitaratnakoṣa* of the Buddhist scholar Vidyākara (fl. eleventh century) (Ingalls 1965, verse 1713 on p. 445) and the *Kāvyaprakāśa* of the Kashmiri rhetorician Mammaṭa Bhaṭṭa (fl. eleventh century) (Jhā [1925] 1967, verse VII.187 on p. 195). Its presence in the colophon appears to be the manuscript's scribe's dedication to Nityānanda.

2.2.3.4 Salient features of Br_B (ff. 37v–42v)

In preparing a critical edition of the *golādhyāya*, I observed the following palaeographic, structural, and orthographic features of Br_B on ff. 37v–42v.

1 Br_B consists of 132 verses (including the *maṅgalācaraṇa*, sectional *ārambhas* and *śeṣas*, and the *puṣpikā* verse) written over ff. 37v–42v. The distribution of the irregularly numbered verses on the different folia is as follows:

   a) f. 37v: verses 1a–3b;
   b) f. 38r: verses 3b–15b and f. 38v: verses 15b–29a;
   c) f. 39r: verses 29a–45a and f. 39v: verses 45a–58c;
   d) f. 40r: verses 58c–71b and f. 40v: verses 71b–84b;
   e) f. 41r: verses 84b–36c and f. 41v: verses 36c–114d;
   f) f. 42r: verses 114d–130b and f. 42v: verses 130b–132 (colophon).

2 Table C.1 (in Appendix C) lists the order of the verses in Br_B vis-à-vis the numbering seen in the critical edition (in Chapter 3).

3 Some of the more common orthographic features of Br_B include the following:

   - *anusvāras* are used as the standard notation for all allophonic nasalisation occurring at morpheme boundaries;
   - *visargas* are shown for allophonic pausa as well as appropriate case-endings of words;
   - *virāma* for consonant-breaks in non-ligated conjuncts and across *pāda* breaks is generally explicit, e.g., *ṣaṯ ṣaṯ likhya* or *syāṯ* (at the end of a *pāda*);

- *avagraha*s are explicitly indicated, e.g., *'sita sitadala*;
- the morpheme boundaries are generally ligated (with the corresponding consonantal change based on euphonic rules), e.g., *udagvivardhane*;
- the gap-filler (*pāda-saṃdhi-cihna*)  (resembling the *daṇḍa* in the Siddhamātṛkā script) is placed at the end of certain lines to indicate continuity between successive lines of the text, e.g., at the end of line 2 on f. 39r  *vyaṣṭā-* before  *śadantonmita* at the beginning of line 3 on f. 39r;
- the doubled consonants *kka*, *tta*, and *dda* have distinct ligatures; however, the doubled consonant *tta* and its singular form *ta* often appear quite similar;
- *cha* is substituted for *ccha* and *tśa*; *ddha* for *dha*, *vva*/*yya* for *va*/*ya* respectively (when succeeding the approximant consonant *r*); and *ṣṭa* for *ṣṭha*; and
- the letters *dva* and *ddha*, *sa* and *ma*, *mṛ* and *ṣṭa*, and *pa* and *ya* appear identical in some instances.

4 The digital traces of select graphemes from Br_B (ff. 37v–42v) alongside their equivalent Devanāgarī glyphs are noted below. The equivalence is based on my editorial reading of the text. The verse numbers, however, follow what is attested on the manuscript folia.

 as ooo (verse 8 on f. 38r)

 as चक्ष् (verse 10 on f. 38r)

 as ooo (verse 11 on f. 38r)

 as च्छ (verse 18 on f. 38v)

 as च (verse 21 on f. 38v)

 as घ (verse 24 on f. 38v)

 as ष्या (verse 24 on f. 38v)

 as द्यु (verse 24 on f. 38v)

 as ध्य (verse 39 on f. 39r)

 as क्षु (verse 39 on f. 39r)

 as ट्य (verse 47 on f. 39v)

 as ष्णु (verse 57 on f. 39v)

 as ष्ण (verse 58 on f. 39v)

 as हु (verse 75 on f. 40v)

 as र्द्ध (verse 76 on f. 40v)

 as ड्या (verse 30 on f. 41r)

 as र्द्ध (verse 40 on f. 41v)

2.2.4 *Np: National Archives Nepal, NAK 5.7255 (NGMCP Microfilm Reel № B 354/15)*

Catalogue description in the NGMCP (Catalogue of Indic manuscripts)[5]

Title	*Siddhāntarājagaṇita*
Language	Sanskrit
Script(s)	Devanagari
Subject(s)	Jyotiṣa
Year	–
Material(s)	not specified
Condition	–
Size	28.3 × 10.0 cm
Folio count	96
Old Record ID	64633
Microfilm	B 354/15
NAK Accession Number	5/7255
Remarks	–
MyCoRe Object ID	aaingmcp_ngmcpdocument_00061190
Last update	2015-06-22 : 04:38:23

The manuscript was microfilmed by the Nepal-German Manuscript Preservation Project at the National Archives Nepal (Kathmandu) and is catalogued by the Nepalese-German Manuscript Cataloguing Project (NGMCP 2002–16) hosted by the University of Hamburg. The microfilm reel is accessible at the Orient and East Asia Reading Room (*Orient- und Ostasienlesesaal*) of the Berlin State Library (*Staatsbibliothek zu Berlin*).

The title page of the microfilm copy identifies the work as *Siddhāntarājagaṇita* without mentioning its author. Table 2.5 describes the metadata of the manuscript (as attested on the title page of the microfilm copy). The cataloguer states the short title of this work as *Siddhāntarājagaṇita* + Komm. However, there is no commentary (*kommentar*) to the base text found of microfilmed manuscript folia.

The manuscript consists of three parts, namely, Np_{α}, Np_{β}, and Np_{γ}, each part containing a separate chapter of the work and differently foliated. All folia have the same marginal padding and each folio has nine lines of text on its recto and verso. Overall, the manuscript appears to be written by a single scribal hand and is generally complete. However, there are a few instances of lacunae within verses (indicated by the Devanāgarī head-strokes or *śirorekhā*), e.g., 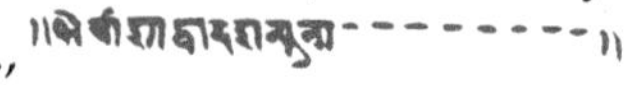 on line 2 on f. 42r of Np_{β}. There are aligned blank spaces on certain folia (e.g., f. 38v of Np_{β}) with the handwritten text surrounding it. Several folia have detailed diagrams alongside text and some folia are entirely occupied by diagrams or numerical tables. Each part has its own benedictory verse (*maṅgalācaraṇas*), inceptive and conclusive statements

Table 2.5 Metadata of Np

Short Title	*Siddhāntarājagaṇita* + Komm.
Title	*Siddhāntarājagaṇita*
Manuscript Number	5.7255
Subject	*Jyauṭīṣa* 1290
Number of leaves	96
Size in cm	28.3 × 10
Reel Number	B$\frac{354}{15}$
Date of filming	9 October 1972
Script	Devanāgarī
Remarks	Paper Ff. 1–10; 1–9; 1-77

(*ārambhas* and *śeṣas*) for each section, colophons (*puṣpikās*), and verse-numbering.

2.2.4.1 *Folio-based divisions of Np*

1 Np_{α} (10 pages, ff. 1v–10v) includes the chapter on instruments (*yantrādhyāya*). The verso folia have *si*∘ *rā*∘ *yaṃ* (with f. 1v having the extended form, *siddhāṃ*∘ *rā*∘ *yaṃ*) along with the page number (written below), both placed at the top-left corner on each folio. The word *rāmaḥ* appears along with the page number (written below) at the bottom-right corner on each folio verso. On the right side of f. 2r, a blank space occupies most of the right-half of the folio between the lines two and eight with text wrapped around it to the left. f. 6r has a few scribal corrections in the top margin.

2 Np_{β} (77 pages, ff. 1v–77v) includes the chapter on computations (*gaṇitādhyāya*). The verso folia have *si*∘ *rā* along with the page number (written below), both placed at the top-left corner of each folio. The word *rāmaḥ* appears along with the page number (written below) at the bottom-right corner on each folio verso.

A few verso folia have irregular words at their top-left corners: namely, *siddhāṃ*∘ *ja* (on f. 1v), *si*∘ *rāja* (on f. 2v), *siddhāṃ*∘ *rāja* (on f. 3v), *siddhāṃ*∘ *rā* (on f. 4v), *si*∘ *nta*∘ *rā* (on f. 74v), *si*∘ *nta*∘ *ja* (on ff. 75v and 76v), and *siddhāntarāja* (on f. 77v).

On ff. 18v, 38v, 48r, 58v, 59r, and 65v, the handwritten text appears positioned alongside reserved white spaces, presumably to draw geometrical diagrams at a later stage (although none are found). f. 38r is complete blank. There are, however, different kinds of geometrical diagrams (of varying sizes) interspersed throughout this manuscript. Most of them are surrounded by handwritten text (ff. 17r, 18r–19v, and 30v), but some of them completely occupy the face of the folio (ff. 29r–30r, 33r, 37v, and 37r).

ff. 15r–16r have Nityānanda's table of Sine values extending across three sides of two folia. The table blocks are drawn with (approximate) dimension of 14 × 4 cm on f. 15r, 15.5 × 4.5 cm on f. 15v, and 15.75 × 4.25 cm on f. 16r (with a marginal column of 0.75 × 4.25 cm extending into the left margin of the table block). The table in each table block (on separate recto and verso folia) is partitioned into two grid-like blocks one above the other. Each grid-like block has 2 rows × 30 columns in which the Sine values corresponding to the arguments are tabulated. A digital trace of the table block on f. 16r is shown above.

Some geometrical diagrams appear as explanatory drawings accompanying versified text describing various calculations and concepts. For example, the digital trace from f. 17r (shown here to the right) depicts how Sine (*jyā*) and Cosine (*koṭijyā*) of the arc-lengths (*cāpas*) can be understood in terms of the base (*koṭi*), the perpendicular (*bāhu*), and the hypotenuse (*karṇa*) of inscribed right triangles in a circle of radius equal to the sine of ninety degrees (*sinus totus* or *trijyā*).[6]

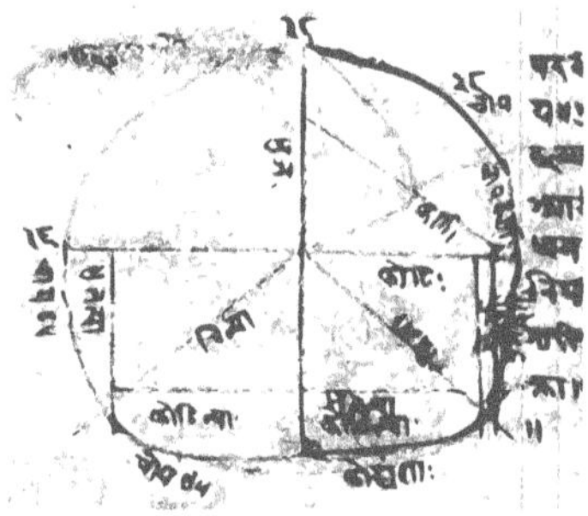

There are some diagrams where vertices are labelled (using Sanskrit alphabets) to indicate line segments and arcs, e.g., the digital trace from f. 18v shown here to the right. These geometrical elements are then referred to by their corresponding letters in the versified text accompanying these diagrams. The diagrams and texts are often placed side by side on the same folio.

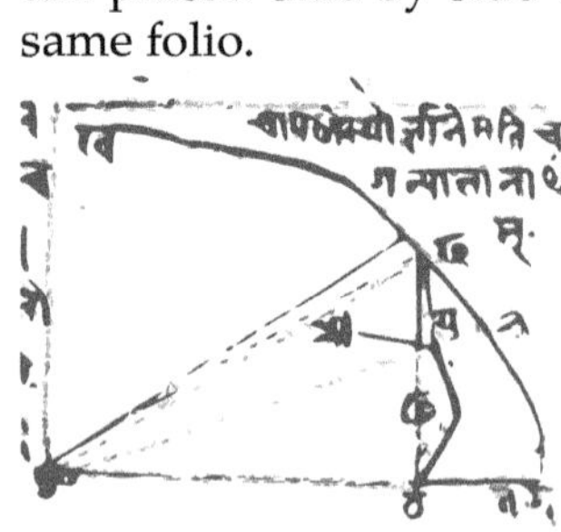

Some diagrams appear to be kinematic representations of the motion of planets, e.g., the digital trace of the diagrams on ff. 29v–30r shown below depicts the epicyclic and eccentric orbits of planets from a geocentric perspective.

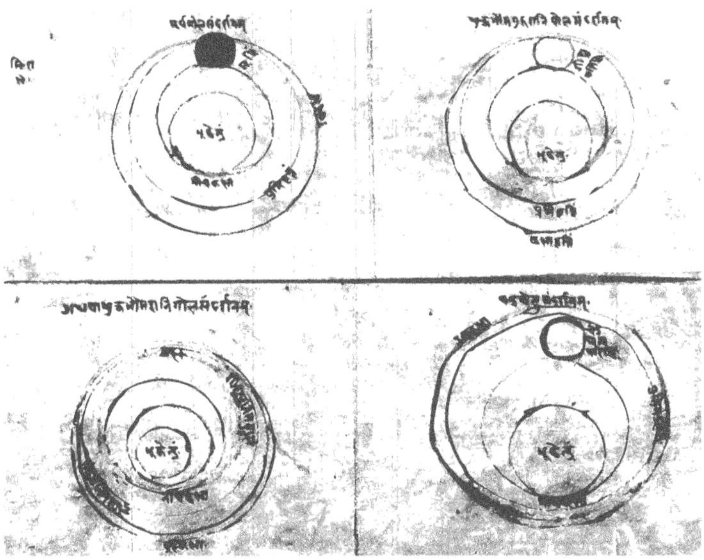

There are also figures (*bhaṅgīs*) that label the different astronomical terms in planetary models, and, at the same time, offer a visual representation of the geometry underlying the working (*ānayana*) of astronomical phenomenon, e.g., the digital trace from f. 33r seen below has two figures for explaining the working of the Sun and the planets.

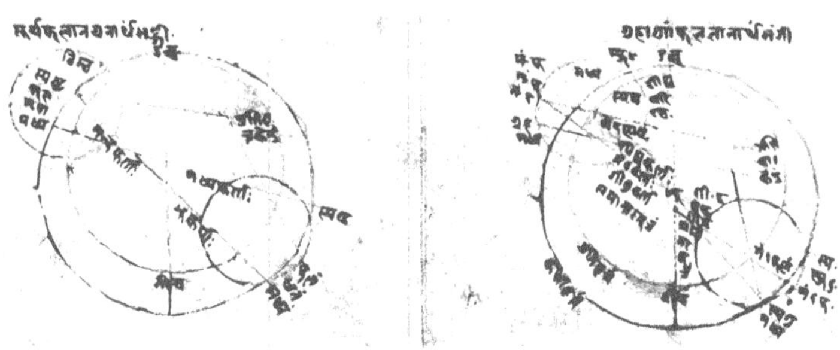

Scribal corrections and emendations are seen on a few folia of Np_{β}, mostly in the top or bottom margins (and most prominently on ff. 44v, 45rv, 49r, 50r, and 51r). Also, ff. 52v, 54r–54v, and 56v have minor marginal corrections.

3 Np$_{\gamma}$ (9 pages, ff. 1v–9v) includes the chapter on spheres (*golādhyāya*). The verso folio have *si◦ rā◦ go* (with f. 9v having *siddhā◦ rā◦ go*) along with the page number (written below), both placed at the top-left corner of each folio. The word *rāmaḥ* along with the page number (written below) appears at the bottom-right corner of each folio verso. There are no blank spaces or any marginal corrections on any these folia.

2.2.4.2 *General comments on Np*

As the three chapters on computations (in Np$_{\beta}$), on spheres (in Np$_{\gamma}$), and on instruments (in Np$_{\alpha}$) are separately foliated (with their own independent verse numbering), I suspect the appearance of the chapter on instruments (in Np$_{\alpha}$) at the beginning of the manuscript copy to be a microfilming artefact. The microfilm reel also bears duplicate images of f. 77v (from Np$_{\beta}$) and f. 9v (from Np$_{\gamma}$) concatenated together on one slide. The presence of detailed geometrical diagrams (in Np$_{\beta}$) alongside text (and on separate folia) is a notable feature of this manuscript. The overall legibility of the microfilm copy is fair; however, there are particular folia (mainly recto sides) where the writing is smudged with ink blots. Most strikingly, ff. 31v, 38v, 45r, 47r, 52r, 55r, 57r, 67r, and 68r of Np$_{\beta}$ and ff. 4r and 7r of Np$_{\gamma}$ have considerable smudging. There are also some ink smears along the margins on some folia, e.g., the top margin along f. 38v of Np$_{\beta}$.

A peculiarity of this manuscript is that the inceptive (and conclusive) statements of each section (*adhyāyas*)in a chapter are separated from the verses by visible inline white spaces, e.g., the ending of the section on the sphere of asterisms (*bhagola*) and the beginning of the section on the oblique celestial sphere (*khagola*) in the chapter on spheres (*golādhyāya*) from f. 5v of Np$_{\gamma}$ (digital trace shown below).

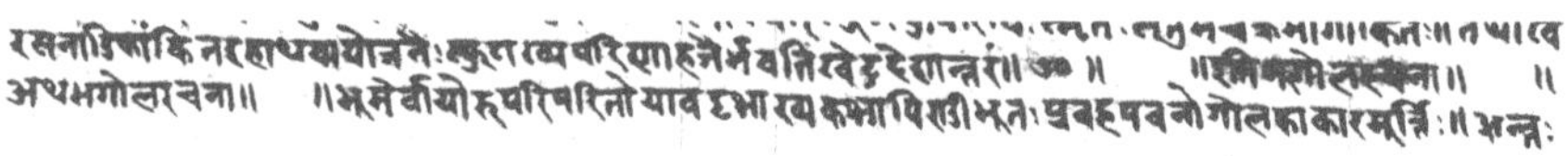

The use of || (double-*daṇḍas* or *dīrgha-virāma*) surrounding inceptive (and conclusive) statements that are then set apart from the surrounding text by indented white spaces is unique to this particular manuscript among the others included in this study.

2.2.4.3 *Colophons of Np*

Parts Np$_{\alpha}$ and Np$_{\beta}$ contain two (slightly) distinct colophons:

1 On f. 10v of Np$_{\alpha}$, the colophon reads

> *iti śrī-sakala-gaṇaka-cakra-cūḍāmaṇi-miśra-nityānanda-kṛtaḥ sarvasiddhāntarājaḥ sampurṇaḥ* || || *śubham* ||

In this manner, the *Sarvasiddhāntarāja* composed by Śrī Nityānanda Miśra, the most excellent jewel amongst all astronomers, is complete. [May all things be] auspicious.

2 On f. 77v of Np_β, the colophon reads

iti śrī-sakala-gaṇaka-śiromaṇi-marīci-nīrājita-caraṇa-kamala-yugala-śrī-devadattputra-nityānanda-viracite siddhāntarāje gaṇitādhyāyaḥ samāptaḥ || || śubhaṃ bhūyāt ||

In this manner, the chapter on computations (*gaṇitādhyāya*) in the *Siddhāntarāja* composed by Śrī Nityānanda, the son of Śrī Devadatta [whose] two lotus feet are illuminated by the radiance of the crest-jewel of all astronomers, has been completed. May all things be auspicious.

Np_γ has no colophon: it simply ends with the benediction *śubhaṃ* '[may all thing be] auspicious' on f. 9v.

The identification of Nityānanda Miśra as the author of *Sarvasiddhāntarāja* in Np_α concurs with what is stated in the colophon of Bn.I. In the colophon on f. 77v of Np_β, the honorific accorded to Nityānanda's father Devadatta as the man whose 'two lotus feet are illuminated by the radiance of the crest-jewel of all astronomers' implies all astronomers bow before Devadatta's feet in reverence of his scholarship. The construct *sakala-gaṇaka-śiromaṇi-marīci-nīrājita-caraṇa-kamala-yugala* is very similar to epithet *sakala-sāmanta-cakra-cuḍāmaṇi-marīci-mañjarī-nīrājita-caraṇa-kamal*[*a*] 'the lotus of whose feet are brightened by clustered light-beams from the crown jewels of the whole circle of his vassals' (Mishra 2016, p. 5) given to Gopāla in Kṛṣṇa Miśra's *Prabodhacandrodaya* I. 5.[7]

2.2.4.4 *Salient features of* Np_γ *(ff. 1v–9v)*

In preparing a critical edition of the *golādhyāya*, I observed the following palaeographic, structural, and orthographic features of Np_γ on ff. 1v–9v were noticed.

1 Np_γ consists of 133 verses (including the *maṅgalācaraṇa*, sectional *ārambhas* and *śeṣas*, and the *puṣpikā* verse) written over ff. 1v–9v. The distribution of the irregularly numbered verses on the different folia is as follows:

- f. 1v: verses 1a–6d;
- f. 2r: verses 6d–13d and f. 2v: verses 14a–22d;
- f. 3r: verses 22d–31c and f. 3v: verses 31c–39b;

- f. 4r: verses 39b–48c and f. 4v: verses 48c–56d;
- f. 5r: verses 56d–64a and f. 5v: verses 64a–71c;
- f. 6r: verses 71c–78a and f. 6v: verses 78a–85d;
- f. 7r: verses 86a–95a and f. 7v: verses 95a–104b;
- f. 8r: verses 104c–113a and f. 8v: verses 113a–122a;
- f. 9r: verses 122a–130d and f. 9v: verses 131a–132b and colophon (unnumbered).

2 Table C.1 (in Appendix C) lists the order of the verses in Np_{γ} vis-à-vis the numbering seen in the critical edition (in Chapter 3).

3 Some of the more common orthographic features of Np_{γ} include the following:

- *anusvāras* are used as (i) the standard notation for the allophone of /m/ at a morpheme boundary, e.g., *dina<u>ṃ</u>*; (ii) a conditioned alterant for all postvocalic nasals within morphemes when succeeded by a fricative, e.g., *a<u>ṃ</u>śa*; and, occasionally, (iii) an allophonic nasal for /ŋ/ (the guttural nasal *ṅa*), e.g., *la<u>ṃ</u>kā* for *la<u>ṅ</u>kā*. In most other cases, appropriate homorganic nasals before plosives are used, e.g., *vada<u>n</u>ti*, or *vedā<u>ṅ</u>gama*;
- the final *visargas* are often erroneously combined to the beginning of the following word using the vowel mark (*mātrā*) ो or the inherent-vowel-suppressed dental sibilant *s* (in its unligated form with a *virāma* or in a ligature);
- *avagrahas* are not explicitly indicated, e.g., *vaḍavānal<u>ovāk</u>* for *vaḍavāna<u>lo' vā</u>k*;
- *virāma* for consonant-breaks in non-ligated conjuncts and across *pāda* breaks is generally explicit, e.g., *śaśidi<u>k</u> sumeruḥ* or *vaḍavānalova<u>k</u>* (at the end of a *pāda*);
- the morpheme boundaries are generally unligated (with the *virāma*), e.g., *uda<u>k</u> vivardhane*; item the gap-filler (*pāda-saṃdhi-cihna*)  (resembling the *daṇḍa* in the Siddhamātṛkā script) is placed at the end of certain lines to indicate continuity between successive lines of the text, e.g., at the end of line 2 on f. 1v गण *gaṇa*- before कोज्ञानन् *kojānan* at the beginning of line 3 on f. 1v;
- tri-conjuncts are often written without the first consonant, e.g., *ta<u>tstha</u>ṃ* is written as *ta<u>stha</u>ṃ*;
- the doubled consonants *tta*, *dda*, *kka*, and *ccha* have distinct ligatures;

- the letters *ma*, *pa*, and *ya* appear identical in some instances; and
- the final consonant-mark for *ra* (*ra-kāra*) in some ligatures or conjunct consonants (e.g., *v<u>ra</u>*) appears different from the Nāgarī diagonal stroke (व्र) and more like its Bengali/Nepali equivalent (ड्र). For example, महेन्द्र appears as महेन्द्र.

4 The digital traces of select graphemes from Np_γ (ff. 1v–9v) alongside their equivalent Devanāgarī glyphs are noted below. The equivalence is based on my editorial reading of the text. The verse numbers, however, follow what is attested on the manuscript folia.

 as ह (verse 5 on f. 1v)	 as प्र (verse 60 on f. 5r)
 as a variant *pada-saṃdhi-cihna* (verse 51 on f. 4v)	 as ल्लं (verse 68 on f. 5v)
 as ह्न (verse 56 on f. 4v)	 as द् (verse 125 on f. 9r)

2.2.5 *Pb: Punjab University Library II 3870 (Pingree Collection Box № C7)*

Catalogue description in the SSS (SSS 1941, p. 233):

Title	*romakasiddhānta* (*golādhyāye*)
Author or commentator	–
Leaves	4
Granthas (Book)	–
Age	–
Remarks	*asam*[*pūrṇa*]
Accession Number	3870

The cover page of the manuscript identifies the work as *golādhyāya romakasiddhānta* (in Nāgarī), presumably by an earlier cataloguer. It bears the (accession?) number 1014 (in Indo-Arabic numerals) written by the same hand. Along with this, it also contains another title, accession number, and remark in (transliterated) English and in Prof. David Pingree's handwriting.

Sarvasiddhāntarāja of Nityānanda
golādhyāya [from Romakasiddhānta]
4A PUL II ~~4066~~ 3870
~~Totally, not Romakas~~
~~Romakas, but very~~
~~interesting~~

Table 2.6 Metadata of Pb

Box Number	C-7
Barcode	3 1236 07181 3805
Location/ Repository Abbreviation	PUL II
Call Number/ Shelf Mark	3870
Title	Goladhyaya
Author (if discernible)	Nityananda
Language	Sanskrit
Old Box Number	21

This manuscript copy belongs to the John Hay Library of Brown University as part of their David E. Pingree collection.[8] It was made available to me by Dr Kim Plofker who catalogued the photocopies of manuscripts in the Pingree collection of John Hay Library (as a preliminary list). Table 2.6 describes the metadata of this manuscript copy as attested in Dr Plofker's list.

The available copy of the manuscript only contains the chapter on spheres (*golādhyāya*) from Nityānanda's *Sarvasiddhāntarāja*. The Nāgarī text is continuous over four folia; however, the text stops abruptly at the beginning of verse 91a on f. 4r. The dimensions of the folia are not discernible from the digital copy. Nevertheless, each folio appears to uniformly marginated. F. 4r contains 13 lines of text; f. 1v, 19 lines; ff. 2rv and 3v, 20 lines each; and f. 3r contains 21 lines of text. There are no blank spaces in this manuscript and no diagrams or drawings are found on any of the folia. On the basis of the benedictory verse (*maṅgalācaraṇa*), inceptive and conclusive statements (*ārambha*s and *śeṣa*s)for each section, and the verse numbers beginning at one, I characterise this fragmentary manuscript as part Pb_{α}.

2.2.5.1 Description of Pb

Pb_{α} (4 pages, ff. 1v–4r) includes the chapter on spheres (*gaṇitādhyāya*). The verso folia have *golādhyā∘* along with the page number (written below), both placed at the top-left corner on each folio. Also, f. 1v has *romaka*, f. 2v has *romakaśiddhāṃta*, and f. 3v has *romakaśiddhādm∘* written in Nāgarī by a third hand (different from the handwriting of the cataloguer or the scribe) at the top-left corner, adjacent or below the page number. The page number also appears at the bottom-right corner on each folio verso. On f. 1v, the words *romakaśi*[.]*ta* appear (in a third hand) at the bottom-right corner above the page number.

The scribal penmanship is relatively clear with no ink smears or smudges. Only a single marginal correction is seen in the left margin of f. 3v. The underlined (accession?) number '1014' appears in the middle of the top margin, right above the first line of text, on f. 1v. This seems to correspond to the inceptive statement (*ārambha*) *atha golādhyāyo vyākhyāyte* also underlined (presumably by the earlier cataloguer).

2.2.5.2 *Salient features of Pb$_\alpha$ (ff. 1v–4r)*

In preparing a critical edition of the *golādhyāya*, I observed the following palaeographic, structural, and orthographic features of Pb$_\alpha$ on ff. 1v–4rv.

1 Pb$_\alpha$ consists of 90 complete verses (including the *maṅgalācaraṇa* and sectional *ārambhas* and *śeṣas*)and an incomplete verse 91 written over ff. 1v–4r. The distribution of the irregularly numbered verses on the different folia is as follows:

- f. 1v: verses 1a–12c;
- f. 2r: verses 12c–29d and f. 2v: verses 29d–48c;
- f. 3r: verses 48c–65c and f. 3v: verses 65c–80a;
- f. 4r: verses 80a–91a (incomplete).

2 Table C.1 (in Appendix C) lists the order of the verses in Pb$_\alpha$ vis-à-vis the numbering seen in the critical edition (in Chapter 3).

3 Some of the more common orthographic features of Pb$_\alpha$ include the following:

- *anusvāras* are used as the standard notation for all allophonic nasalisation occurring at morpheme boundaries;
- *visargas* are shown for allophonic pausa as well as appropriate case-endings of words;
- *avagrahas* are not explicitly indicated, e.g., *vaḍavānalovāk* for *vaḍavānalo ' vāk*;
- *virāma* for consonant-breaks in non-ligated conjuncts and across *pada* breaks is generally explicit, e.g., *ṣat̲ ṣat̲ vilikhya* or *syāt̲* (at the end of a *pada*);
- the morpheme boundaries are generally ligated (with the corresponding consonantal change based on euphonic rules), e.g., *udagvivardhane;*
- the gap-filler (*pāda-saṃdhi-cihna*) (resembling the *daṇḍa* in the Siddhamātṛkā script) is placed at the end of certain lines to indicate continuity between successive lines of the text, e.g., at the end of line 13 on f. 2r 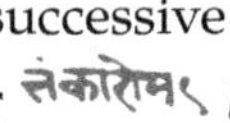 *laṃkāroma-* before 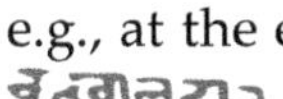 *kakhāladāta* at the beginning of line 14 on f. 2r;

- tri-conjuncts are written explicitly in a ligated form with all three participating consonants, e.g., *tatsthaṃ*; however, the (optional) doubling of the consonants of the same class in tri-conjuncts is omitted, e.g., *syādyujyākhya* for *syāddyujyākhya*;
- the doubled consonants *tta* and *kka* have distinct ligatures;
- the fricative palatal consonant *śa* is routinely replaced by its dental equivalent *sa*;
- the conjunct *bdha* is often written as *dhba* (or *dhva*);
- *cha* is substituted for *ccha*, *ddha* for *dha* (when succeeding the approximant consonant *r*), and *ṣṭa* for *ṣṭha*; and
- the letters *ma*, *pa*, and *ya* appear identical in some instances.

4 The digital traces of select graphemes from Pb_{α} (ff. 1v–4r) alongside their equivalent Devanāgarī glyphs are noted below. The equivalence is based on my editorial reading of the text. The verse numbers, however, follow what is attested on the manuscript folia.

 as ब्द्र (verse 15 on f. 2r)

 as स्यु (verse 40b on f. 2v)

 as ध्रु (verse 16b on f. 2r)

 as ऋ (verse 63a on f. 3r)

 as क्ष्म (verse 18b on f. 2r)

 as द्धा (verse 72b on f. 3v)

 as ह (verse 36a on f. 2v)

 as द्यु (verse 82b on f. 4r)

2.2.6 *Pm: Poona Mandlik Jyotisha 15/BL 368*

Catalogue description in the CESS:

> Poona, Mandlik. Jyotisha 15. 54 ff. Copied from a Jayapura manuscript, allegedly in Saṃ. 1696 = A.D. 1639 (the date of composition).[9]
>
> (CESS A5 1994, p. 184)

The original manuscript is held in the collection of Fergusson College Library in Pune (India). The digital copy provided to me identifies the work as *Sarvasiddhāntarāja* with an accession number 368, 15/BL. The cover page of this manuscript bears the cataloguer's remarks (in Marāṭhī) that reads

> *sarvasiddhāntarāja*
> *(hī prata mūla jayapūra (rājapūta saṃsthānāṃtīla) ethīla*
> *junī bastīṃtīla śahāṇāra lakṣmīnātha śāstrī drāviḍa*
> *yaṃcyā pustkāvaruna utaruna ghetalī.*

mūla bukā sarikheṃ lihileleṃ asūna tyācī patreṃ 62
āheta; darapatrācī lāṃbī 11 iṃca , ruṃdī 7½ iṃca : kāgada deśī.
kāṃhīṃ patreṃ kiḍīṃnī kharāba kelelīṃ āheta.

Sarvasiddhāntarāja
This copy of the original text has been made from the book belonging to Paṇḍita Lakṣmīnātha Śāstrī Drāviḍa [resident at] the Old Colony (Junī Bastī) in Jayapura, at the Rājaputa Saṃsthāna. As the original text is written like a book, it has 62 pages. Each page is 11 inches in length and 7½ inches in width. [Written on] country paper. Some pages have been damaged by insects.

The last page of the manuscript also bears a remark (in Marāṭhī) in the cataloguer's hand:

hī prata viśvanātha nārayaṇa maṃdalika
yāṇīṃ caitra māsa śā◦ śāke 1804 madhyeṃ
utaravilī

This copy was acquired by Viśvanātha Nārāyaṇa Maṇḍalika in Caitra month of Śālivāhana Śāka 1804 [i.e., around March–April 1882 CE].

On the same page, immediately below this remark, is the signature of Viśvanātha Maṇḍalika (a digital trace of the signature is shown to the right).

Table 2.7 describes the metadata of the manuscript (based on the cataloguer's remark and colophon).

The manuscript contains the Nāgarī text written by a single scribe and is complete. Based on the cataloguer's remark, the folia of this manuscript resemble the dimensions of a book, roughly, 27.9 × 19.1 cm. Each side of a folio contains text in two text-blocks stacked one above the other.[10] All folia have left and right bounding lines, and each text-block is ruled to include nine lines of text in Nāgarī. There are a few blank spaces in this manuscript with justified-text wrapped around them. Several folia include detailed geometrical diagrams. Overall, the manuscript is continuous over 54 pages without any chapter breaks in the handwritten text. The presence of benedictory verses (*maṅgalācaraṇa*s), inceptive and conclusive statements (*ārambha*s and *śeṣa*s)for each section, colophons (*puṣpikā*s), and changes in verse numbering suggest a tripartite content-based division of Pm, namely, parts Pm_A, Pm_B, and Pm_C. These are described below.

Table 2.7 Metadata of Pm

Library	Fergusson College Library, Pune (India)
Collection	Mandlik Jyotish Collection
Title	*sarvasiddhāntarāja*
Author	*nityānanda*
Accession Number	368
Series Number	15/BL
Number of folia	54
Number of verses	2104
Lines per page	18 (in two text-blocks of 9 lines each)
Folia Size (in inches)	11 × 7.5
Script	Devanāgarī
Material	Paper
Completeness	Complete
Date of composition	Saṃ. 1696 (1639 CE)
Date of acquisition	Śāka 1804 (1882 CE)

2.2.6.1 Content-based divisions of Pm

1 Pm_A (44 pages, ff. 1v–44r) includes the chapter on computations (*gaṇitādhyāya*). Page numbers are written at the top-left corners of verso folia in the left margin.

Marginal and interlinear additions (letters and words) are seen on several folia, sometimes indicated by an insertion or correction mark (*kākapada*), e.g., *bhūpaiḥ* in the left margin of f. 2r to be inserted at the end of verse 12, i.e., ◦*rathādirūḍhai.h*$_\wedge$ || *12* ||; or the interlinear addition *pā*$^{ṭa}_{\wedge}$ in verse 25 of f. 2r. Often, the corrections are changes to a letter presumably copied incorrectly in the first instance, e.g., ~~*sṛ*~~ *ṇaṃ*, with *ṛ* in the bottom margin on f. 7v, suggesting a correction to *ṛṇaṃ*.

Occasionally, a part of text is cancelled, e.g., on f. 34v, as seen in the digital trace .

In a few other places, parts of the text are underlined and some alternative text (comment?) on them is written below, e.g., on f. 13v, s seen in the digital trace 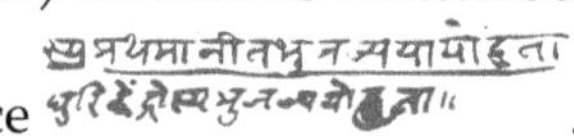.

Prominent foxing brown stains are seen on ff. 22v, 30v, 41v, and 42v. Although rare, erroneous letters are sometimes erased with white (chalk?), e.g., 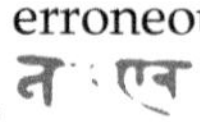 on f. 28r. There is only one instance of a correction to the text being illegibly smudged on f. 39v (in the top margin) 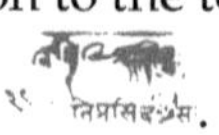.

There are underlined and semi-bracketed portion (with *bhūta-saṁkhyā* word-numerals and associated number) of ff. 12v–13r; an excerpted digital trace from f. 13r is shown here to the right.

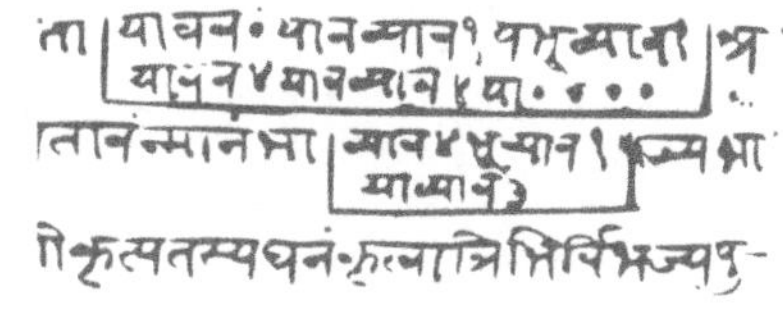

There are geometrical diagrams positioned alongside text on the following folia: ff. 9v, 10r, 10v, 11v (bottom text-block), 12r, 13v–15v, and 19v. Like the diagrams and figures found in Np_β, these geometrical diagrams are extremely detailed and serve as explanatory or representational aids in understanding astronomical models and trigonometric methods.

For example, the digital trace from f. 9v (shown to the right here) shows how Sine (*jyā*) and Cosine (*koṭijyā*) of the arc-lengths (*cāpa*s) are understood in terms of the parts of inscribed right triangles. Compare the diagram from f. 17r on Np_β on page 76.

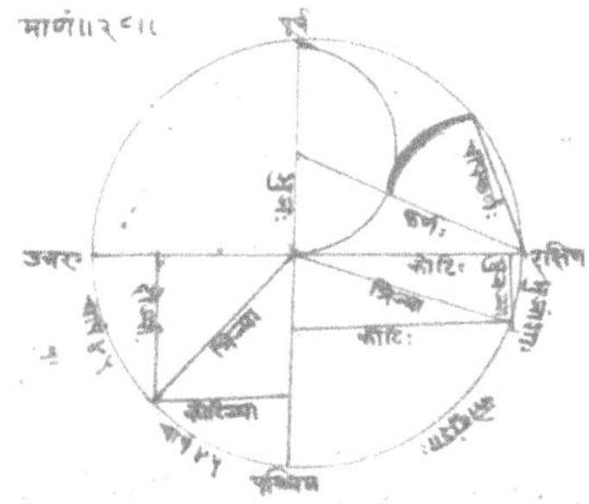

The top text-block of ff. 11v, 20r, 20v, and 21v and the bottom text-block of ff. 11r, 19v, and 21r include diagrams without any text. These diagrams are labelled and commented, often representing the geometry of the planetary model being described.

For example, the digital trace from f. 20v on the left shows the working (*ānayana*) of the Sun and the planets. Compare this with the figure from f. 33r of Np_β on page 77.

F. 18 includes incomplete diagrams, resembling kinematic representations of the motion of planets, on its recto and verso sides (with no text). A digital trace of a partially incomplete geometrical diagram from f. 21v is shown below:

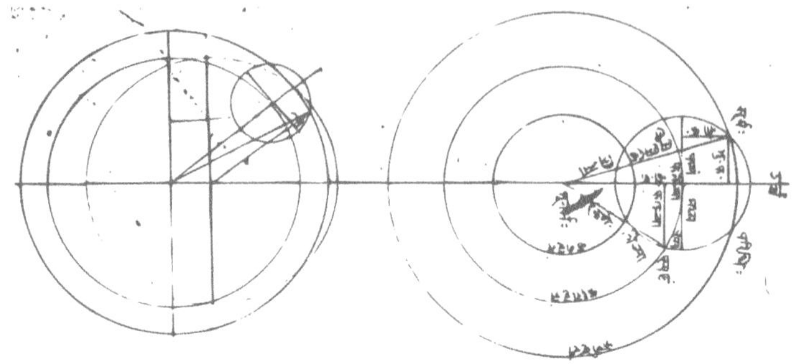

Ff. 25v–26v, 27v, 28r, 35v–37r, and 41v–43v contain blank spaces (with no diagrams or drawings) surrounded by text wrapped around them, whereas ff. 32r and 33r have vacant spaces with ruled lines in the text-block.

2 Pm_B (5 pages, ff. 44r–48v) includes the chapter on spheres (*golādhyāya*). Page numbers are written in the top-left corners of verso folia in the left margin. There are no blank spaces or any diagrams on any these folia. Like Pm_A, frequent marginal and interlinear corrections are found on all folia of Pm_B.

3 Pm_C (7 pages, ff. 48v–54v) includes the chapter on instruments (*yantrādhyāya*). Page numbers are written in the top-left corners of verso folia in the left margin (with the exception of f. 54v). There is a blank space with text wrapped around it on f. 50r (top text-block). No diagrams are found on any of these folia. Along with several marginal and interlinear corrections on different folia, ink smears are seen on ff. 49v and 53rv. The text ends in the top text-block of f. 54v with the colophon (described below). The words *saṃkhyāśloka 2104* 'number of verses (*śloka*) 2104' are written at the end of colophon (on the last line of f. 54v).

2.2.6.2 *General comments on Pm*

The three chapters on computations (in Pm_A), on spheres (in Pm_B), and on instruments (in Pm_C) appear as a continuous text with no word breaks. The handwriting is clear and distinct; however, there are several marginal and interlinear additions. These frequent corrections suggest that the text was carefully proofread by a scrupulous scribe. For the most part, the text is free of ink smudges or smears and the graphemes are highly regular with very few palaeographic peculiarities. The presence of blank spaces (with justified-text wrapped around them) hints at a similar structure in an intermediary hyparchetypal manuscript from which this manuscript was presumably copied.

In the digital copy of the manuscript, the bounding lines to the left and right sides of ff. 1v–6r are red; thereafter from f. 6v onwards, these bounding lines appear grey. The marginal corrections often seem to have been written in a lighter ink than the dark black ink with which the text is written. The ruled lines in each text-block are also very faint. The geometrical diagrams are well formed suggesting the use of a straight rule and a compass as aids, and some diagrams (on f. 14rv) are drawn in red ink.

The verses of the three chapters are numbered individually, and moreover, the different sections of the chapter on computations (in Pm_A) have different verse numbers. The half-verse and change-of-verse breaks often lack punctuation marks (like the *daṇḍa*); the change in verse number at the end of a verse is simply appended at the end of last word of the line (*pāda*).

2.2.6.3 *Colophon on f. 54v*

Pm_C includes an extended colophon on f. 54v (lines 5–7) that reads

> *anyoktaṃ vṛttaṃ || ye nama kecid iha naḥ prathayaṃty avajñāṃ jānaṃti te kim api tān prati naiṣa yatnaḥ || ~~ya~~tpatsya* *u* (in the right margin)
>
> *tesmi mama kopi samānadharmākālo hy ayaṃ niravadhir vipulā ca pṛthvī 1 || || ~~saṃpūrṇajuja graṃtha~~*
>
> *saṃkhyāśloka 2104*

This verse is identical to what is found in the colophon of Br_C (see page 71 for an English translation of this verse). The presence of this colophon in Br_C and Pm_C suggests a familial relation between these two manuscripts even though Br_C is devoid of blank spaces or diagrams. As will be seen subsequently in Section 2.2.7, this same colophon is also found in Rr_C.

2.2.6.4 *Salient features of* Pm_B *(ff. 44r–48v)*

In preparing a critical edition of the *golādhyāya*, I observed the following palaeographic, structural, and orthographic features of Pm_B on ff. 44r–48v.

1 Pm_B consists of 133 verses (including the *maṅgalācaraṇa*, sectional *ārambhas* and *śeṣas*, and the *puṣpikā* verse) written over ff. 44r–48v. The distribution of the irregularly numbered verses on the different folia is as follows:

- f. 44r: verses 1a–8b and f. 44v: verses 8b–19a;
- f. 45r: verses 19a–inceptive statement (before verse 33a) and f. 45v: verses inceptive statement (before verse 31a)–49c;

- f. 46r: verses 49c–62a and f. 46v: verses 62a–74b;
- f. 47r: verses 74c–88b and f. 47v: verses 88c–105d;
- f. 48r: verses 106a–120d and f. 48v: verses 120d–132 (colophon).

2 Table C.1 (in Appendix C) lists the order of the verses in Pm_B vis-à-vis the numbering seen in the critical edition (in Chapter 3).

3 Some of the more common orthographic features of Pm_B include the following:

- *anusvāra*s are used as the standard notation for all allophonic nasalisation occurring at morpheme boundaries;
- *visarga*s are shown for allophonic pausa as well as appropriate case-endings of words;
- *virāma* for consonant-breaks in non-ligated conjuncts and across *pāda* breaks is generally explicit, e.g., *ṣaṭ ṣaṭ vilikhya* or *syāṭ* (at the end of a *pāda*);
- *avagraha*s are explicitly indicated, e.g., *'sita sitadala*;
- the morpheme boundaries are generally ligated (with the corresponding consonantal change based on euphonic rules), e.g., *udagvivardhane*;
- the gap-filler (*pāda-saṃdhi-cihna*) ÷ is placed at the end of certain lines to indicate lexical continuity between successive lines of the text, e.g., at the end of line 3 on f. 46r  *vi-* before 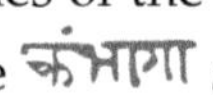 *ṣkaṃbhāgā* at the beginning of line 4 on f. 46r;
- the *anusvāra*s are often corrected with a diagonal mark striking it out, e.g.,  *kuṃ* in verse 70b on f. 46v to indicate deletion;
- the doubled consonants *kka*, *tta*, and *dda* have distinct ligatures; however, the doubled consonant *tta* and its singular form *ta* often appear quite similar;
- *cha* is substituted for *ccha*, *ddha* for *dha*, *vva*/*yya* for *va*/*ya* respectively (when succeeding the approximant consonant *r*), and *ṣṭa* for *ṣṭha*;
- the consonant conjuncts *dga*, *bdha*, and *hna* are frequently written (in reverse) as *gda*, *dhba*, and *nha* respectively; and
- the letters *dva* and *ddha*, *sa* and *ma*, *sva* and *kha*, and *pa* and *ya* appear identical in several instances.

4 The digital traces of select graphemes from Pm_B (ff. 44r–48v) alongside their equivalent Devanāgarī glyphs are noted below. The equivalence is based on my editorial reading of the text. The verse numbers, however, follow what is attested on the manuscript folia.

 as ऋ (verse 61 on f. 46r)	 as लं (verse 86 on f. 47r)
 as क्सु (verse 65 on f. 46v)	 as ऽ (verse 9 on f. 48r)
 as ग्ध्रु (verse 70 on f. 46v)	 as ध्दं (verse 126 on f. 48v)

2.2.7 *Rr: Rajasthan Oriental Research Institute (Alwar) 2619*

Catalogue description in the CESS:

> Alwar 2005. (CESS A3 1976a, p. 174)
>
> RORI (Alwar) 2619 = *Alwar 2005. 61 ff. Copied on Thursday 10 śuklapakṣa of Kārttika in Saṃ. 1903 = 29 October 1846. (CESS A5 1994, p. 184)

The cover page of the manuscript identifies the work as *Siddhāntarāja* of Nityānanda (written in transliterated Latin in black ink in the handwriting of Prof. David Pingree). It has the shelf-mark 'RORI – Alwar # 2619' (top-left corner) and 'Pingree' (top-right corner) both written in red ink (in a different hand). The (older?) accession number 2677 is stamped on the page, and the same number is written in Devanāgarī (by a cataloguer at the Rajasthan Oriental Research Institute in Alwar?).

The cover page, identified as f. 1r of the manuscript as it includes bounding lines to the left and right (similar to the rest of the manuscript), includes the statement (in the scribe's own handwriting) in the middle in Nāgarī:

> *siddhāntarājaḥ pa· 61*
> *śrīman mahārājādhirāja mahārāvarājāśrī*
> *savāī vinaya siṃha deva varmaṇāṃ pustakam adaḥ*
>
> *Siddhāntarāja* 61 pages
> This book belongs in the princely protection of the eminent king of kings, the Great King[11] Śrī Savāī Vinay Siṃha.[12]

Below the scribal incipit on f. 1r, there is an impression of a seal (*mohara*). The image to the right is its digital trace from f. 1r. While the writing on the relief of the seal is unclear, I suspect this to be the seal of the royal house (Kachwaha Rajput) of Alwar, perhaps belonging to the King Savāī Vinay Singh's government.

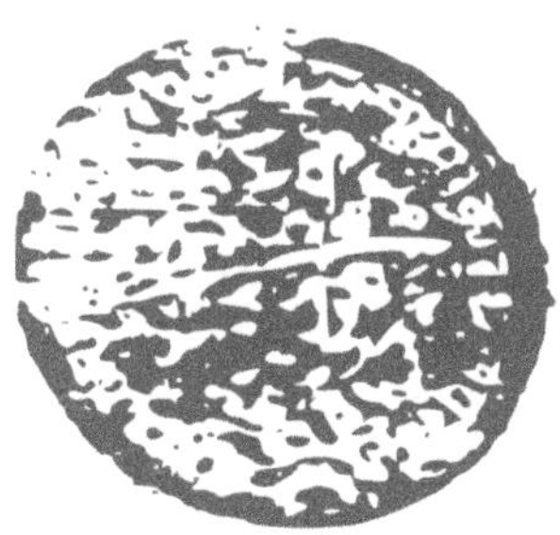

An identical seal impression is also seen on the last f. 61v of this manuscript. The image to the left is its digital trace from f. 61v. Also, the words *sarkāra alavara* 'State of Alwar' are written after the colophon (in the scribe's handwriting) towards the middle of f. 61v.

Ff. 1r and 61v also have two stamp impressions of the *rajasthāna prācya vidyā pratiṣṭhāna* (Rajasthan Oriental Research Institute). The colophon on f. 61v mentions (in vernacular) *saṃ 1903 kārttika śukla 10 guruvāsare pustaga likhī* '[this] book was written on Thursday, the 10th [*tithi*] of the bright [fortnight] of Kārttika in Saṃ. 1903'. (i.e., Thursday, 29 October 1846 CE).

Like Pb, this manuscript copy also belongs to the John Hay Library of Brown University as part of their David E. Pingree collection.[13] It was made available to me by Dr Kim Plofker who has catalogued the manuscript photocopies in the Pingree collection of John Hay Library (as a preliminary list). Table 2.8 describes the metadata of this manuscript copy as attested in Dr Plofker's list.

The manuscript contains the Nāgarī text written on hand-made paper; however, the dimensions of the folia are not clear from the digital photocopy provided by the library. All folia have equidistant left and right double bounding lines and are also padded to the top and bottom uniformly. There are fourteen lines of text on the recto and verso sides of each folio. There are no aligned blank spaces in this manuscript, and several folia have geometrical drawings with justified-text wrapped around them. A few folia are entirely filled with diagrams. Overall, the manuscript is complete and continuous over 61 pages without any chapter breaks in the handwritten text. The presence of benedictory verses (*maṅgalācaraṇa*s), inceptive and conclusive statements (*ārambha*s and *śeṣa*s)for each section, colophons (*puṣpikā*s), and changes in verse

Table 2.8 Metadata of Rr

Box Number	C-25X
Barcode	3 1236 07181 3979
Location/ Repository Abbreviation	Alwar (RORI)
Call Number/ Shelf Mark	2619
Title	Siddhantaraja
Author (if discernible)	Nityananda
Language	Sanskrit
Old Box Number	21

numbering suggest a tripartite content-based division of Rr, namely, parts Rr_A, Rr_B, and Rr_C. These are described below.

2.2.7.1 *Content-based divisions of Rr*

1 Rr_A (48 pages, ff. 1v–49v) includes the chapter on computations (*gaṇitādhyāya*). Page numbers are written in the top-left corner of verso folia. Ff. 1v, 2v, and 30v have $\overset{\textit{ramaḥ}}{1}$, $\overset{\textit{rama}}{2}$, and $\overset{\textit{ramaḥ}}{30}$ written at the bottom-right corner respectively. Marginal corrections (letters and words) are seen on several folia with a variety of in-line correction marks (placed over the letters), e.g., *ya* in the left margin on f. 3r to be corrected for $\overset{|||}{\textit{pa}}$ on line 6; or *tyanatvataḥsatyamatra* in the top margin of f. 4r to be inserted in the middle of verse 55, i.e., ◦*vyabbhicāraṃ*$_\wedge$ on line 2. The handwriting is without ink smears or smudges, although there are a few marginal ink stains seen on some folia, e.g., the bottom margin of f. 37v. There is one instance where the erroneous part of a verse is erased:  at the beginning of verse 17 on f. 24v. Also, at the beginning of verse 70 on f. 37v, a single dittographic error is identified and the repeated words are enclosed within parenthesis: .

There are underlined and semi-bracketed portion (with *bhūtasaṁkhyā* word-numerals and associated number) of f. 14rv; an excerpted digital trace from f. 14r is shown here to the right.

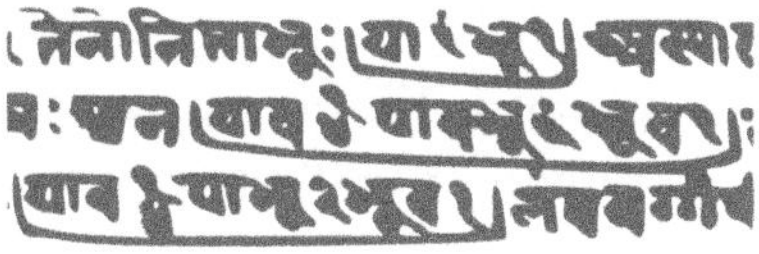

There are several different geometrical diagrams interspersed throughout this manuscript on different folia, namely, on ff. 11v–13v, 15r–17r, 20rv, 21v–23v, 27v–30v, 35v, 36v, 39r–40v, 41v, 47r, 48r, and 49r. The diagrams are mostly situated within the text-block surrounded by text wrapped around it. Occasionally, the diagrams encroach into the margins, e.g., on f. 15v. In some cases, the diagrams are completely placed in the margin, e.g., those on f. 30r. F. 23r is the only folio to include geometrical diagrams without any text.

Like the diagrams and figures seen in Np_β and Pm_A, these geometrical diagrams are extremely detailed and serve as explanatory or representational aids in understanding astronomical models and trigonometric methods.

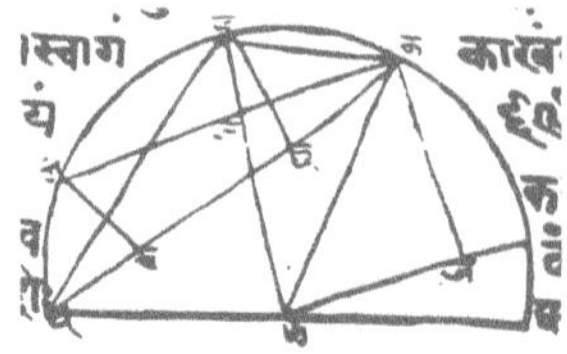

While most diagrams are labelled (using Sanskrit alphabets), there are also instances of unlabelled (or incomplete) diagrams. The digital traces from f. 13r (left) and f. 13v (right) seen above are two examples of unlabelled and labelled diagrams respectively. Like Np_β, the versified text accompanying these diagrams, often found side by side on the same folio, refers to the geometrical elements by the letters assigned to them in the diagrams.

The digital trace from f. 11v (shown here to the right) depicts how Sine (*jyā*) and Cosine (*koṭijyā*) of the arc-lengths (*cāpa*s) are to be understood in terms of the parts of inscribed right triangles, and how the distribution of measures of arc helps the calculation of Sine values. Compare the diagrams from f. 17r on Np_β on page 76 and f. 9v on Pm_B on page 91.

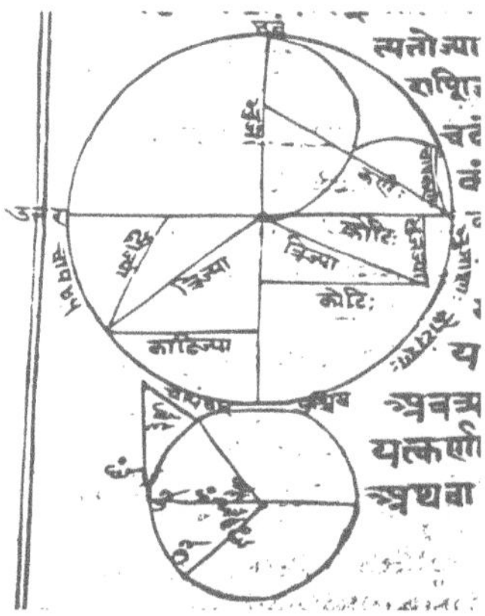

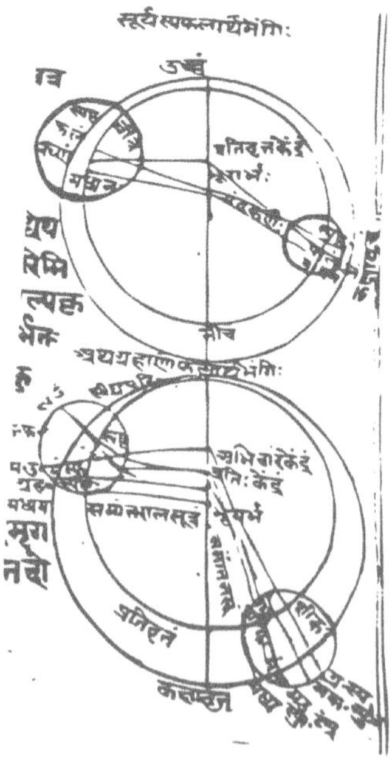

Figures (*bhaṅgī*s) showing the different astronomical terms in planetary models, and, at the same time, offering a visual representation of the geometry underlying the working (*ānayana*) of astronomical phenomenon are also seen in Rr_A; e.g., the digital trace from f. 21v (shown here to the left) has two figures for explaining the working of the Sun and the planets. These figures are sometimes incomplete or unlabelled, e.g., the ones on f. 23rv.

Throughout Rr_A, there are geometrical drawings (*chedaka*s or *chedyaka*s) representing the projection of the eccentric and epicyclic orbits of the planets, e.g., in the digital traces from ff. 15v to 16r shown below.

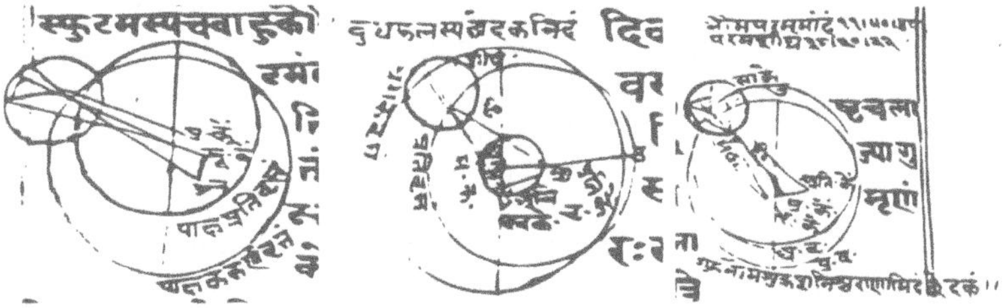

There are also diagrams depicting planetary orbits as concentric shaded spheres centred on the Earth. Different tones of ink are used to indicate the different orbital spheres in these diagrams (*gola-darśana*), e.g., in the excerpted digital trace from f. 20r seen below.

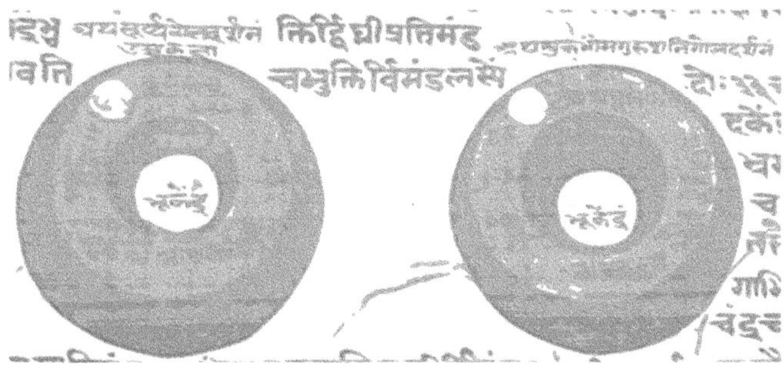

There are diagrams depicting the spherical triangles formed by intersecting great circles on the celestial sphere on ff. 28v–30v; a digital trace from f. 28v is shown here to the right.

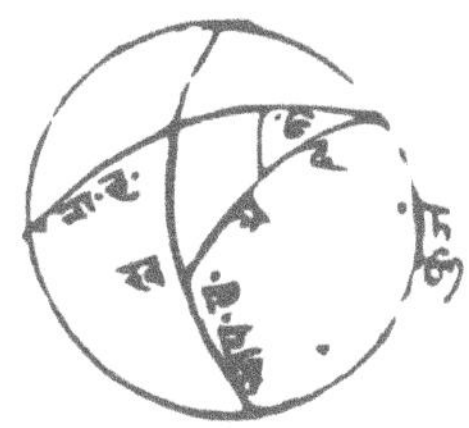

On f. 35v, we find a remarkable diagram (digital trace seen to the left) presenting the sizes of the orbits (*kakṣa*s)of the planets (measured in *yojana*s)in a geocentric model of concentric spheres extending outwards up to the sphere of constellations (*nakṣatragola*).

2 Rr_B (7 pages, ff. 49v–55r) includes the chapter on spheres (*golādhyāya*). Page numbers are written in the top-left corner of verso folia. A single marginal addition *koṇeṣu 2* is seen in the right margin on f. 53r (corresponding to line 13 of the main text). There are no blank spaces or geometrical diagrams on any these folia. Overall, the penmanship is clear and distinct with very few erasures or emendations. At the beginning of verse 34 on f. 51r, a dittographic error is identified and the repeated words are enclosed within parenthesis: 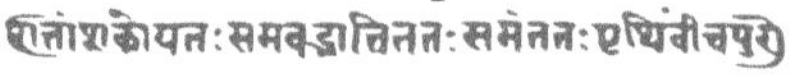.

3 Rr_C (7 pages, ff. 55r–61v) includes the chapter on instruments (*yantrādhyāya*). The verso folio have *si∘* with the page number (written below), both placed at the top-left corner of each folio. F. 56v is an exception: it only includes the page number in the top-left corner. Ff. 60v and 61v have $\overset{\textit{ramaḥ}}{60}$ and $\overset{\textit{rama}}{61}$ written at the bottom-right corner respectively. There are no blank spaces, but geometrical diagrams are found on ff. 55v, 56r, and 61r. The diagrams on ff. 55v and 56r extend into the right margins of the folia. Like Rr_A and Rr_B, the handwriting is distinct with a few interlinear corrections, e.g., in line 10 of f. 58r where 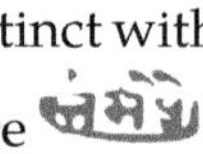 shows a string of letters being circled off, or in line 8 of f. 52v where the supra-linear corrective mark $\overset{||}{\textit{ra}}$ indicates deletion of the letter *ra*.

Rr_C is the only manuscript that includes geometrical drawings in the *yantrādhyāya*. As an example, the digital trace of the drawing from f. 62r is shown below.

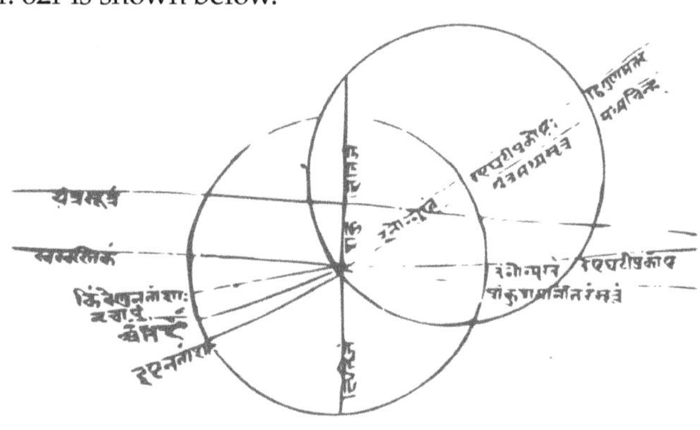

2.2.7.2 General comments on Rr

The three chapters on computations (in Rr_A), on spheres (in Rr_B), and on instruments (in Rr_C) appear as a continuous text with no word breaks. The handwriting is clear and distinct; however, there are marginal and interlinear corrections. These emendations appear to be in the same scribal hand; however, the difference in the thickness of the strokes (in the main and marginal handwriting) suggests the use of two different pens (*lekhanas*): the marginal emendations have thinner strokes compared to the main text. A digital trace of an excerpt from f. 6v is shown below.

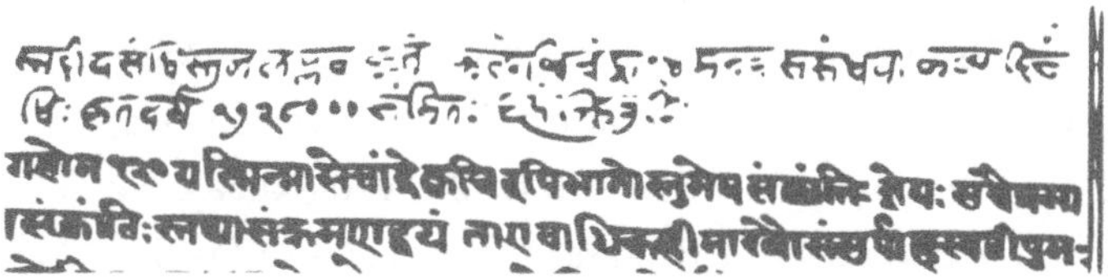

The presence of geometrical drawings alongside the text suggests that this manuscript was copied from a parent manuscript that included diagrams. Np_{β} and Pm_{A} both contain diagrams (in the chapter on computations); however, Rr_{C} is unique in containing diagrams (in the chapter on instruments).

For the most part, the text is free of ink smudges or smears and the graphemes are highly regular with very few palaeographic peculiarities. The verses of the three chapters are numbered individually, and moreover, the different sections of the chapter on computations (in Rr_{A}) have different verse numbers. The half-verse and change-of-verse breaks often lack punctuation marks (like the *daṇḍa*); the change in verse number at the end of a verse is simply appended at the end of last word of the line (*pāda*).

2.2.7.3 *Colophon on f. 61v*

Rr_{C} includes an extended colophon on f. 61v that reads

> *saṃ 1903 kārttikaśukla 10 guruvāsare pustaga likhī* ||
>
> *śubhaṃ bhavatu* ||
>
> *anyoktaṃ vṛttaṃ* || *ye nama kecid iha naḥ prathavaty avajñāṃ jānaṃti te kim api tān prati naiṣa yatnaḥ yatpatsya tesmi mama kopi samānadharmākālo hy ayaṃ niravadhir vipulā ca pṛthvī* || 1 || ||

This verse is identical to what is found in the colophon of Pm_{C} (see page 89) and Br_{C} (see page 71). The presence of this colophon in Br_{C}, Pm_{C}, and Rr_{C} suggests a familial relation between these three manuscripts even though Br_{C} is devoid of blank spaces or diagrams.

2.2.7.4 *Salient features of Rr_{B} (ff. 49v–55r)*

In preparing a critical edition of the *golādhyāya*, I observed the following palaeographic, structural, and orthographic features of Rr_{B} on ff. 49v–55r.

1 $\mathrm{Rr_B}$ consists of 132 verses (including the *maṅgalācaraṇa*, sectional *ārambhas* and *śeṣas*, and the *puṣpikā* verse) written over ff. 49v–55r. The distribution of the irregularly numbered verses on the different folia is as follows:

- f. 49v: verses 1a–4b;
- f. 50r: verses 4c–14b and f. 50v: verses 14b–25b;
- f. 51r: verses 25b–40a and f. 51v: verses 40a–54a;
- f. 52r: verses 54a–64d and f. 52v: verses 65a–75a;
- f. 53r: verses 75a–87a and f. 53v: verses 87a–37c;
- f. 54r: verses 37c–13b and f. 54v: verses 13b–127a;
- f. 55r: verses 127a–132 (colophon).

2 Table C.1 (in Appendix C) lists the order of the verses in $\mathrm{Rr_B}$ vis-à-vis the numbering seen in the critical edition (in Chapter 3).

3 Some of the more common orthographic features of $\mathrm{Rr_B}$ include the following:

- *anusvāras* are used as the standard notation for all allophonic nasalisation occurring at morpheme boundaries;
- *visargas* are shown for allophonic pausa as well as appropriate case-endings of words;
- *virāma* for consonant-breaks in non-ligated conjuncts and across *pāda* breaks is generally explicit, e.g., *ṣaṯ ṣaṯ vilikhya* or *syāṯ* (at the end of a *pāda*);
- *avagrahas* are explicitly indicated, e.g., *'rkato*;
- the morpheme boundaries are generally ligated (with the corresponding consonantal change based on euphonic rules), e.g., *udagvivardhane*;
- the gap-filler (*pāda-saṃdhi-cihna*)  (resembling the *daṇḍa* in the Siddhamātṛkā script) is placed at the end of certain lines to indicate lexical continuity between successive lines of the text, e.g., at the end of line 9 on f. 59r 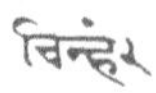 *cinhaṃ-* before 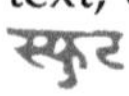 *sphuṭa* at the beginning of line 10 on f. 59r;
- tri-conjuncts are written explicitly in a ligated form with all three participating consonants, e.g., *tatsthaṃ*; however, the (optional) doubling of the consonants of the same class in tri-conjuncts is omitted, e.g., *syādyujyākhya* for *syāddyujyākhya*;
- the doubled consonants *tta* and *kka* have distinct ligatures;
- the consonant conjuncts *dga, bdha,* and *hna* are frequently written (in reverse) as *gda, dhba* (or *dhva*), and *nha* respectively;

- *ta* is substituted for *tta*, *cha* is substituted for *ccha*, *ddha* and *yya* for *dha* and *ya* respectively (when succeeding the approximant consonant *r*), and *ṣṭa* for *ṣṭha*;
- the letters *ma*, *pa*, and *ya* appear identical in some instances; and
- the final consonant-mark for *ra* (*rakāra*) in some ligatures or conjunct consonants (e.g., *p<u>ra</u>*) appears different from the Nāgarī diagonal stroke (प्र) and more like its Bengali/Nepali equivalent (প্র). For example, <u>प्रा</u>क्य appears as প্রাক্য.

4 The digital traces of select graphemes from Rr_B (ff. 49v–55r) alongside their equivalent Devanāgarī glyphs are noted below. The equivalence is based on my editorial reading of the text. The verse numbers, however, follow what is attested on the manuscript folia.

घ्रा as घ्रा (verse 11 on f. 50r)	क्प्र as क्प्र (verse 45 on f. 51v)
घो as द्यो (verse 12 on f. 50r)	द्र as द्र (verse 47 on f. 51v)
घ्रि as घ्रि (verse 20 on f. 50v)	द्वृ as द्वृ (verse 29 on f. 55r)

2.2.8 *Sc: Scindia Oriental Institute 9409*

Catalogue description in the CESS and SOI:

> SOI 9366. (CESS A3 1976a, p. 174)
>
> 905, 9409, *sarvasiddhāntarāja sāroddhāra* (SOI 1983–85, p. 27)[14]

The cover page of the manuscript identifies the work as *Sarvasiddhāntarāja Saroddhāra* authored by Nityānanda Miśra in Saṃ. 1696 (1639 CE). Table 2.9 describes the metadata of the manuscript (as attested on the cover page). The name of the scribe or the date on copying is not mentioned on the cover page. The cataloguer's remark (on the cover page) indicates pages 180, 181, 182, and 183 have been numbered on the same page.

The manuscript appears to have been completed by a secondary scribe. There is a difference in penmanship on certain folia.

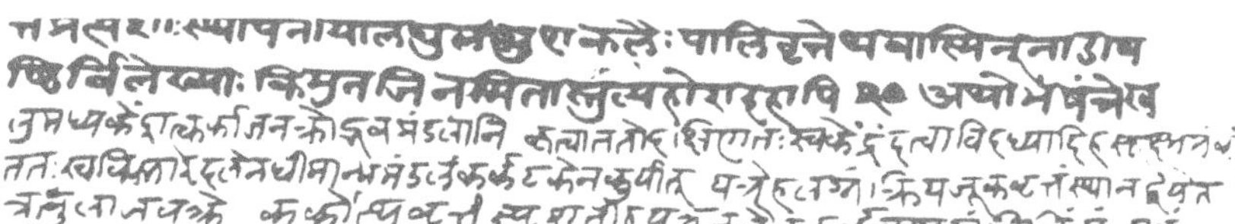

Table 2.9 Metadata of Sc

Library	The Scindia Oriental Institute, Ujjain. Manuscript Collection
Title	*sarvasiddhāntarāja sāroddhāra*
Author	*nityānandamiśra*
Accession Number	9409
Subject	*jyotiṣa*
Number of folia	193
Lines per page	7
Letters per line	31 (approximate)
Folia Size (in inches)	10.5 × 4.5
Folia Size (in cm)	26 × 11
Language	Sanskrit
Script	Devanāgarī
Condition	Excellent
Completeness	Complete
Date of composition	Saṃ. 1696 (1639 CE)

For example, the beginning *atho bhapatre kha* of verse 21 on f. 180v (excerpted digital trace shown above) is followed by *lu madhyakeṃdrā*◦... ◦*evaṃ* (in the bottom margin) in a different handwriting. Subsequently, f. 184r resumes with the words *mṛgasyāpi*◦... in the original hand.

The manuscript contains Nāgarī text written on hand-made paper with black ink. There is extensive use of yellow pigments to indicate scribal corrections (erasures) throughout this manuscript. Also, select verses (incipit and colophon), important breaks (*pada* breaks in verses and sectional breaks within chapters), and verse numbers are rubricated. Seven lines of text are written on each folio with equal marginal padding on all sides. There are a few aligned blank spaces in this manuscript with text wrapped around them; however, there are no diagrams or drawings on any of the folia. Overall, the manuscript is continuous over 193 pages without any chapter breaks in the handwritten text. Like some of the other manuscripts, the presence of benedictory verses (*maṅgalācaraṇa*s), inceptive and conclusive statements (*ārambha*s and *śeṣa*s)for each section, colophons (*puṣpikā*s), and changes in verse numbering suggest a tripartite content-based division of Sc, namely, parts Sc_A, Sc_B, and Sc_C. These are described below.

2.2.8.1 *Content-based divisions of MS Sc*

1. Sc_A (155 pages, ff. 1v–155v) includes the chapter on computations (*gaṇitādhyāya*). Page numbers are written at the top-left corners

of verso folia. The stacked numbers $\overset{9}{8}$ appear on the recto sides of ff. 1–6 and 76–132, in the top-right corner in Nāgarī. In some instances, e.g., on ff. 77r and 78r, this number appears to be written in the Moḍī script (historically used for writing Marāṭhī) as .

Marginal corrections (in the form of vocalic marks, consonant ligatures, words and sentence corrections) are seen on several folia. A variety of corrective or insertion marks are used to indicate a marginal emendation, e.g., *ñchā* in the bottom margin on f. 114v as a correction for $\underset{=}{\textit{ghṣya}}$ on line 7. Occasionally, the marginal emendations are followed by the line number on which the correction applies, e.g., *yo 4* in the right margin on f. 86r indicates the letter *yo* is to be inserted in between the words *taccāpa*$_{\wedge}$*reka*∘ on line 4. The line numbers associated with the marginal corrections are labelled symmetrically 1, 2, 3, 4, 3, 2, and 1 for the seven lines of text on each folio.

The verse numbers, (sectional) inceptive and conclusive statements, and some arithmetical operations (indicated by words like *varga*, *mūla*, etc.) are rubricated. There are frequent and extensive scribal emendations and erasures on these folia. The erasures appear to be painted over the original text (in black) with a yellow-coloured (turmeric) paste, e.g., 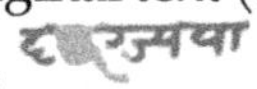 on line 1 of f. 122v. In some places, a part of original text is blackened out, e.g., on line 4 on f. 146v. F. 34r begins with two overhead bars (*śirorekhā*s) at the beginning of line 1: .

Prominent blank spaces with text wrapped around it are found on ff. 33r, 34v, 36r, 37r, 38r, 39v, 70v, 72v, 84v, 109v, 130v, and 132v. Ff. 59v–61r, 66v, and 69r–70r are completely blank and devoid of any text.

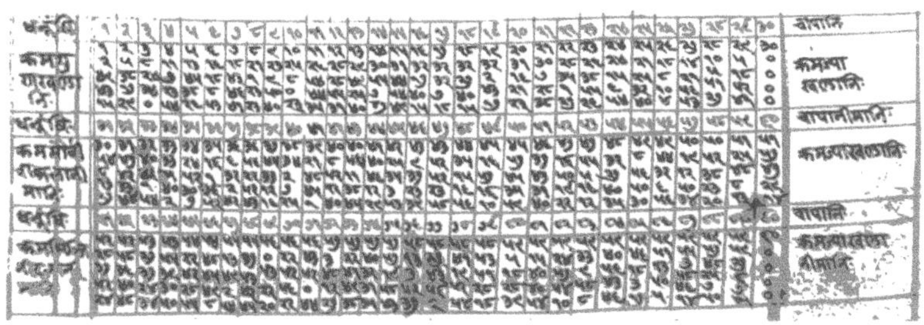

Ff. 30r has Nityānanda's table of Sine values. The table block is partitioned into three grids, placed one above the other. Each grid has 2 rows × 32 columns in which the Sine values corresponding to the arguments are tabulated. The leftmost and rightmost columns include labels for the entries tabulated in the two rows: in each grid the top entry is the argument and the bottom entry is the Sine value (up to the fifth sexagesimal place) stacked vertically. The grid rule and the arguments are written in red ink, while the Sine values and labels are in black. A digital trace of the table block on f. 16r is shown above.

2 Sc_B (20 pages, ff. 155v–174r) includes the chapter on spheres (*golādhyāya*). Page numbers are written at the top-left corners on verso folia. There are no blank spaces on any of these folia. Like Sc_A, there are frequent marginal and interlinear emendations and yellow-stained erasures. The verse numbering and sectional breaks are also rubricated for emphasis.

3 Sc_C (23 pages, ff. 174r–196r) includes the chapter on instruments (*yantrādhyāya*). Page numbers are written at the top-left corners on verso folia. F. 180v includes the folia numbers 181, 182, and 183 (written in a second hand) on the top-right corner. The bottom margin of f. 180v includes the missing text (written by the second hand) that connects verse 21a (line 7 of f. 180v) and verse 23b (unmarked, line 1 of f. 183v).[15] In the top margin of f. 180v, we find the following statement (in Sanskrit) by the secondary scribe in connection to the three ff. 181r–183v:

> *patratrayasya dvir ullekhatvād avaśiṣṭam adho ' bhilikhya patre trayāṅkam aṅkitam agre caturśīty uttara[ṃ] śatatam patre pāṭhaḥ* ||
>
> Due to the doubled description of the three pages, having written the remaining three [page] numbers below [the page number] on [this] page, the reading [continues] on the next page numbered one hundred and eighty four.

The f. 187rv appears to have been dyed yellow before writing; the foreground text (in black), however, is clearly legible. To the left and right of ff. 187r–188v there are red bounding lines demarcating the text from the margins. Overall, there are no blank spaces on any of these folia. Like Sc_A and Sc_B, there are frequent marginal and interlinear emendations and yellow-stained erasures. The verse numbering and sectional breaks are also rubricated for emphasis.

2.2.8.2 *General comments on Sc*

The arrangement of the three chapters on computations (in Sc_A), on spheres (in Sc_B), and on instruments (in Sc_C) as a continuous text with no word breaks. The peripheral edges of some folia are slightly frayed and have water stains; however, the damage does not affect the handwritten text. Occasional ink smudges can be seen on a few folia, e.g., in the top margin of f. 190v, but in general the penmanship of the main scribe is clean, distinct, and legible. The presence of copious marginal and interlinear corrections, scribal emendations and erasures, rubricated text for emphasis, and statement of omission of repeated text (on f. 180v) confirms the text was carefully proofread by a scrupulous scribe. While there are no diagrams in this manuscript, the blank spaces (with justified-text wrapped around them) suggest this manuscript was perhaps copied from an intermediary hyparchetypal manuscript that had diagrams (or similar blank spaces for drawing them).

The verses of the three chapters are numbered individually, and moreover, the different sections of all three chapters have different verse numbers. The numbering is, however, highly irregular and inconsistent. In several instances, verse numbers are altogether absent for large sections of the text. The half-verse and change-of-verse breaks, when indicated, often lack punctuation marks (like the *daṇḍa*); the change in verse number at the end of a verse is simply appended at the end of last word of the line (*pāda*).

2.2.8.3 *Colophon on f. 54v*

Sc_C has the colophon on f. 196r that reads

> *iti śrīmiśranityānandakṛtaṃ sarvasiddhāntarājaṃ samāptaṃ ra[n]ganātha dāsitasya* ||

> Thus is the complete *Sarvasiddhāntarāja* composed by Nityānanda Miśra. Belonging to [what] is given by Raṅganātha.

The colophon suggest a relation of this manuscript to a copy belonging to Raṅganātha. If the word *dāsita* is perhaps a morphemisation of *dīkṣita* then this would mean the manuscript belongs to Raṅganātha Dīkṣita. Even as this name appears to be the same as the owner (and possibly, the primary scribe) of Bn.I (see Section 2.2.1), the palaeography of these two manuscript is quite different. The handwriting styles of the primary and secondary scribes in these two manuscripts are inter-comparably different. Despite a clear palaeography, the presence of copious orthographic and scribal errors in the text suggests the primary scribe was a novice. I suspect the manuscript was assiduously corrected by a second more experienced scribe at a later time in its transmission history.

2.2.8.4 *Salient features of* Sc_B *(ff. 155v–174r)*

In preparing a critical edition of the *golādhyāya*, I observed the following palaeographic, structural, and orthographic features of Sc_B on ff. 155v–174r.

1 Sc_B consists of 137 verses (including the *maṅgalācaraṇa*, sectional *ārambhas* and *śeṣas*, and the *puṣpikā* verse) written over ff. 155v–174r. The distribution of the irregularly numbered verses on the different folia is as follows:[16]

- f. 155v: verses 1a–<2a>;
- f. 156r: verses <2a>–4b and f. 156v: verses 4b–inceptive statement (before verse <7>);
- f. 157r: verses inceptive statement (before verse <7>)–3b and f. 157v: verses 3c–<13b>;
- f. 158r: verses <13b>–<17b> and f. 158v: verses <17b>–<20d>;
- f. 159r: verses <20d>–<24b> and f. 159v: verses <24c>–<28b>;
- f. 160r: verses <28b>–<32d> and f. 160v: verses <33a>–2c;
- f. 161r: verses 2c–6c and f. 161v: verses 6c–11a;
- f. 162r: verses 11a–<49b> and f. 162v: verses <49c>–<53b>;
- f. 163r: verses <53b>–21c and f. 163v: verses 21c–24d;
- f. 164r: verses 25a–27d and f. 164v: verses 27d–31c;
- f. 165r: verses 31d–35b and f. 165v: verses 35b–2a;
- f. 166r: verses 2a–4d and f. 166v: verses 4d–9a;
- f. 167r: verses 9a–12c and f. 167v: verses 12c–2b;
- f. 168r: verses 2b–6b and f. 168v: verses 6c–<95b>;
- f. 169r: verses <95b>–14d and f. 169v: verses 14d–19a;
- f. 170r: verses 19a–21d and f. 170v: verses <107a>–25d;
- f. 171r: verses 25d–30a and f. 171v: verses 30a–<119a>;
- f. 172r: verses <119a>–37d and f. 172v: verses 37d–42a;
- f. 173r: verses 42a–<131d> and f. 173v: verses <131d>–48c;
- f. 174r: verses 48c–colophon (unnumbered).

2 Table C.1 (in Appendix C) lists the order of the verses in Sc_B vis-à-vis the numbering seen in the critical edition (in Chapter 3).

3 Some of the more common orthographic features of Sc_B include the following:

- *anusvāras* are used as the standard notation for all allophonic nasalisation occurring at morpheme boundaries;
- *visargas* are shown for allophonic pausa as well as appropriate case-endings of words;

- *virāma* for consonant-breaks in non-ligated conjuncts and across *pāda* breaks is generally explicit, e.g., *udaḵpara* or *syāṯ* (at the end of a *pāda*);
- *avagrahas* are not explicitly indicated, e.g., *sevaṃtebhimatārtha* for *sevaṃte' bhimatārtha;*
- the morpheme boundaries are generally unligated (with the *virāma*), e.g., *udaḵvivardhane;*
- tri-conjuncts are written explicitly in a ligated form with all three participating consonants, e.g., *tatsthaṃ;*
- the doubled consonant *tta* has a distinct ligature; however, *dda, da,* and *dṛ* are mutually indistinct.
- the consonant conjuncts *dga, bdha,* and *hna* are frequently written (in reverse) as *gda, dhba* (or *dhva*), and *nha* respectively;
- *ta* is substituted for *tta, cha* is substituted for *ccha, ddha* for *dha* (when succeeding the approximant consonant *r*), and *ṣṭa* for *ṣṭha;*
- the letters *pa, gha,* and *dya,* and, again, *tha* and *ya* appear identical in some instances; and
- The vocalic mark for *ā* (in the handwriting of a second scribe) is indicated as a hook above the consonant, e.g.,  *bhaga* corrected to *bhāga* on line 7 of f. 173r.

4 The digital traces of select graphemes from Sc_B (ff. 155v–174r) alongside their equivalent Devanāgarī glyphs are noted below. The equivalence is based on my editorial reading of the text. The verse numbers, however, follow what is attested on the manuscript folia (if applicable; otherwise, they follow the verse numbering from the critical edition).

 as क्र (verse 3 on f. 157r)

 as ड्ड (verse <17> on f. 158r)

 as ड्डु (verse <20> on f. 158v)

 as ष्णो (verse 25 on f. 164r)

 as र्द्धे (verse 7 on f. 166v)

 as द (verse 7 on f. 168v)

 as द्धे (verse 21 on f. 170r)

 as ह्यं (verse 33 on f. 171v)

2.3 Stemma of the manuscript witnesses

A *stemma codicum* (stemma in short) is a diagram (formally, a cladogram) indicating the relationship between different extant manuscripts of a text. The guiding principle behind this form of textual criticism is the belief that a commonality in the pattern of errors across different manuscripts indicates a common origin. Thus, manuscripts that share a common pattern of features (e.g., errors, omissions, additions, or emendations) are presumed to derive from earlier hyparchetype manuscripts, which, in turn, are presumed to derive from an archetypal manuscript that is a progenitor closest to the original text.[17] In this section, I propose a stemma of the eight manuscript witnesses of the *golādhyāya* of Nityānanda's *Sarvasiddhāntarāja* consulted in this study, and explain its recension on the basis of my observations on these manuscripts (described in Section 2.2).

2.3.1 *Remark on manuscript labels*

In describing the different parts of the eight manuscripts in Section 2.2, I used affix-labels to indicate the content-based division (e.g., Bn.I_B or Br_B) or folio-based division (e.g., Np_γ or Pb_α) of the manuscripts. In what follows here (and in the critical edition), I simply use the manuscript sigla without affixes, e.g., Bn.I for Bn.I_B, or Np for Np_γ. Henceforth, all references to the manuscripts pertain to those parts in them that contain the *golādhyāya*.

Figure 2.1 is a cladogram that depicts an approximate genealogy of the eight manuscripts of the *golādhyāya* of Nityānanda's *Sarvasiddhāntarāja*. The manuscripts derive from an archetypal manuscript α, considered the closest to the protograph, via two intermediary branches β and γ. The branch β includes the manuscripts Bn.I, Np, Pb, and Sc (cluster 1)

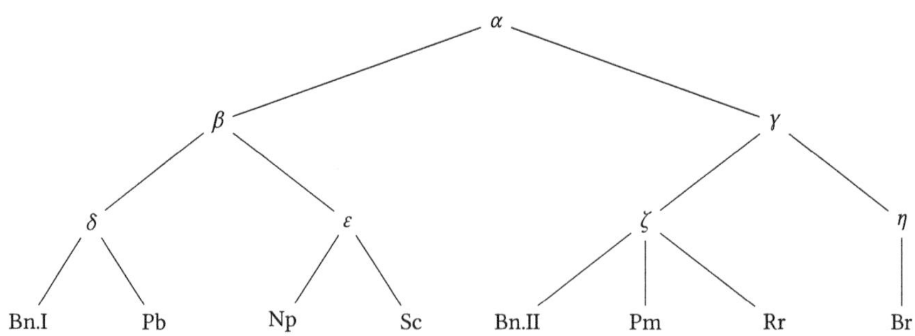

Figure 2.1 A tentative *stemma codicum* of the eight manuscripts of the *golādhyāya* of Nityānanda's *Sarvasiddhāntarāja*.

while the branch γ includes Bn.II, Br, Pm, and Rr (cluster 2). The division of the manuscripts into two branches is based on the following observations.

1 Cluster 1 with Bn.I, Np, Pb, and Sc follows a near-similar structuring of the 135 verses in the *golādhyāya* (see Table C.1 in Appendix C). While the numbering of the verses in this set of manuscripts differs, such differences are mainly due to scribal irregularities; for example, verse 31 is marked as '31' in Bn.I, whereas it is unlabelled in Np. This creates a offset of 1 between the numbering of the verses in these two manuscripts, until Bn.I erroneously (and as a duplicate) labels verse 83 as '82'. Following this, the remaining verses in these two manuscripts are identically numbered till we get to verse 129/130. Pb (incomplete) has the most regular numbering of verses until verse 90 where it terminates, whereas Sc has the most irregular verse numbering with several verses unnumbered.

   Moreover, as Table C.1 indicates, the arrangement of *pāda*s (quarters) of the stanzas in these manuscripts follows a near-identical pattern throughout the chapter. Towards the end, Bn.I arranges verses 129–133 slightly differently compared to Np and Sc; however, all three manuscripts end identically with an unnumbered colophon (*puṣpikā*). More significantly, the first variant reading of verse 73.1 is attested exclusively in this set of manuscripts, and verse 116 is only found in Bn.I, Np, and Sc.

   The orthography of the manuscripts in cluster 1 are highly regular in their use of punctuation breaks (*daṇḍa*s), and the palaeography of the ligatures and glyphs is clear and distinct. Overall, Bn.I, Np, and Pb contain the least number of prosodic and orthographic errors when compared to all other manuscripts.

2 Cluster 2 with Bn.II, Br, Pm, Rr groups the verses differently to what is proposed in the edition. Occasionally, there are aberrations in writing the verse numbers; however, the sequence and structure of the verses in these manuscripts are identical for the entire chapter. Verses 96–105 are erroneously labelled '31–40' in all four manuscripts. Also, the chapter-colophon (*puṣpikā*) in these manuscripts is numbered '132'. In contrast to the manuscripts in cluster 1, the second variant reading of verse 73.2 is found in this set of manuscripts, and also, verse 116 is conspicuously absent in these manuscripts.

   The orthography of the manuscripts in cluster 2 are similar in many respects; for example, they are missing punctuation breaks (*daṇḍa*s), identical grammatical mistakes, and repetition of doubled-consonants and vowel marks.

2.3.2 *Manuscripts derived from hyparchetype β*

The manuscripts in cluster 1 that derive from hyparchetype *β* appear to originate from two sub-branches; namely, the closely related sub-hyparchetypes *δ* and *ε*. I propose this sub-division on the basis of the following features in the manuscripts in cluster 1.

1 The chronology of the manuscripts in cluster 1 suggests Bn.I (copied in 1747) as the oldest manuscript; however, looking at the penmanship of Np, I suspect Np to be a copy from around the same period (or perhaps, slightly younger). The microfilm image of Np, however, does not indicate any date of copying. Similarly, Pb and Sc are also undated but the penmanship in these manuscripts suggests Pb is closer to Bn.I, and Sc, being the youngest among this set of four manuscripts, appears closer to Np.

2 The *apparatus criticus* (in Chapter 3) also indicates Np and Sc are closer in relation to one another than Bn.I and Pb; for example, in both Np and Sc, the compound *kāvyālaṃkṛtiśilpaśāstrakuśalo* (in verse 4) appears as *kāvyālaṃkṛtiśilpaśāstranipuṇo*; the Sanskritised name of the city of Gaṅgadujda *āgaṅgadujdādatha* (in verse 25) appears as *āgaṃgadajdādatha*; the word *kudale* (in verse 40) appears as *śakale*; or the word *dharaṇīpuṭasthāḥ* (in verse 43) appears as *pṛthivīpuṭasthāḥ*. Similarly, Bn.I and Pb also include variants with an orthography unique to these two manuscripts, e.g., *nāma* (in verse 15) appears as *vātra* and *sakale'sti* is spelled as *sakalosti*.

3 Np and Sc both includes well-marked sectional breaks; e.g., the prosaic statement *atha bhūgolabhagolakhagoladṛggolāstattatsthavastūni ca procyante* (f. 2r in Np and ff. 156v–157r in Sc) is distinctly indicated in both manuscripts.

4 Np and Sc often omit the numerals following *bhūtasaṁkhyā* or descriptive numerals, e.g., 1|15 after the word *sapādanāḍī*, lit. 60 with one-quarter (in verse 38) and 23|30 after the word *sārdhatrayoviṃśati*, lit. 23 with one-half (in verse 76) are absent in both Np and Sc.

5 There are, however, other indications that suggest that manuscripts Np and Bn.I are near-related. For example, Bn.I and Np both include a top-left marginal abbreviation of *Sarvasiddhāntarāja* (*si∘rā* in Bn.I and *si∘rā∘*go in Np) and a bottom-right marginal invocation (*rāma∘* in Bn.I and *rāmaḥ* in Np), along with the page number, *infrascripti* in both places. This suggests that the sub-hyparchetypes *δ* and *ε* were perhaps closely related to one another.

2.3.3 *Manuscripts derived from hyparchetype* γ

The manuscripts in cluster 2 that are derived from hyparchetype γ appear to originate from two sub-branches; namely, the closely related sub-hyparchetypes ζ and η. I propose this sub-division on the basis of the following features in the manuscripts in cluster 2.

1. The chronology of the manuscripts in cluster 2 suggests Bn.II (copied in 1839) is the oldest manuscript compared to Rr (copied in 1846) and Br (copied in 1884). Pm does not mention a date of copying; however, the cataloguer's remark mentions the manuscript was acquired in 1882 implying a *terminus ante quem* of 1882.

2. The *apparatus criticus* (in Chapter 3) indicates Bn.II, Pm, and Rr are closer in relation to one another than Br. The variants share a common orthography in Bn.II Pm, and Rr, for example, *tribhajyā* (in verse 11) is spelled as *tribhajyaṃ; natam* (in verse 30) is spelled as *tanaṃ; gacchantī* (in verse 59) is spelled as *gachaṃti*; or *samupaitya-hardale* (in verse 107) is spelled as *samupaitihardale.*

3. There are, however, some instances where Br shares the same variant reading as Pm, and Rr; e.g., the variant *yāvadṛkṣā-khyakakṣā* (instead of *vyomakakṣāvadhisthaḥ* in verse 72) is seen in all three manuscripts (and also Np), or the word *viṣu-vavṛttabhamaṇḍalamadhyadgaṃ* (instead of *viṣuvamaṇḍalakhecara-madhyadgaṃ* in verse 117) is seen in all four manuscripts of cluster 2. This suggests that the sub-hyparchetypes ζ and η were perhaps also closely related to one another.

Additionally, in the other part of these manuscripts that include the chapter on computations, diagrams are seen in Np, Pm, and Rr; blank spaces (with justified-text wrapped around them) are found in Bn.I, Bn.II, occasionally Pm, and Sc; and no diagrams or blank spaces are seen in Br. The distribution of manuscripts with or without diagrams/blank spaces is roughly equal, suggesting the presence of diagrams (or blank spaces) in both intermediary hyparchetype manuscripts β and γ.

2.4 Editorial notes

I discuss below the orthographic standards, transcription/transliteration schemes, and typographic conventions adopted in this book, in particular, the structure and format of the critical edition of the *golādhyāya* in Nityānanda's *Sarvasiddhāntarāja* (presented in Chapter 3).

2.4.1 *Remarks on Sanskrit orthography*

I have silently emended some common scribal oversights in the Nāgarī palaeography of the eight manuscripts of the *golādhyāya* in Nityānanda's *Sarvasiddhāntarāja*. These include:

1 the use of over-letter diacritic ◌ं (*anusvāra*) for all conjoined nasal consonants;

2 the omission of the diacritical marks ◌ः (*visarga*) for the terminal aspirate, ◌् (*virāma*) for inherent-vowel suppression, and ऽ (*avagraha*) for prodelision of *a*/*ā*;

3 the irregular use or omission of punctuation marks like the ॥ (*double-daṇḍa*);

4 the irregular use of vernacular Nāgarī letters, e.g., य़ for the Sanskrit य, and (corrected or uncorrected) ill-formed vocalic marks, e.g., णी, मी, ति etc.;

5 the irregular use of doubled consonant, e.g., द्ध in अर्द्ध or र्य्य in कुर्य्यात् (after a vowel-suppressed *r*-consonant) or across line (*pāda*) breaks in a stanza;

6 the irregular use of conjunct consonant ligatures following external euphonic rules, e.g., ग्भ or क्भ for क् + भ, ड्ढ or ट्ढ for ट् + ढ, etc.;

7 the change in consonant glyphs based on optional euphonic rules, e.g., शि changed to च्छि following the terminal *t*-consonant in an antecedent word;

8 the reversal of conjunct-consonant pairs, e.g., ध्ब/ध्व for ब्ध/व्ध, न्ह for ह्न, य्द for द्य, etc.;

9 commonly confused consonants like ब and व, प and य, म and स, ष and ख, etc., and ligatures like ष्ट for ष्ठ, त for त्त, त्व for त्त्व, त्क for त्त्क, etc.

However, those orthographic irregularities (mainly, morphosyntactic errors) that affect the reading of the text are duly noted in the *apparatus criticus*. Moreover, certain errors like haplography (inadvertent omission) and dittography (inadvertent repetition) that indicate scribal practices are also noted in the apparatus.

2.4.2 *Transcription and transliteration schemes*

I adopt the following transcription/transliteration schemes to render Arabic, Persian, and Sanskrit characters into the Roman (Latin) script.

- Arabic and Persian texts are transcribed with the EI3 standard of phonetic transcription in Brill's *Encyclopedia of Islam*, third edition (Fleet et al. 2007–20).
- Sanskrit text is transliterated following the International Alphabet of Sanskrit Transliteration (IAST) scheme. Commonly attested words of Indian origin (e.g., Hindu, Brahmin, Mughal, Varanasi) are presented without diacritics.

2.4.3 *Typographic conventions*

2.4.3.1 *Apparatus criticus in Chapter 3*

1 The manuscript witnesses are identified by their corresponding sigla; see Table 2.1.

2 Folio breaks in the edited text are indicated by a raised ⌈, slightly above the line, and the corresponding folio numbers and manuscript sigla are indicated in the margins.

3 The lemma (the edited *base text*) is separated from its variant readings (from different manuscripts) by a single right square bracket].

4 Fragments of Sanskrit compounds in both the lemma and the variant are indicated with ∘ at the breakpoint. For example,

> विघ्नौघविध्वंसकृत् (verse 1) is abbreviated as विघ्नौघ∘ in the lemma, or कोणगमण्डलं (verse 6) is abbreviated as कोणग∘ in the lemma.

An extended string of words (typically, one or more *pāda*s)is abbreviated by inserting ∘…∘ between an antecedent and subsequent syllable (*akṣara*s) in the string. For example,

> क्षितेर्गोलः क्वम्बुज्वलनपवनाकाशनिचितः (verse 7b) is abbreviated as क्षितेर्गोलः∘…∘ निचितः in the lemma.

5 Different manuscripts with the same variant reading corresponding to a lemma are separated by commas, while different variant readings corresponding to the same lemma are separated by semicolons. For example,

> नाम] वात्र Bn.I, Pb; चात्र Np, Sc (verse 15).

6 The characters (i.e., the consonant-marks in conjunct-consonants, vocalic marks, or entire syllables) or words on a manuscript that are erased or marked for deletion by the scribe(s) are indicated by enclosing them in double brackets ⟦ ⟧.[18] For example,

शैवः] ⟦शैवः⟧ शैवः ...Pb;... (verse 1)

indicates the first शैवः is a correction (dittography) by the scribe of Pb, marked for deletion.

When the consonant-marks in conjunct-consonants or the vocalic-marks are erased (by the scribe) and effectively a new character (syllable) is implied, the emended syllable is enclosed in double brackets ⟦ ⟧ followed by the corrected syllable enclosed in a pair of hyphens. For example,

निराधारोऽजस्रं] निरा⟦र्धा⟧-धा-राजस्रं ...Pm (verse 7)

indicates the consonant-mark for *r* (*repha*) in the internal syllable र्धा is erased by the scribe of Pm, and hence, the syllable is effectively emended to धा.

7 Scribal corrections/additions (vocalic marks or syllables) that are above or below a line, without the inline-characters (syllables) being erased, are simply indicated by vertically stacked syllables (as seen on a manuscript). For example,

सकला०]...स^क^ला० ...Sc (verse 3)

suggests क is written above the word सला to indicate a scribal emendation.

8 In some instances, the scribal additions (above or below a line) are explicitly indicated by a caret (*kākapada*) ˅, ˄, or ˅˄. For example,

तस्मादु०] त˅ ^स्मा^दु ० ...Pm (verse 40).

When the scribal additions are in the margins, and a *kākapada* is used to indicate the point of correction in the text, the marginal addition is enclosed in a pair of hyphens. For example,

क्रमादथ] क्रमा˅थ -द- ...Pm;...(verse 49).

9 Variant readings with marginal additions (by the scribe) across line breaks (in a manuscript) are marked by ⎿⎾, and the marginal addition is again enclosed in a pair of hyphens. For example,

०द्यदाद्यं] ०द्यम ⎿⎾ द्यं -ा- ...Bn.II,...(verse 44)

where the variant reading in Bn.II ends with the characters ०द्यम at the end of a line (on f. 71v) and begins with द्यं on the next line, and, correspondingly, has the vocalic mark ा (for *ā*) in the (right) margin at the line break.

10 In some manuscripts, the transposition of characters (syllables) in a word is indicated by over-line numbers. For example,

॰ककुभं] $\overset{२}{\text{कु}}\overset{१}{\text{क}}$भं ...Pb (verse 72).

11 In a few instances, characters (words or syllables) that I consider superfluous or erroneous are indicated by enclosing them in curly brackets { } (as an editorial emendation). For example,

मिथ्येति] {असंलग्न१} मिथ्येति ...Bn.II (verse 1).

12 Illegible characters (syllables) are indicated by [-#-], where '#' is the number of unrecognisable characters. In instances where I have proposed an editorial reconstruction of an illegible character, my proposition is enclosed in angular brackets ⟨ ⟩. For example,

स्तस्मा॰] स्त[-1-] स्मा॰ ... ⟨द्रा⟩ Rr (verse 2)

where the variant reading in Rr has one unrecognisable character in the compound word, which I propose to be द्रा.

13 Table 2.10 lists some common Latin expressions used in the apparatus with their English translations.

Table 2.10 List of common Latin expressions used in the *apparatus criticus*

add.	*addidit*, added or supplementary
alterum	the second entry of two repeated characters (numbers, syllables, or words)
corr.	*correctio*, correction
delere	*delere*, spurious text (according to a modern editor)
desunt (sing. *deest*)	missing characters (numbers, syllables, or words)
et	and
ex	from
finis imperfecta	incomplete ending
fort.	*fortasse*, perhaps (according to a modern editor)
idem	the same as something previously mentioned
ign.	*ignotus*, unknown
in linea confrac.	*in linea confractus*, at the line break
in marg. dext.	*in margine dextra*, in the right margin
in marg. inf.	*in margine inferius*, in the bottom margin
in marg. sins.	*in margine sinsitra*, in the left margin

(*Continued*)

Table 2.10 (Continued)

in marg. sup.	*in margine superius*, in the top margin
in ras.	*in rasura*, on top of an erasure
ins.	*inseruit*, inserted
intra lineam	written in line
litt.	*littera* (pl. *litterae*) letters
om.	*omisit*, omitted
prius	the first entry of two repeated characters (numbers, syllables, or words)
per errorum	by error
ras.	*rasura*, erasure
secl.	*seclusit*, secluded or bracketed text (by the original scribe)
sign.	*signum* (pl. *signa*), signs or marks
subscr.	*subscripta*, written below the line
suprascr.	*suprascripta*, written above the line
transp.	*transposuit*, transposed characters (numbers, syllables, or words)
ut videtur	as it appears (used to indicate inconsistent spellings in the manuscript witnesses)
vox (pl. *voces*)	words

2.4.3.2 *Edited Sanskrit text and its English translation in Chapter 4*

1 For syntactic clarity, any additional words (in the English translation) to the original Sanskrit text are enclosed in a pair of square brackets [ ].

2 The sexagesimal numbers are separated at the fractional break by a single-*daṇḍa* in Devanāgarī and by a semicolon in the English translations, e.g., १।१५ (in verse 34) and 1;15.

3 The verse numbers are indicated in inner margins *inter lineum* with the English translations of the corresponding Sanskrit verses.

Notes

1 An opisthograph refers to an early document, e.g., an ancient manuscript, parchment, tablet, or book, which contains writing on both its front (*recto*) and back (*verso*) sides.

2 In Sanskrit literature, the final words of a chapter that describe the subject discussed in the chapter are called *puṣpikā* (i.e., colophon). The *puṣpikā*s help identify breaks between the contents of a work and are often introduced by the signifying particle *iti* 'in this manner'—similar in function to the use of the word *explicit* 'here ends' in medieval Latin works.

3 In Sanskrit palaeography, corrective marks (*kākapadas* or *haṃsapadas*)can be indicated above, below, or in between the letters using a variety of signs, e.g.,

a caron ˅ or double vertical bars ॥ above the letter(s); a caret ∧ or insertion caret ⁁ below the letter(s); a stacked wedge ⩛ or saltire × in between letters. Scribes are also known to modify these marks in various ways, e.g., a dotted caron above the letter(s) or a dotted stacked wedge in between letters.

4 The unedited version on f. 47v of Br reads *anyokta vṛttaṃ ye nama kecid iha naḥ trathayaty avajñāṃ jānaṃti te kim api tān prati naiṣa yatna* || *utpatsyatesmi mama kopi samānadharmākālo hy aya niravadhir vipulā ca pṛthvī* 1 || *mitī baiśāṣasūdī 10 saṃvat 1941 kā*.

5 https://catalogue.ngmcp.uni-hamburg.de/receive/aaingmcp_ngmcpdocument_00061190. Last accessed on 11th June 2022.

6 Montelle, Ramasubramanian and Dhammaloka (2016) study Nityānanda's geometrical method of computing Sines in his *Sarvasiddhāntarāja* in detail. Interested reader can find a technical analysis of this figure (in conjunction with the text) for computing the Sines of ninety, thirty, and eighteen degrees on pp. 16–23 of their paper.

7 The Sanskrit philosophical drama *Prabodhacandrodaya* 'Rise of the Wisdom Moon' of Kṛṣṇa Miśra (fl. eleventh–twelfth century, Mithilā) was a widely celebrated work with several Sanskrit commentaries written on it. It promulgated the philosophy of Advaita Vedānta (mixed with the doctrine of devotion or *bhakti* towards Lord Viṣṇu) via the allegorical story of Gopāla asking for the *Prabodhacandrodaya* to be performed after defeating the Cedis and restoring the great Candella King Kīrtivarman (r. 1060–1100) to power (Mishra 2016, pp. xvii–xxxii). On the basis of this famous epithetic address to Gopāla by the narrator, Gopāla has been differently argued to be the supreme Lord Viṣṇu or, more conceivably, a great general under the service of the King Kīrtivarman (R. K. Dikshit 1976, pp. 112–114).

8 https://library.brown.edu/collatoz/info.php?id=302. Last accessed on 11th June 2022.

9 The cataloguer's remark on the cover page mentions a Jayapura manuscript from which the present manuscript was copied; however, there is no mention of any date of copying.

10 Each text-block is captured as a separate image in the digital copy of the manuscript provided to me by the Fergusson College Library. For the critical edition, I have considered two consecutive images as constituting a recto or verso side of a folio.

11 Mahārāva (or Mahārāo) *rājā* was the title of the Kings of Alwar until April 1889.

12 Savāī Vinay Singh was the third king of the princely state of Alwar from 1815 to 1857.

13 https://library.brown.edu/collatoz/info.php?id=302. Last accessed on 11th June 2022.

14 While CESS catalogues the manuscript of Nityānanda's *Sarvasiddhāntarāja* from the Scindia Oriental Institute (SOI) at Vikram University in Ujjain as 'SOI 9366', the SOI catalogue registers '9366' as the accession number for (a part of) another work, the *Siddhāntasundara* (*bījagaṇitādhyāya*) of Jñānarāja composed in 1503. The Sanskrit manuscript collection at the Scindia Oriental Institute only includes one copy of Nityānanda's *Sarvasiddhāntarāja*, and hence I suspect the reference number in CESS to be in error.

15 I have verified the continuity of the text by cross-referencing other manuscripts that contain the *yantrādhyāya*, e.g., Rr_C: f. 57rv, verses 54–56 and Np_α: f. 4v, verses 18–20.

16 For an unnumbered verse at a folio break, a corresponding verse number from the critical edition is indicated in angle brackets.

17 See Trovato (2014) for an excellent overview of the history of stemmatics as a method of textual criticism.

18 Typically, scribes indicate emendations by using correction marks (*signe de renvoi*) and erasures by applying coloured pastes—white (*saphedā*) or yellow (*haratāla*) in colour—over the characters. In some instances, corrections/deletions are indicated with strikethrough lines over the characters.

3 Critical Edition

f. 67v Bn.I
f. 68v Bn.II
f. 37v Br
f. 1v Np
f. 1v Pb
f. 44r Pm
f. 49v Rr
f. 155v Sc

⌜अथ गोलाध्यायो व्याख्यायते ।

तं वन्दे गणनायकं सुरगणा यत्पादपद्मद्वयं
सेवन्तेऽभिमतार्थसाधनविधौ विघ्नौघविध्वंसकृत् ।
यं ⌜सांख्याः पुरुषं प्रधानमपि वा शैवाः शिवं मेनिरे f. 69r Bn.II
नित्यं ब्रह्म चिदात्म सर्वमपरं मिथ्येति वेदान्तिनः ॥ १ ॥

यत्किंचिद्गणितं ग्रहस्य गण⌜को जानन्विना वासनां f. 156r Sc
पृष्टः सन्नपरेण गोलविदुषा प्रश्नप्रपञ्चोक्तिभिः ।
किं ब्रूते प्रति तं तदुत्तरमयं गोलागमाज्ञो यत-
स्तस्माद्गोलविचारचारुरचनां वक्ष्ये सतां प्रीतये ॥ २ ॥

---

1 अथ गोलाध्यायो व्याख्यायते] इतिपूर्ववृत्तिः Bn.I, Br, Sc; इतिर्ववृति Bn.II, Rr; श्रीगुरुगणेशाभ्यान्नमः Np; इतिसर्ववृत्तिः Pm
2 यत्पाद॰] यत्पत्द॰ Br
3 विघ्नौघ॰] विघ्नोच्च॰ Bn.II, Pm, Rr
4 शैवाः] ⟦शैवाः⟧शैवाः *corr. ras.* Pb; शेवा Br
5 ब्रह्म] ब्रह्मपरं Sc
5 चिदात्म सर्वम॰] चिदात्वर्वम॰ Bn.I; चिदात्ससर्वम॰ Pb; चिदात्म⟦सर्व⟧म॰ *corr. ras.* Sc
5 मिथ्येति] {असंलग्न१}मिथ्येति *intra lineam, delere vox et ign. sign.* Bn.II
5 वेदा॰] वेद॰ Pb
6 यत्किं॰] य⟦ा⟧त्किं॰ *corr. ras.* Pb
6 ॰चिद्गणितं] ॰चिहुणितं Bn.I, Bn.II
6 जानन्विना] जानान्विता Bn.II, Rr; ⟦जान⟧जानन्विना *corr. ras.* Pb
7 ॰विदुषा] ॰विडषा Pm
8 गोला॰] गो{॥}ला॰ *intra lineam, delere ign. litt.* Bn.II
9 स्तस्मा॰] स्त[-1-]स्मा॰ *intra lineam, fort.* ⟨द्गा⟩ Rr
9 ॰चारुरचनां] ॰चारुरुरचनां Bn.II
9 सतां] शतां Pb
9 २] १ Bn.II, Br, Pm, Rr; *om.* Sc

DOI: 10.4324/9780429506680-3

रात्रौ सद्मविचित्रचित्ररचनं दीपं विना किं यथा
त्रैलोक्यं सकलार्थपूर्णमपि किं भास्वत्प्रकाशं विना ।
वैदग्ध्येन विना प्रगल्भतरुणी सौन्दर्ययुक्तापि किं
तद्वद्गोलविवर्जितो गणितवित्प्रश्नोत्तरे किं विभुः ॥ ३ ॥

[a] f. 68r Bn.I [b] f. 38r Br [c] f. 156v Sc
पाटीकुट्टकबीजगोलनिपुणो वेदाङ्गपारङ्गमः
काव्यालंकृति⌈[a]शिल्पशास्त्रकुशलो यो न्यायशास्त्रादिवि⌈[b, c]त् ।
सिद्धान्तेऽत्र समस्तवस्तुसहिते तस्याधिकारो भवे-
च्चेदेवं न यथा कथंचिदभिधामात्रं प्रसिद्धिं नयेत् ॥ ४ ॥

कीदृग्भूगोलसंस्था कथय कथमहो चक्रसंज्ञो भगोलः
कीदृग्भूयः खगोलः स्थिर इह सततं कीदृशो दृष्टिगोलः ।
f. 50r Rr ⌈तत्तत्स्थं वस्तु कीदृक्सकलमपि यथा भूमिगोलेऽब्धिशैल-
द्वीपाद्यं तद्यथा वा विषुववृतिमुखं चक्रसंज्ञे भगोले ॥ ५ ॥

पूर्वापरं दक्षिणसौम्यवृत्तं भूजाभिधं कोणगमण्डलं च ।
f. 2r Np उन्मण्डलाद्यं च खगोलसंस्थं तथैव दृग्गोलगतं ⌈यथा वा ॥ ६ ॥

---

1 सद्म॰] सप्र॰ Pm
1 ॰चित्ररचनं] ॰रचनं Bn.II
2 सकला॰] सकल॰⌟-ा- *add. in marg. dext. in linea confrac.* Bn.II; स(क)ला॰ *add. suprascr.* Sc
3 वैदग्ध्येन] वैदग्धेन Np, Pb, Sc
4 तद्वद्गोल॰] तद्वगोल॰ Pm
4 गणित॰] गणि॰ Pb
4 ॰श्नोत्तरे] ॰श्नोत्तत्तरे Br
4 ३] २ Bn.II, Br, Pm, Rr
5 ॰कुट्टकबीज॰] ॰कुट्टकविज॰ Pm
5 वेदाङ्ग॰] वेदांत॰ Bn.II, Br, Pm, Rr
6 ॰कृति॰] ॰कृत॰ Bn.II, Rr
6 ॰शिल्प॰] ॰शल्प॰ Bn.II, Rr
6 ॰कुशलो] ॰निपुणो Np, Sc
6 ॰शास्त्रादिवित्] ॰शास्त्रार्थवित् Np, Sc
7 तस्या॰] तस्य॰ Bn.I
7 ॰धिकारो] ॰धिका⟦लो⟧रो *corr. ras.* Bn.II
8 च्चेदेवं] च्चेदेयं Pm
8 यथा] (य)कथा *corr. suprascr.* Sc
8 नयेत्] न्ययेत् Pb
8 ४] ३ Bn.II, Br, Pm, Rr
9 कथय] *om.* Bn.II; कथ Br, Pm, Rr; कथय⟦कथय⟧ *corr. ras.* Sc
9 भगोलः] भगोला Sc
10 कीदृग्भूयः] कादोभूयः Sc
11 तत्तत्स्थं] तत्तहथं Br; तत्रस्थं Sc
11 भूमिगोले॰] भूमिगोले⟦ा⟧॰ *corr. ras.* Pb; भूगोले॰ Rr
12 ॰मुखं] ॰सुखं Br
12 चक्रसंज्ञे] चक्रसं Bn.I
12 ५] ४ Bn.II, Br, Pm, Rr
13 पूर्वापरं] पूर्वावरं Br
13 ॰सौम्यवृत्तं] ॰सौम्यत्तं Br
13 कोणग॰] गोजग॰ Sc
14 खगोल॰] स्वगोल॰ Br
14 ॰संस्थं] ॰मंस्थं Br; ॰संज्ञं Np
14 दृग्गोल॰] दृगगेल॰ Bn.II
14 ॰गतं] ॰तं Br
14 वा] च Np, Rr
14 ६] *om.* Bn.II, Rr; ५ Br, Pm

अथ भूगोलभगोलखगोलदृग्गोलास्तत्तत्स्थवस्तूनि च ⌈प्रोच्यन्ते । f. 157r Sc

निराधारोऽजस्रं भवति निजशक्त्या स्थिरतरः
क्षितेर्गोलः क्वम्बुज्वलनप⌈वनाकाशनिचितः । f. 69v Bn.II
यथाऽयोगोलोऽन्तर्वियति परितश्चुंबकमये
गृहे ग्रामरामाचलजलधिदेवासुरनृभृत् ॥ ७ ॥

उत्तुङ्गं शैलशृङ्गं खगुण ३० परिमितैर्योजनैर्दृश्यते चे-
दर्वागेवाध्वमध्ये कथय ⌈च कियती शैलशृङ्गोन्नतिः सा । f. 44v Pm
किं वा शृङ्गोच्छ्रयो मे भवति सविदितो दृश्यते योजनैः कै-
रेवं गोलाकृतिश्चेद्भवति वसुमती त्वन्मते मां प्रचक्ष्व ॥ ८ ॥

तदुत्तरस्य गणितम् ।

नृपादशैलान्तरमार्गयोजनै ३० र्भचक्रभागा ३६० गुणिता हृताः पुनः ।
⌈वसुंधरायाः परिणाहयोजनै ६००० स्तदाप्तचापं विवरांशका मताः ॥ ९ ॥ f. 157v Sc

---

1 भूगोलभगोल॰] भूगोल॰ Rr
1 ॰दृग्गोला॰] ॰दग्गोला॰ Br; ॰दृगोल॰ Pm
1 ॰त्तत्स्थ॰] ॰त्तस्थ॰ Bn.I, Pm; ॰त्ततूथ॰ Rr
1 ॰वस्तूनि] ॰वस्तूनि⟦ज⟧ *corr. ras.* Bn.II
2 निराधारोऽजस्रं] निरा⟦र्धा⟧-धा-राजस्रं *corr. ras.* Pm
2 स्थिरतरः] स्थिरतरं Rr
3 क्षितेर्गोलः॰…॰निचितः ॥] *desunt* Pb
3 क्वम्बु॰] कंब्रु॰ Pm
3 ॰पवना॰] ॰पपवना॰ Bn.II
4 ॰श्चुंबक॰] ॰श्चुंब्यक॰ Br; ॰श्चंबक॰ Rr
5 गृहे] गृह Np; ग्रहे Pb
5 ग्राम॰] ग्रामा॰ Pm
5 ॰रामाचल॰] ॰रा(मा)चल॰ *add. suprascr.* Sc
5 ॰जल॰] ॰जुल॰ Rr
5 ॰नृभृत्] ॰नृभूत् Bn.II, Rr; ॰निभृत् Np, Sc
5 ७] ६ Bn.II, Br, Pm, Rr; *om.* Sc
6 उत्तुङ्गं] उत्तंगं⟦ा⟧ *corr. ras.* Bn.II; उत्तंगं Rr
6 ३०] *om.* Sc
6 ॰र्दृश्यते] ॰र्दश्यते Sc
7 दर्वागे॰] दर्वावो॰ Br; दर्वीवो॰ Rr
7 ॰वाध्व॰] ॰काध्व॰ Np
7 कियती] विकयती Rr
7 ॰शृङ्गोन्नतिः] ॰श्रृंगंन्नति Bn.II, Rr
8 शृङ्गोच्छ्रयोमे] श्रृंगछ्र्यो Bn.II, Rr; श्रृंगोधूयो (छ्र) *corr. suprascr.* Sc
9 रेवं] रवं Bn.II
9 ॰श्चेद्भवति] ॰श्चेद्भभवति Pb
9 वसुमती] वसुप्रती Bn.II; वसु(म)ती *add. suprascr.* Pm
9 ८] ७ Bn.II, Br, Pm, Rr; २ Sc
10 तदुत्तरस्य] तदुत्तर Np, Sc
11 नृपाद॰] नद॰ Pm
11 ॰योजनै] ॰जनै Bn.I
11 ॰भागा] ॰⟦र्भा⟧-भा-गा *corr. ras.* Sc
11 ३६०] *om.* Np, Sc
11 ३६० गुणिता] गुणि ३६० ता Pb
11 पुनः] द्युनः Rr
12 ६०००] ६ Bn.II; ६ ∴ *ut videtur* Br; ६००० (६०००) *idem suprascr.* Rr
12 विवरांशका] विवगंशका Br
12 ९] *om.* Bn.I, Pm; ८ Bn.II, Br, Rr; ३ Sc

तत्कोटिज्योना त्रिभज्या ६० विधेया शेषं भूव्यासार्धमानेन निघ्नम् ।
f. 68v Bn.I तत्कोटिज्यामानभक्तं च लब्धं शृङ्गौन्नत्यं योजनैः प्रस्फुटं ⌈स्यात् ॥ १० ॥

भूव्यासार्द्धं शैलशृङ्गोच्छ्रयाढ्यं हारः कल्पोऽथोन्नतिस्त्रिज्यकाघ्नी ।
हारेणाप्ता तद्विहीना त्रिभज्या तत्कोटिज्या कल्पनीयात्र तज्ज्ञैः ॥ ११ ॥

तत्कोटिज्यावर्गहीनात्त्रिमौर्व्या वर्गान्मूलं तस्य चापस्य भागाः ।
f. 2r Pb चक्रांशाप्ता ३६० भूपरिध्युन्मि⌈ति ६००० घ्ना ज्ञेयः सोऽध्वा शैलपुंपादमध्ये ॥ १२ ॥

f. 158r Sc अथो कुमध्ये यदि वास्ति लङ्का प्राच्यां त⌈दा स्याद्यमकोटिरेव ।
अवाचि वेद्यो वडवानलाख्यो मेरुः सदोदक्पररोमकं च ॥ १३ ॥

f. 2v Np ⌈चेद्भूमिमध्ये यमकोटिपूः स्यात्प्राच्यां तदा सिद्धपुरी विचिन्त्या ।
प्रत्यक्चलं काथ यथा कुमध्ये सिद्धाभिधाना नगरी प्रकल्प्या ॥ १४ ॥

प्राच्यां तदा रोमकपत्तनं हि पश्चात्स्थिता स्याद्यमकोटिरेषा ।
[a] f. 70r Bn.II [b] f. 50v Rr ⌈[a]चेद्रोमकं नाम कुमध्यसंस्थं प्राच्यां हि लङ्का ⌈[b]परतश्च सिद्धा ॥ १५ ॥

---

1 ॰मानेन] ॰मानेन {१२} *intra lineam, delere ign. litt.* Pm; ॰न Np
2 योजनैः] योजैः Rr
2 १०] ९ Bn.II, Br, Pm, Rr; ४ Sc
3 ॰शृङ्गोच्छ्रयाढ्यं] ॰श्रृंगो⟦धू⟧याथं -च्छ्र- *corr. in marg. sins.* Sc
3 कल्पोऽथो॰] कल्पोथौ॰ Bn.II; कल्प्येथो॰ Pb; कल्पे -ा- थो॰ *add. intra lineam* Sc
4 ॰हीना] ॰होना Br
4 त्रिभज्या] त्रिभज्यं Bn.II, Pm, Rr
4 तत्कोटिज्या] *om.* Bn.I
4 तज्ज्ञैः] तज्ञैः Bn.I, Pb; त^ज्^ज्ञैः *add. suprascr.* Sc
4 ११] १० Bn.II, Br, Pm, Rr; ५ Sc
5 ॰हीनात्त्रि॰] ॰नात्त्रि॰ Bn.I; ॰हीनात्रि॰ Bn.II, Br, Pm, Rr
6 ॰शाप्ता ३६०] ॰शा ३६० प्ता Bn.II, Br, Pm, Np, Rr
6 ॰ध्युन्मिति ६०००] ॰ध्यु६००० न्मिति Np
6 ६०००] ६०० Bn.II, Rr; ६००० Sc
6 सोऽध्वा] शौध्वा Bn.II
6 १२] ११ Bn.II, Br, Pm, Rr; ६ Sc
7 वास्ति] चास्ति Bn.I, Np, Pb, Sc; वास्तिकं Bn.II
8 ॰पर॰] ॰परे॰ Sc
8 १३] १२ Bn.II, Br, Pm, Rr; *om.* Sc
9 ॰कोटिपूः] ॰कोटिप्तः Bn.I; ॰कोटिभूः Sc, Pm
9 स्यात्प्राच्यां] स्यात्माच्यां Sc
9 तदा] ⟦द⟧तदा *corr. ras.* Pm
9 सिद्धपुरी] सि^द्ध^पुरी *add. suprascr.* Sc
9 विचिन्त्या] प्रकल्प्या Bn.II, Br, Pm, Rr
10 प्रत्यक्च॰] प्रक्च॰ Pb
10 यथा] दा Bn.I; यदा Pb; ⟦यथा⟧यदा *corr. ras.* Sc
10 ॰धाना] ॰धा Br
10 नगरी] नकरी Np
10 १४] *om.* Bn.II, Br, Pm, Rr, Sc
11 ॰त्स्थिता॰] ॰तूस्थि⟦ति⟧ता॰ *corr. ras.* Pm
11 ॰स्याद्यम॰] ॰स्यायम॰ Bn.I, Bn.II, Br, Pb, Pm, Rr
11 ॰रेषा] ॰रेखा Bn.I; ॰रेखा १३ Bn.II, Rr; ॰रेषा १३ Br, Pm
12 चेद्रोमकं] चेद्रोम⟦को⟧-का- *corr. ras.* Bn.II
12 नाम] वात्र Bn.I, Pb; चात्र Np, Sc
12 १५] *om.* Bn.II, Br, Pm, Rr, Sc

एतासु सर्वासु नृणां पुरीषु मेरुः सदोदग्वडवानलोऽवाक् ।
मूर्ध्नि स्थितं भ्राम्यति कालचक्रं ध्रुवद्वयं तिष्ठति भूजलग्नम् ॥ १६ ॥

परस्परं षड्वरणीस्थलानि भूपा[a]दभागैः ९० [b]सततं स्थितानि । [a] f. 38v Br [b] f. 158v Sc
एतानि भूमेः परिणाहतुर्यभागेन १५०० भूयः परिकल्पितानि ॥ १७ ॥

मेरुर्यदा तिष्ठति भूमिमध्ये प्राच्यां यमाख्यापररोमकं हि ।
सिद्धोत्तरा दक्षिणगा च लङ्का सौम्यो ध्रुवः खेऽङ्घ्रितले च याम्यः ॥ १८॥

कुमध्यगश्चेद्वडवानलाख्यः प्राक्पश्चिमे स्तो यमरोमकाख्ये ।
लङ्कोत्तरस्था यमदिक्च सिद्धा याम्यो ध्रुवः खेऽङ्घ्रितले च सौम्यः ॥ १९ ॥

भूजासक्तमिवाभितः सुरगणा मेरुस्थिता वैषवं
वृत्तं गच्छदसव्यतोऽष्टककुभां मार्गेण पश्यन्ति हि । f. 69r Bn.I
सव्यं व्युत्क्रमतो दिशां वडवगा दैत्याः सदा प्राङ्मुखा-
स्तद्वत्तेन धराखिला विजयते गोलाकृतिः सर्वतः ॥ २० ॥ f. 159r Sc

---

1 सर्वासु] र्वस्सौ Bn.I
1 नृणां] नृ ५ णां Sc
1 पुरीषु] पुरासु Bn.I
1 ◦नलोऽवाक्] ◦नलोग्वाक् १४ Bn.II, Pm, Rr; ◦नलो वाक् १४ Br
2 मूर्ध्नि] मूर्धि Bn.II
2 तिष्ठति] ति Bn.II
2 १६] *om.* Bn.II, Br, Pm, Rr, Sc
3 षड्वरणी◦] षट्सरणी◦ Np
3 भूपाद◦] भूपाद्द◦ Np
3 ९०] १० Pm
3 स्थितानि] स्थितानि १५ Bn.II, Br, Pm, Rr
4 एतानि] ⟦एतानि⟧एतानि *corr. ras.* Sc
4 ◦तुर्य◦] ◦तूर्य◦ Pb
4 ◦कल्पितानि] ◦कल्पेतानि Pb
4 १७] *om.* Bn.II, Br, Pm, Rr, Sc
5 ◦पर◦] ◦परे◦ Sc
5 हि] हि १६ Bn.II, Br, Pm, Rr
6 ◦गा] ◦ग Bn.I
6 सौम्यो◦] सौम्यौ Bn.II
6 खेऽङ्घ्रितले] खेंघ्रितल Bn.I; खेंघ्रिले त *add. suprascr.* Pm; खेंघ्रितलै Sc
6 याम्यः] याम्या Pb
6 १८] *om.* Bn.II, Br, Pm, Rr, Sc
7 यम◦] दय◦ Np, Sc
7 ◦काख्ये] ◦काख्ये १७ Bn.II, Br, Pm, Rr
8 यमदिक्च] यमदिक्क Bn.II, Br, Pm, Rr; यम⟦को⟧दिक् च *corr. ras.* Sc
8 सौम्यः] याम्यः Sc
8 १९] १⟦७⟧ *corr. ras.* Bn.II; १७ Br, Pm, Rr; *om.* Sc
9 भूजासक्त◦] भूजाशक्ति◦ Bn.II; भूजासक्ति◦ Np; भूवासक्त◦ Sc
9 ◦वाभितः] ◦नामिताः Sc
9 वैषवं] वैषुवं Np, Sc
10 ◦सव्य◦] ◦शव्य◦ Pb; ◦दव्य◦ Sc
10 हि] ज्ञि Br
11 ◦क्रमतो] ◦क्रमतौ Sc
12 धरा◦] ⟦र⟧धरा◦ *corr. ras.* Pm
12 २०] १८ Bn.II, Br, Pm, Rr; *om.* Sc

f. 45r Pm यत्र क्षितौ तिष्ठति मान⌈वोऽसौ तस्मात्कुपादान्तरितं समन्तात् ।
वृत्तं तिरश्चीनमवैहि भूजं खं मूर्ध्नि भूम्यर्धगतं ३००० पदाधः ॥ २१ ॥

तथा हि लङ्का क्षितिपृष्ठमध्ये तद्भूजगा एव सुमेरुपूर्वाः ।
f. 3r Np कुबेरदिक्तोऽङ्घ्रितलस्थिता च सिद्धापुरी तद्व⌈दतोऽपरत्र ॥ २२ ॥

रोमकाभिधपुराच्च पश्चिमे खालदातपुरमस्ति वारिणि ।
नेत्रलोचन २२ लवैरतः पुरो गङ्गदुज्दमिह भार्धभागकैः १८० ॥ २३ ॥

लङ्कारोमकखालदातनगरं संस्पृश्य यन्मण्डलं
f. 70v Bn.II यायाद्भूपरिधिः ६००० स एव गणकैरुक्तोऽत्र ⌈मध्याभिधः ।
f. 159v Sc ⌈तत्रस्थो विषयो निरक्ष इह ये लोका वसन्ति ध्रुवौ
ते पश्यन्त्यवनीजगौ च विषुवद्वृत्तं स्वमूर्ध्नि भ्रमत् ॥ २४ ॥

आगङ्गदुज्दादथ खालदातं यावत्पुनर्मेरुमुदक्स्थितं च ।
एतावती भूमिरिह त्रिकोणा पादोन्मिता तिष्ठति वारिमुक्ता ॥ २५ ॥

---

1 यत्र] यत्रा Bn.II, Rr
1 तिष्ठति] तिष्ठाति Bn.I
1 तस्मात्कु॰] तस्मा⟦त्⟧त्कु॰ *corr. ras.* Sc
2 वृत्तं] वृत्तां Pm
2 तिरश्चीन॰] तिरस्चीन॰ Pb; तिराश्चीन॰ Pm
2 भूम्यर्धगतं ३०००] भूम्यर्द्धग ३००० तं Bn.I, Bn.II, Br, Pb, Pm, Rr; भूम्यर्द्ध ३००० गतं Np
2 पदाधः] यदाधः Sc
2 २१] १९ Bn.II, Br, Pm, Rr; *om.* Sc
3 तथा] त⟦ा⟧था *corr. ras.* Sc
3 एव] ⟦ऐ⟧-ए-व *corr. ras.* Sc
4 ॰दिक्तोऽङ्घ्रि॰] ॰दिक्तोंलि॰ Bn.I; ॰दिक्तोंह्रि॰ Br, Pb, Rr; ॰दिकूतांह्रि॰ Sc
4 ॰पुरी] ॰पुरो Bn.II
4 तद्वदतो॰] तद्वदमो॰ Bn.II, Np; तद्धदतो॰ Rr
4 २२] २० Bn.II, Br, Pm, Rr; *om.* Sc
5 खालदात॰] खालद॰ Bn.I
6 २२] *om.* Sc
6 लवैरतः] लवैरितः Bn.II, Rr; लवेरितः Pm
6 गङ्गदुज्द॰] गङ्गदज्द॰ Np
6 २३] २१ Bn.II, Br, Pm, Rr; *om.* Sc
7 संस्पृश्य] संस्पश्य Bn.II
8 यायाद्भू॰] या^ा^यद्भू॰ *add. suprascr.* Bn.II
8 ॰परिधिः] ॰र्यारधिः Br
8 गणकै॰] कणकै॰ Bn.II, Br, Pm, Rr; ग⟦णितो⟧णकै॰ *corr. ras.* Sc
10 पश्यन्त्य॰] पश्यात्य॰ Sc
10 ॰नीजगौ] ॰नीजगो Bn.II
10 विषुव॰] विषुव⟦ा⟧॰ *corr. ras.* Pm
10 २४] २२ Bn.II, Br, Pm, Rr; *om.* Sc
11 आगङ्गदुज्दा॰] आगंगदज्दा॰ Np, Sc; आगंगदुत्दा॰ Pm, Rr
11 यावत्पु॰] या^व^त्पु॰ *add. suprascr.* Pm; य^ा^वत्पु॰ *add. suprascr.* Sc
11 ॰मेरुमुद॰] ॰मेरुमु^द^॰ *add. suprascr.* Pm; ॰मेरुरुद॰ Sc
12 त्रिकोणा] त्रिकौणा Bn.II; त्रिकोणो Pm
12 पादो॰] पाद⟦ौ⟧-ो-॰ *corr. ras.* Sc
12 २५] २३ Bn.II, Br, Pm, Rr; *om.* Sc

षष्ट्या घटीनां द्युनिशव्यवस्था यावद्ध्रुवेत्तावदिहापि लोकाः ।
तिष्ठन्ति नीहारमयाद्रिभीत्या तत्रापि वस्तुं विभवो न सन्ति ॥ २६ ॥

पश्यन्तु काश्मीरमुखप्रदेशान्हिमागमे दुर्ग⌈तरान्नराणाम् । f. 51r Rr
सदैव यत्र प्रपतेद्धिमानी शक्नोति को गन्तुमिहोत्तरेषु ॥ २७ ॥

लङ्काकाञ्चीश्वेतशैलोज्झयिन्यो रोहीताख्यं ⌈सांनिहित्यं सुमेरुः । f. 160r Sc
रेखा भूमेर्मध्यनाम्नी निरुक्ता तस्यां किंचिन्ना⌈स्ति देशान्तरं हि ॥ २८ ॥ f. 69v Bn.I

निरक्षदेशाच्चलितः पुमान्यथा कुबेरकाष्ठां च तथोन्नतं ध्रुवम् ।
विलोकयेन्नभ्रमपीह वैषवं नभस्तलाद्याम्यविभागसंस्थितम् ⌈॥ २९ ॥ f. 2v Pb

यथा यथा गच्छति दक्षिणां पुमांस्तथा तथा पश्यति सध्रुवं नतम् ।
अतो ध्रुवो यत्र शिरःस्थितो भवेन्निरक्षतो भूचरणान्तरे तु सः ॥ ३० ॥

---

1 घटीनां] घटिनां Pm
1 द्युनिश॰] द्युतश॰ Rr
2 नीहार॰] नीसारि॰ Bn.II
2 ॰भीत्या] ॰नीत्या Pm
2 तत्रापि] तत्रपि Bn.I
2 २६] २४ Bn.II, Br, Pm, Rr; *om.* Sc
3 पश्यन्तु] पश्यन्ति Bn.II
3 काश्मीर॰] कश्मीर॰ Bn.I, Br, Pm, Rr; कस्मीर॰ Bn.II; कास्मीर॰ Np
3 ॰मुख॰] ॰सुख॰ Br
3 दुर्गत॰] दुर्गात्त॰ Bn.II, Br, Pm
4 प्रपतेद्धि॰] प्रयतेद्धि॰ Np; प्रददेद्धि॰ Sc
4 शक्नोति] शक्तोति Bn.I, Pm, Rr; शक्तोते Bn.II
4 २७] २५ Bn.II, Br, Pm, Rr; *om.* Sc
5 ॰काञ्ची॰] ॰कातौ॰ Bn.I; ॰कांति॰ Bn.II, Rr; ॰कांती॰ Br, Pb, Pm
5 ॰श्वेत॰] ॰श्वे⟦ते⟧-त-॰ *corr. ras.* Sc
5 ॰शैलोज्झ॰] ॰शैज्झ॰ Bn.I; ॰शैलोज॰ Bn.II, Rr
5 ॰यिन्यो] ॰यिज्यो Br
5 सांनिहित्यं] सान्निहत्यं Bn.II, Rr
5 सुमेरुः] सुमेहः Bn.II
6 ॰र्मध्य॰] ॰मध्ये॰ Pm
6 ॰नाम्नी] ॰मास्ता Np, Sc
6 तस्यां] स्तस्यां Np
6 २८] २६ Bn.II, Br, Pm, Rr; *om.* Sc
7 ॰च्चलितः] ॰च्चलेतः Bn.II
7 कुबेर॰] कुवे⟦व⟧र॰ *corr. ras.* Bn.II
8 विलोक॰] विलोम॰ Bn.II, Rr
8 ॰न्नभ्रमपीह] ॰न्नर्म्मयीह Bn.I; ॰न्नभ्रमयीह Bn.II, Pb, Rr; ॰न्नमृमयीह Pm; ॰न्नमपीह (भ्र) *add. suprascr.* Sc
8 वैषवं] वेंषवं Bn.II; वैष Pm; वैषुवं Np, Sc
8 नभस्तला॰] ⌊भस्तला॰ -न- *add. in marg. sins. in linea confrac.* Pm; नभस्थला॰ Pb; भभस्तला॰ Sc
8 ॰द्याम्यवि॰] ॰द्यम्वि॰ Np
8 २९] *om.* Bn.II, Pm, Rr, Sc; २७ Br
9 *prius* यथा] ⟦त⟧-य-था *corr. in ras.* Sc
9 गच्छति] गक्षति Bn.I; गछतिति Br
9 तथा] स्तथा Bn.II, Rr; तथा तथा Pb
9 सध्रुवं] स्वध्रुवं Sc
9 नतम्] तनं Bn.II, Pm, Rr
10 अतो] अत Pb
10 ॰न्निरक्षतो] ॰न्निरक्षती Bn.II; ॰निरक्षतो Pb
10 ३०] २८ Bn.II, Br, Pm, Rr; *om.* Sc

[a] f. 39r Br
[b] f. 3v Np यत्र यत्र वसति क्षमोप⌈[a]रि प्राणभृद्धरणिगर्भ⌈[b]कर्षितः ।
तत्र तत्र किल मन्यते स्वयं भूपरिष्ठमिव चेतसा स्वकम् ॥ ३१ ॥

भूचतुर्थलवकान्तरे यथा तिर्यगेव खलु मन्यते स्थलम् ।
भूदलान्तरमधःस्थितं स्वतोऽन्यत्र तद्वदपि कल्पनान्यथा ॥ ३२ ॥

f. 160v Sc ⌈निरक्षतः सौम्यविभागसंस्थे यत्र प्रदेशे परमद्युमानम् ।
व्यष्टांशदन्तोन्मितनाडिकाभि ३१|५२|३० स्तत्राद्यखण्डस्य भवेत्प्रवेशः ॥ ३३ ॥

f. 71r Bn.II ⌈सपादनाडी १|१५ सहिते च तस्मिन्द्वितीयखण्डस्य तथा प्रवेशः ।
एवं तृतीयादिकखण्डकानां ज्ञेयाः प्रवेशाः किल सप्त यावत् ॥ ३४ ॥

f. 45v Pm अथ पुराणमतेन द्वीपसमुद्रपर्वतखण्डादीनां संस्थो⌈च्यते ।

परिधेस्तु शतांशको यतः समवद्धाति ततः समन्ततः ।
पृथिवीं च पुराणपण्डिताः सकलामेव समां समूचिरे ॥ ३५ ॥

भूपृष्ठस्पृक्दृष्टिसूत्रं विलग्नं व्योम्ना तिर्यग्यत्र तद्दृष्टिभूजम् ।
f. 161r Sc चक्राका⌈रं सर्वतो दृश्यते यल्लोकालोकं पर्वतं तं वदन्ति ॥ ३६ ॥

---

1 क्षमोपरि] क्षमोपभ Sc
1 प्राणभृ॰] प्राणभू॰ Np, Sc
1 ॰द्धरणि॰] ॰द्धराणि॰ Bn.II, Rr
1 ॰कर्षितः] ॰कर्णितः Sc
2 स्वयं] स्थलं Sc
2 भूपरिष्ठ॰…॰स्वकम्] *desunt* Sc
2 स्वकम्] श्वतकं Np
2 ३१] २८ Bn.II; २९ Br, Pm, Rr; *om.* Np, Sc
3 भूचतुर्थ॰…॰स्थलम्] *desunt* Sc
4 ॰ऽन्यत्र तद्वदपि] ॰न्यत्रादपि Bn.II
4 ३२] *om.* Bn.II, Rr, Sc; ३० Br, Pm, Rr; ३१ Np
5 ॰विभाग॰] ॰विभाग⟦ा⟧॰ *corr. ras.* Pm
5 ॰संस्थे] ॰संस्थिे *ut videtur* Bn.I
5 ॰द्युमानम्] ॰द्यमानं Bn.II
6 व्यष्टां॰] व्यष्टं॰ Bn.II, Rr
6 ३१|५२|३०] ३१ ५२|५२|३० Rr
6 ॰प्रवेशः] ॰प्रवे⟦शे⟧-श- *corr. ras.* Sc
6 ३३] ३१ Bn.II, Br, Pm, Rr; ३२ Np
7 सपाद॰] पपाद॰ Pm
7 १|१५] *om.* Np, Sc
7 सहिते] सहित्वे Pb; रहिते Sc
7 तस्मिन्द्वि॰] तस्मिन्द्व॰ Bn.I; स्मिनद्वि॰ (त) *add. suprascr.* Bn.II
7 तथा] तस्य तथा Bn.I; भवेत् Np, Sc
7 प्रवेशः] प्रवेशेः Pm
8 एवं] एकं Bn.II, Br, Pm, Rr
8 ३४] ३२ Bn.II, Br, Pm, Rr; ३३ Np; *om.* Sc
9 पुराण॰] पुरा॰ Pm
9 ॰समुद्र॰] ॰समुग्द्र॰ Bn.II
10 शतांशको] शतांशक्ये Br; शकांशतो Pm
11 पृथिवीं] पृथवी Bn.II
11 समूचिरे] समुचिरे Bn.II
11 ३५] *om.* Bn.II, Rr; ३३ Br, Pm; ३४ Np; १ Sc
12 ॰स्पृक्दृष्टिसूत्रं] ॰स्पृण्दृष्टिसूत्रं Bn.II; स्पृग्दृष्टसूत्रं Pb; ॰स्पृग्(शतांशको यतः समवद्धाति ततः समन्ततः पृथिवीं च पुरा)द्दष्टिसूत्रं Rr
12 तद्दृष्टि॰] तद्दष्टि॰ Pb; दृष्टि॰ Pm; तदष्टि॰ Rr
13 चक्रा॰] चक्र॰ Bn.I
13 सर्वतो] *om.* Sc
13 य] यत्रयत्त Np, Sc
13 ३६] ३४ Bn.II, Br, Pm, Rr; ३५ Np; २ Sc

दृश्यादृश्यगिरिं केचिन्नभःकक्षां जगुर्बुधाः ।
तेन किं खचरा नित्यमुदयास्तौ प्रकुर्वते ॥ ३७ ॥

क्षितिजमेव भवेदुदयाचलश्चलति ⌈येन कुतोऽपि न नित्यशः । f. 70r Bn.I
स्वविषयेऽल्पतरं हि यदन्तरं पतति तच्चलनेऽपि खगोदये ॥ ३८ ॥

सुवर्णशैलं विपुलोन्नतं रविस्तदीय कट्यां परितः सदा भ्रमन् ।
कथं न कुर्याद्द्युनिशोर्व्यवस्थितिं तदेकपार्श्वे वसतां सुपर्वणाम् ॥ ३९ ॥

क्षाराम्बुधिः क्षितिकटीस्थितमेखलेव व्यासोऽस्य योज⌈नशतं मुनिभिः प्रदिष्टः । f. 4r Np
तस्मादुदीचिकुदले सकलेऽ⌊ स्ति जम्बूद्वीपं परत्र कुदले तुपराणि षट्स्युः॥४०॥ f. 161v Sc

क्षाराब्धिः पयसां सिन्धुस्ततो दधिघृताम्बुधी ।
इक्षुमद्यार्णवौ स्वादुसिन्धुः सप्ताब्धयोऽत्र वै ॥ ४१ ॥

---

2 खचरा] खचग Br; ख(च)रो *add. suprascr.* Sc
2 ◦र्वते] ◦र्वत Bn.I
2 ३७] ३५ Bn.II, Br, Pm, Rr; ३६ Np; ३ Sc
3 ◦दयाचल] ◦दया˅ -चल- *add. in marg. dext.* Pm
3 कुतोऽपि] कुपितो Bn.II, Rr; कुजोपि Np, Sc
3 नित्यशः] नित्यस Bn.I
4 हि यदन्तरं] हि यदन्तरं हि यदन्तरं Br; (२ १) हिदंयतरं *corr. transp.* Pm
4 पतति] पति Bn.I; यतति Br
4 ३८] ३६ Bn.II, Br, Pm, Rr; ३७ Np; ४ Sc
5 विपुलो◦] विपलो◦ Pb
5 कट्यां] कद्यां Bn.I, Br
6 ◦द्द्युनिशोर्व्य◦] ◦द्युनिशोर्व्य◦ Bn.I, Br, Pm; ◦द्द्युनिशंव्य◦ Pb
6 ◦स्थितिं] ◦स्थितं Np
6 तदेक◦] पदेक◦ Bn.II, Rr
6 वसतां] वसतं Bn.II, Rr
6 सुपर्वणाम्] सुप{ष}र्वणां *intra lineam, delere ign. litt.* Pb
6 ३९] ३७ Bn.II, Br, Pm, Rr; ३८ Np; ५ Sc
7 क्षाराम्बुधिः] क्षाराम्बुभिः Np
7 ◦कटीस्थित◦] ◦तटीत◦ Bn.I
7 ◦खलेव] ◦त्वलेव Br; ◦खले˅ -व- *add. in marg. dext.* Pm
7 व्यासो◦] व्य -ा- सो◦ *add. intra lineam* Sc
7 ◦शतं] ◦संतं Bn.I; ◦सतं Bn.II, Rr; ◦शतं १०० Np
8 तस्मादु◦] त˅ (स्मा) दु ◦ *add. suprascr.* Pm
8 ◦दीचि] ◦रीचि Rr
8 ◦कुदले] ◦कुले (द) *add. subscr.* Pm; ◦शकले Np, Sc
8 ◦लेऽस्ति] ◦लोस्ति Bn.I, Pb
8 कुदले] शकले Np, Sc
8 ४०] ४० *add. in marg. sup.* Bn.I; ३८ Bn.II, Br, Pm, Rr; ३९ Np; ६ Sc
9 सिन्धुस्ततो] सिंधूःस्ततो Bn.I
9 ◦ताम्बुधी] ◦ताम्बुधिः Np; ◦तांबुधिः Sc
10 ◦मद्यार्णवौ] ◦मद्यार्ण⟦स⟧वौ *corr. ras.* Sc
10 स्वादु◦] स्वा⟦द्यु⟧दु◦ *corr. ras.* Bn.II
10 ४१] ३९ Bn.II, Br, Pm, Rr; ४० Np; ७ Sc

f. 51v Rr तेषां मध्ये क्षा|रसिन्धुर्महीयान्व्यक्षो देशः प्रोच्यते सोऽयमेव ।
तस्मादल्पाल्पप्रमाणाः परे स्युः स्वल्पोऽत्यन्तं सप्तमः स्वादुसिन्धुः ॥ ४२ ॥

स्वादूदकान्तर्वडवानलोऽस्ति वसन्ति दैत्या इह मेरुमूले ।
f. 71v Bn.II पातालक्लोका धरणीपुट|स्थाः प्रकाशिताः सर्पमणिप्रकाशैः ॥ ४३ ॥

द्वयोर्द्वयोरन्तरगं समुद्रयोरुदाहृतं द्वीपमुदारबुद्धिभिः ।
महद्यदाद्यं परमल्पमल्पकं ततोऽस्ति षड्वीपमिलादलं परम् ॥ ४४ ॥

f. 162r Sc शा|कं च शाल्मलं कौशं क्रौञ्चं गोमेदपुष्करे ।
सप्तमं जाम्बवं द्वीपं भूगोलार्धे व्यवस्थितम् ॥ ४५ ॥

लङ्कापुरीतोऽपि कुबेरदिक्स्थो हिमाचलो नाम नगाधिराजः ।
प्राक्पश्चिमाम्भोनिधिसीमदैर्घ्यस्ततोऽप्युदीच्यामिह हेमकूटः ॥ ४६ ॥

f. 39v Br त|द्वत्परं निषधशैल उदीचिभागे सिद्धात्पुराद्गिरिरुदक्प्रभवेत्पुरावत् ।
शृङ्गी च शुक्ल इति नीलगिरिः क्रमेण तेषां द्वयोर्विवरदेशगतं च वर्षम् ॥ ४७ ॥

---

1 ०हीयान्व्यक्षो] ०हीजाध्याक्षो Bn.I; ०हियाच्यक्षो Pm
2 ०ऽत्यन्तं] ०ल्पन्तं Np
2 ४२] ४० Bn.II, Br, Pm, Rr; ४१ Np; ८ Sc
3 ०ऽस्ति] ०सौ Np, Sc
3 दैत्या] देत्या Bn.II; लोका Np, Sc
3 मेरुमूले] मेवमूले (रु) *corr. suprascr.* Sc
4 ०लोका धरणि०] ०लोकाः पृथिवी० Np, Sc
4 ४३] ४१ Bn.II, Br, Pm, Rr; ४२ Np; ९ Sc
5 ०योरुदा०] ०योरुद्रा० Pm
6 ०द्यदाद्यं] ०द्यम⌐द्यं -ा- *add. in marg. dext. in linea confrac.* Bn.II; ०द्यमाद्यं Rr
6 ४४] ४२ Bn.II, Br, Pm, Rr; ४३ Np; १० Sc
7 शाल्मलं] शल्मकं Bn.I; शाल्मकं Pb; ⟦शो⟧-शा-ल्मलं *corr. ras.* Pm
7 कौशं] कौचं Br
7 क्रौञ्चं] क्रोचं Bn.II; *om.* Br
7 ०पुष्करे] ०पुष्करः Bn.I
8 ०गोलार्ध०] ०गोलर्द्धि० Br
8 ४५] ४३ Bn.II, Br, Pm, Rr; ४४ Np; ११ Sc
9 लङ्कापुरी०] लंकापुरो० Pm
9 ०दिक्स्थो] ०दिक्स्था Bn.I
10 प्रक्पश्चिमाम्बो०] शश्चिमांभो० Bn.II
10 ०सीम०] ०साम० Bn.I, Np
10 ०तोऽप्युदी०] ०तोदी० Bn.I; ०तो⟦षु⟧दी० -प्यु- *corr. in marg. dext.* Pm
10 ०च्यामिह] ०व्यामिह Bn.II; ०च्यामपि Np
10 ४६] ४४ Bn.II, Br, Pm, Rr; ४५ Np; ११ Sc
11 निषध०] निषिध० Bn.II
11 उदीचि०] उदीच्य० Bn.II
11 ०त्पुराद्गिरि०] ०त्पुराहिरि० Bn.I; ०त्पुद्गिरि० (रा) *add. suprascr.* Pm; ०त्पुराज्दिरि० Sc
12 शृङ्गी] शृङ्गो Bn.II
12 द्वयोर्वि०] द्वयोवि० Bn.I; द्वषोर्वि० Br
12 ०र्विवर०] ०र्विव० Bn.II
12 ०देशगतं] ०देश⟦ा⟧तं (रा) *corr. ras., et add. suprascr.* Pm
12 वर्षम्] वर्षः Bn.II, Rr, Pm; वर्षाः Br
12 ४७] ४५ Bn.II, Br, Pm, Rr; ४६ Np; १२ Sc

भारतं किन्नरं तस्माद्धरिवर्षं यथोत्तरम् ।
लङ्कातोऽ⌈[a]थ पुरात्सिद्धा⌈[b]त्कुरुवर्षं तथोत्तरम् ॥ ४८ ॥

[a] f. 70v Bn.I
[b] f. 3r Pb

हिरण्मयं रम्यकनामखण्डकं सुमाल्यवद्गन्धगिरी क्रमादथ ।
⌈[a]यमादिकोट्यन्तपुराच्च रोम⌈[b]कादुदक्ककुप्स्थौ च पुराणसन्मतौ ॥ ४९ ॥

[a] f. 162v Sc
[b] f. 4v Np

निषधपर्वतनीलनगावधी भवत उत्तरदक्षिणगामिनौ ।
जलधिमाल्यवदन्तरवर्ति यत्तदिह भद्रतुरङ्गमखण्डकम् ॥ ५० ॥

गन्धमादनपयोधिमध्यगं यज्ञगुस्तदिति केतुमालकम् ।
नीलगन्धनिषधाद्रिमा⌈ल्यवन्मध्यसंस्थितमिलावृतं बुधाः ॥ ५१ ॥

f. 46r Pm

इलावृतान्तर्निवसन्ति निर्जराः सुचारुचामीकरभूतलान्विते ।
इहैव मेरुः किल मध्यसंस्थितः सुरालयः काञ्चनरत्नवान्गिरिः ॥ ५२ ॥

---

1 यथोत्तरम्] यथोत्तरः Bn.I
2 लङ्का॰] लोका॰ Bn.II, Rr
2 पुरात्सिद्धात्कुरुवर्षं] पुरातसिद्धातकुरंवर्षे Bn.II
2 ४८] ४६ Bn.II, Br, Pm, Rr; ४७ Np; १३ Sc
3 ॰नाम॰] ॰नाग॰ Np
3 ॰खण्डकं] ॰खण्डके Br
3 ॰वद्गन्ध॰] ॰वहुधि॰ Bn.II; ॰वहुंधि॰ Rr
3 क्रमादथ] क्रमा^∨^थ -द- *add. in marg. dext.* Pm; क्रमादथा Sc
4 यमादि॰] यमाद्य॰ Bn.II, Rr
4 ॰कोट्यन्त॰] ॰कोट्यान्त॰ Pb
4 रोमकादु॰] रोमका⟦त⟧दु॰ *corr. ras.* Sc
4 ॰दक्ककुप्स्थौ] ॰दक्कुकुप्स्थो Bn.II; ॰दक्कुकुस्थौ Pm, Rr
4 ॰सन्मतौ] ॰सं⟦ ि⟧मतौ *corr. ras.* Sc
4 ४९] ४७ Bn.II, Br, Pm, Rr; ४८ Np; *om.* Sc
5 ॰पर्वत॰] ॰पर्वज॰ Br
5 ॰नगा॰] ॰ननगा॰ Bn.I; ॰नमा॰ Np
5 ॰मिनौ] ॰मिने⅃ -ा- *add. in marg. dext. in linea confrac.* Bn.II
6 यत्तदिह] यत्तहिह Br
6 ॰रङ्गमखण्डकम्] ॰रंगमध्यगं Pm
6 ५०] ४८ Bn.II, Br, Rr; *om.* Pm; ४९ Np; १५ Sc
7 गन्ध॰…॰मध्यगं] *desunt* Pm
7 यज्ञ॰] यज्प्र॰ Br
8 नीलगन्ध॰] नी^∨^गंध॰ -ल- *add. in marg. dext.* Pm
8 ५१] ४९ Bn.II, Br, Pm, Rr; ५० Np; १६ Sc
9 इलावृतान्तर्निवसन्ति] इलावृत्तावत्रवसंति Bn.I, Bn.II, Br, Pm, Rr
9 निर्जराः] नि⟦र्वृ⟧र्जराः *corr. ras.* Pm
9 ॰भूतला॰] ॰तला॰ Bn.II
10 ॰रत्न॰] ॰रत्न॰ Sc
10 ५२] ५० Bn.II, Br, Pm, Rr; ५१ Np; *om.* Sc

f. 163r Sc श्रीमद्विरिञ्चिजननाय वसुंधराब्जे पौराणिका जगुरमुं कि⌈ल कर्णिकेति ।
यस्माद्धरा भवति नाभिसरोरुहं हि पुंसो विराज इतरत्र षडङ्गसंस्था ॥ ५३ ॥

f. 72r Bn.II विष्कंभागा मन्दरो गन्धशैलो वि⌈स्तीर्णाद्रिः पार्श्वभूभृत्क्रमेण ।
तेषु प्रोचुः केतुवृक्षान्मुनीन्द्रा नीपं जम्बूं सद्वटं पिप्पलं च ॥ ५४ ॥

यज्जम्बूफलतोऽगलद्रसमयी जम्बूनदीप्रस्रुता
या या तद्रसमिश्रिता मृदभवत्तत्तत्सुवर्णं स्फुटम् ।
तस्मादुत्तरतः समागतनदीवाहेषु जाम्बूनदं
किंचित्किंचिदवाप्यतेऽपि सिकतामध्ये नरैरुत्तमम् ॥ ५५ ॥

f. 52r Rr विष्कम्भशैलेषु वना⌈नि चैत्ररथं तथा नन्दननामधेयम् ।
f. 163v Sc धृत्याख्यवैभ्राजकसंज्ञि⌈ते च सरांस्यथैतेषु यथा क्रमेण ॥ ५६ ॥

अरुणमानसके च महाह्लदं सितजलं पुनरेषु विलासिभिः ।
[a] f. 71r Bn.I [b] f. 5r Np सुरत⌈[a]केलिसमाकुलमानसाः स्वपतिभिश्च रम⌈[b]न्त्यमराङ्गनाः ॥ ५७ ॥

---

1 ०द्विरिञ्चिजन०] ०द्विरंचिजन० Bn.I, Bn.II, Rr; ०द्विवरं^जन० -चि- *add. in marg. dext.* Pm
1 वसुं०] वसं० Bn.I
1 ०धराब्जे] ०धराजे Np, Sc
1 ०का] ०⟦को⟧-का- *corr. ras.* Np
1 ०रमुं] ०रमं Bn.I; ०रुमुं Bn.II
1 ०केति] ०के⟦तू⟧-ते- *corr. ras.* Pb
2 यस्माद्धरा०] यस्म⌟⌜द्धरा० -ा- *add. in marg. dext. in linea confrac.* Bn.II
2 नाभि०] न-ा-भि० *add. intra lineam* Rr
2 ०संस्था] ०मंस्था Br
2 ५३] ५१ Bn.II, Br, Pm, Rr; ५२ Np; १८ Sc
3 विष्कंभागा मन्दरो] विष्कुंभागा मंदिरो Pb
3 ०शैलो] ०शैलौ Bn.II, Pm
3 विस्तीर्णाद्रिः] विस्तीर्णानिः Bn.II
4 ०मुनीन्द्रा] ०मुमनीद्रा Bn.II
4 ५४] ५२ Bn.II, Br, Pm, Rr; ५३ Np; १९ Sc
5 यज्जम्बू०] यं जम्बू० Np
6 *prius* या] व्याया Bn.I; *om.* Br
6 या या तद्रसमिश्रिता] (या या तद्रसमयी जंनंदीप्रश्रुता) या या तद्रसमिश्रिता *secl., ins. per errorum ex* verse 55a (line 1) Bn.II
6 मृद्भव०] मृदभव० Bn.I, Br, Np; मृदुभव० Bn.II, Rr; मदभव० Pb; मृदभ^व^० *add. suprascr.* Sc
7 ०दुत्तरतः] ०द्धूतरतः Pm
8 ०वाप्यतेऽपि] ०वाप्यतेपिवाप्यतेपि Bn.I
8 ५५] ५३ Bn.II, Br, Pm, Rr; ५४ Np; २० Sc
9 ०रथं तथा] ०रथत्तथा Np
10 ० धृत्याख्य०] ०धृत्याख्यत्या० Br
10 ०भ्राजक०] ०भ्राजित० Np, Sc
10 ०संज्ञिते] ०संज्ञिके Sc
10 सरांस्य०] सरांश्य० Pm
10 ५६] ५४ Bn.II, Br, Pm, Rr; ५५ Np; २१ Sc
11 ०मानसके च] ०मानसकेपि Np, Sc
11 महाह्लदं] महाह्रदं Bn.II, Pm, Rr
11 विलासिभिः] विलासिनः Np, Sc
12 ०समा०] ०सवमा० Bn.II
12 ०भिश्च रमन्त्य०] ०भि^श्च^रमंत्य० *add. suprascr.* Sc
12 ५७] ५५ Bn.II, Br, Pm, Rr; ५६ Np; २२ Sc

शृङ्गत्रयं कनकरत्नमयं सुमेरौ तस्मिन्विरञ्चिहरिशम्भुपुराणि सन्ति ।
इन्द्राग्निकालनिर्ऋताम्बुपवायुसोमेशानां पुराणि तदधोऽपि भवन्ति चाष्टौ ॥ ५८ ॥

स्वर्णाद्रौ पतिता हि विष्णुपदतो जाता चतुर्धा ततो
विष्कम्भाचलमस्तकोपरिसरः प्राप्ता चतस्रो दिशः ।
गच्छन्ती गिरिगह्वरादिषु रटत्कल्लोललोलाम्बुभृ-
त्सीताख्यालकनन्दिका सुरसरिच्चक्षुः सुभद्रा क्रमात् ॥ ५९ ॥

⌈शम्भोर्मस्तकभूषणाय विमला मुक्तावलीवत्स्थिता f. 164r Sc
या सोपानपरम्परेव विदिता स्वर्गोन्मुखानां नृणाम् ।
या विष्णोश्चरणाम्बुजस्य विल⌈सल्लोकत्रये विस्तृता f. 40r Br
सा गङ्गा हृदयस्थिता भवतु मे पापौघविध्वंसिनी ॥ ६० ॥

ऐन्द्रं खण्डमथो कशेरुशकलं स्यात्ताम्रपर्णं ततो
विज्ञेयं च गभस्तिमादनमतः खण्डं कुमारीति च ।
⌈नागं सौम्यमथापि वारुणमतः प्रान्त्यं च गन्धर्वकं f. 72v Bn.II
वर्षेऽस्मिन्नपि भारते नव सदा खण्डानि सन्ति ध्रुवम् ॥ ६१ ॥

---

1 सुमेरौ] सुमेरो Bn.II, Np
1 तस्मिन्विरञ्चि॰] तस्मिनवि⟦व⟧रं॰ *corr. ras.* Pm; स्तस्मिन्विरिंचि॰ Np; ⟦र्त⟧-त-स्मिन्विरिंचि॰ *corr. ras.* Sc
1 ॰पुराणि] ॰पुणि Bn.I
2 इन्द्राग्नि॰] इन्द्रा{ा}ग्नि॰ *ut videtur* Bn.I
2 ॰निर्ऋताम्बु॰] ॰निनिऋतिनिज्ञजंइत्यपितांबु॰ Bn.II; ॰निऋतांबु॰ Pb, Rr; ॰निसृतांयु॰ Pm; निऋताम्बु॰ Np; ॰निऋतोंबु॰ Sc
2 भवन्ति] भ^ति^वं *add. suprascr.* Sc
2 ५८] ५६ Bn.II, Br, Pm, Rr; ५७ Np; २३ Sc
3 हि विष्णु॰] हि^∨^ष्णु॰ -वि ३- *add. in marg. dext.* Sc
3 जाता] व्राता Br
4 विष्कम्भा॰] वि⟦ष्कुं⟧-कुं-भा॰ *corr. ras.* Sc
4 ॰परिसरः] ॰परिसरा Pb
4 प्राप्ता चतस्रो] प्राप्ताच्चतस्रो Pm; संसारिणीस्त्रो Sc
5 गच्छन्ती] गछेती Bn.I; गछंति Bn.II, Pm, Rr
5 ॰गह्वरादिषु] ॰गह्वरगदिषु Pb
6 ॰सरिच्चक्षुः] ॰सरिचक्षुः Pm
6 ५९] ५७ Bn.II, Br, Pm, Rr; ५८ Np; २४ Sc
7 ॰र्मस्तक॰] ॰र्मस्त॰ Bn.II
7 ॰वत्स्थिता] ॰वस्थिता Bn.I, Bn.II, Pm, Rr, Sc
8 ॰परम्परेव] ॰रंपरेव Bn.I; ॰परम्परैव Np
8 नृणाम्] मृणां Bn.I
9 विलसल्लोक॰] विसवल्लोक॰ Bn.I, Bn.II, Br, Pb, Rr; विसवल्लोके॰ Pm
10 हृदयस्थिता] हृ^य^दस्थिता *add. suprascr.* Pm
10 ६०] ५८ Bn.II, Br, Pm, Rr; ५९ Np; २५ Sc
11 ऐन्द्रं] ऐंद्रं⟦ि⟧ *corr. ras.* Bn.II
11 कशेरु॰] कसेरु॰ Bn.I, Np, Sc
11 ॰शकलं] ॰सकलं Np
12 ॰मतः] ॰मथो Np, Sc
13 ॰मतः] ॰मलः Pm
13 प्रान्त्यं] प्रांतं Pb
14 ६१] ५९ Bn.II, Br, Pm, Rr; ६० Np; २६ Sc

विप्रक्षत्रियवैश्यादि व्यवस्था या पृथक्पृथक् ।
f. 164v Sc कुमारीखण्डमध्यस्था दृश्यते ⌈नेतरत्र सा ॥ ६२ ॥

सप्तैव सन्ति कुलभूमिधरा महेन्द्रः शुक्तिस्तथा मलय ऋक्षकपारियात्रौ ।
सह्यः सविन्ध्य इह भारतवर्षमध्ये प्रोक्ताः पुराणविबुधैर्जगति प्रसिद्धाः ॥ ६३ ॥

f. 46v Pm यद्वीपषट्कं ⌈खलु दक्षिणस्थं व्यक्षात्स भूर्लोक इह प्रसिद्धः ।
f. 71v Bn.I सौम्यो भुवर्लोक इतश्च मेरुः स्वर्लोक उक्तश्च ⌈पुराणविद्भिः ॥ ६४ ॥

f. 5v Np ततः परं ⌈व्योम्नि महो जनोऽतस्तपश्च सत्यं किल सप्त लोकाः ।
f. 3v Pb केचिज्जगुर्लोक⌈चतुर्दशत्वमधः स्थिताः सप्त तथोर्ध्वसंस्थाः ॥ ६५ ॥

लङ्कापुरे रविरुदेति यदा तदानीं मध्याह्नसंज्ञसमयो यमकोटिपुर्याम् ।
f. 165r Sc सिद्धाभिधेऽस्तसमयोऽथसदाप्रतीच्यां ⌈स्याद्रोमकेस्फुटतरंरजनीदलं हि ॥ ६६ ॥

---

1 ॰वैश्यादि] ॰विट्भूद्ध Np, Sc
1 ॰व्यवस्था] ॰व्यस्था Bn.I
1 पृथक्पृथक्] पृथक् Bn.II
2 ॰मध्यस्था] ॰मध्येसा Sc
2 दृश्यते] दृश्य⟦ति⟧-ते- *corr. ras.* Sc
2 ६२] ६० Bn.II, Br, Pm, Rr; ६१ Np; २७ Sc
3 सन्ति] संते Bn.II
3 ॰धरा] ॰धर -ा- *add. intra lineam* Sc
3 ॰पारियात्रौ] ॰पारयात्रो Bn.II; ॰पारियात्रौ Sc
4 सविन्ध्य] सर्विध्य Bn.I
4 प्रोक्ताः] प्रे⌟⌈क्ता -ा- *add. in marg. dext. in linea confrac.* Bn.II
4 ॰विबुधै॰] ॰विविधै॰ Pb
4 ६३] ६१ Bn.II, Br, Pm, Rr; ६२ Np; *om.* Sc
5 यद्वीप॰] यव्द्वीप॰ *ut videtur* -दी- *corr. in marg. dext.* Pm
5 ॰षट्कं] ॰खट्कं Bn.II, Rr
5 खलु दक्षिणस्थं] खलक्षिणं Bn.II; खलु^v^क्षिणस्थं -द- *add. in marg. sup.* Pm; खलुक्षिणंस्थं Rr; खलु दक्षि[ण]स्थं *add. suprascr.* Sc
5 व्यक्षात्स] व्यक्षात्म Br; व्यक्ष[ा]त्स *add. suprascr.* Sc
5 इह] इति Np, Sc
6 सौम्यो] से⌟⌈म्यो -ा- *add. in marg. dext. in linea confrac.* Bn.II; सौम्ये Np, Sc
6 पुराण॰] पुरा[ण]॰ *add. suprascr.* Sc
6 ६४] ६२ Bn.II, Br, Pm, Rr; ६३ Np; २९ Sc
7 व्योम्नि] व्यो⟦म्मि⟧ -म्नि- *corr. in marg. sins.* Pm
7 सप्त लोकाः] सप्तं किल सप्तलोकाः Rr
8 केचिज्जगु॰] केचि⟦ज्जु⟧-ज्ज-गु॰ *corr. ras.* Sc
8 सप्त] सत्त Br
8 तथोर्ध्व॰] ततोर्द्ध॰ Bn.II, Rr; तथोर्द्ध॰ Pm
8 ६५] ६३ Bn.II, Br, Pm, Rr; ६४ Np; ३० Sc
9 रविरुदेति] [र]विरुदेति *add. suprascr.* Pm
9 यदा] यक्ष Br
9 ॰ह्नसंज्ञ॰] ॰ह्नमंज्ञ॰ Br
10 स्फुटतरं] स्फु[ट]ततरं *add. suprascr.* Sc
10 ६६] ६४ Bn.II, Br, Pm, Rr; ६५ Np; ३१ Sc

⌈यत्रोदयो भवति भानुमतश्च तत्र पूर्वा दिशाऽस्तमुपयाति पुनः स यत्र । f. 52v Rr
प्रत्यक्तयोस्तिमिकृता खलु दक्षिणोदक्तस्मात्समस्तविषये शशिदिक्सुमेरुः ॥ ६७ ॥

लङ्कानगर्या अथवोज्जयिन्याः श्रीगर्गराटात्कुरुजाङ्गलाद्वा ।
सुमेरुतो वा कुचतुर्थभागे प्राचीप्रतीच्योर्यमरोमके स्तः ॥ ६८ ॥

ताभ्यां न पश्चादथवा पुरस्ताल्लङ्कां विना यत्ककुभो विचित्राः ।
गोलाकृतिं तेन वदन्ति धात्रीं ज्योतिर्विदो दर्पणसंनिभां न ॥ ६९ ॥

भूगोलखण्डस्य शरप्रमाणं निघ्नं धरायाः परिणाह⌈मित्या । f. 165v Sc
तद्भूमिखण्डोपरिगं फलाख्यं क्षेत्रोद्भवं योजनकैः स्फु⌈टं हि ॥ ७० ॥ f. 73r Bn.II

सुमेरुमभितो निजं पुरमवाप्य वृत्तं च य-
त्स्फुटः कुपरिधिः स्मृतः स तु भचक्र ३६० भागाङ्कितः ।
तथा खरस ६० नाडिकाङ्कित इहाथवा योजनैः
स्फुटाख्यपरिणाहजैर्भवति खेटदेशान्तरम् ॥ ७१ ॥

इति भूगोलरचना ।

---

1 तत्र] यत्र Bn.I, Bn.II, Br, Pb, Pm, Rr
1 दिशाऽस्तमुपयाति] दिशास्त⟦म्यु⟧-मु-(प)याति *corr. ras., et add. suprascr.* Sc
2 प्रत्य॰] (प्र)सत्य॰ *corr. suprascr.* Pb
2 प्रत्यक्तयोस्ति॰] प्रत्यक्पुनस्ति॰ Np; प्रत्यक्(न)पुस्ति॰ *add. suprascr.* Sc
2 ६७] ६५ Bn.II, Br, Pm, Rr; ६६ Np; ३२ Sc
3 ॰गर्या अथवो॰] ॰गर्यामथवो॰ Np, Sc
4 वा कुचतुर्थ॰] वाहुचतुर्थ॰ Pm; वाक्चतुर्थ॰ Pb
4 प्राची॰] प्राचो॰ Pm
4 ॰प्रतीच्यो॰] ॰प्रतिच्यी॰ Bn.I; ॰प्रतीच्र्यो॰ Bn.II
4 ६८] ६६ Bn.II, Br, Pm, Rr; ६७ Np; ३३ Sc
5 ॰ल्लङ्कां] ॰थंकां Bn.I
6 तेन] तन Bn.I; तैन Br
6 वदन्ति] वं⟦ा⟧दंति *corr. ras.* Bn.I
6 धात्रीं] धात्र्या Np; धांत्रिं Pm; धात्र्य(ा) *add. suprascr.* Sc
6 ज्योति॰] ज्यो॰ Pm
6 ६९] ६७ Bn.II, Br, Pm, Rr; ६८ Np; ३४ Sc
7 भूगोल॰] भगोल॰ Sc
7 निघ्नं] निघ्नः Br
7 ॰मित्या] ॰मित्यां Bn.I; ॰भित्यां Bn.II; ॰नित्या Br; ॰मित्थां Pm
8 ७०] ६८ Bn.II, Br, Pm, Rr; ६९ Np; ३५ Sc
9 ॰वाप्य वृत्तं] ॰वावृप्यत्तं Bn.I; ॰नाथवृत्तं Sc
10 स] म Bn.II, Rr
10 भचक्र] भक्र Br
10 ३६०] *om.* Np, Sc
10 ३६० भागाङ्कितः] भागां ३६० कितः Bn.I; भाग ३६० कितः Pb
11 ६०] *om.* Bn.II, Br, Np, Pb, Pm, Rr, Sc
11 ॰काङ्कित] ॰कांकितद्र Bn.I
12 ७१] ६९ Bn.II, Br, Pm, Rr; ७० Np; ३६ Sc

अथ भगोलरचना ।

भूमेर्वायोरुपरि परितो व्योमकक्षावधिस्थः
पिण्डीभूतः प्रवहपवनो गोलकाकारमूर्तिः ।
[a] f. 6r Np [b] f. 72r Bn.I अन्तः [a]स्वच्छः स्फटिकघटवन्नी[b]रवत्सावकाशः
शश्वद्भ्राम्यन्नपरककुभं दक्षिणोदग्ध्रुवाभ्याम् ॥ ७२ ॥

f. 166r Sc कुर्वन्मध्ये विषुववलयं पार्श्वजात|द्युवृत्तं
तस्यान्तःस्थं श्लथमिव समं तुङ्गपातादियुक्तम् ।
प्राचीं गच्छन्मृदुगतियथा ज्ञेयमेतद्भचक्रं
तस्यान्तःस्थाः सकलखचराः प्राचि यान्ति स्वभुक्त्या ॥ ७३.१ ॥

कुर्वन्मध्ये विषुववलयं पार्श्वजातद्युवृत्तं
f. 40v Br यस्मिन्नन्तर्जल इव झषास्ता|रकाः प्राक्तरन्ति ।
उड्डीयन्ते युगपदुडवः खे यथा खेचरेन्द्रा-
स्तच्चक्रं वा भवलयमिति प्रोच्यते वा भगोलः ॥ ७३.२ ॥

---

1 भगोल॰] भ⟦ा⟧गोल॰ *corr. ras.* Bn.II
2 व्योमकक्षावधिस्थः] यावदक्षाख्यकक्ष Bn.II; यावद्दक्षाख्यकक्षा Br, Np, Pm, Rr
3 प्रवहपवनो] प्रवह⟦व⟧पवना *corr. ras.* Bn.I
4 स्वच्छः] स्वछुः Bn.II
4 स्फटिक॰] स्फुटिक॰ Bn.II, Pb
4 ॰घटवन्नीर॰] ॰घटवन्निल॰ Np; ॰घ⟦ ि⟧टवन्नीर॰ *corr. ras.* Pm; ॰घटव⟦र⟧न्नील॰ *corr. ras.* Sc
5 ॰द्भ्राम्यन्न॰] ॰द्भ्रम्यन्न॰ Np; ॰द्भ्राम्यत्त॰ Pb
5 ॰न्नपरक॰] ॰न्न⟦र⟧परक॰ *corr. ras.* Rr
5 ॰ककुभं] ॰ कुकभं (२ १) *corr. transp.* Pb
5 दक्षिणोदग्ध्रुवाभ्याम्] दग्धध्रुवाभ्यां Bn.II, Rr
5 ७२] ७० Bn.II, Br, Pm, Rr; ७१ Np; १ Sc
6 विषुवव॰] विषुव (व) *add. suprascr.* Sc
6 ॰वलयं] ॰वलये Np, Sc
6 ॰द्युवृत्तं] ॰युवृत्तं Sc
7 श्लथ॰] श्वथ॰ Bn.I
7 समं] स⁻ *ut videtur, fort.* ⟨भं⟩ Np; सभम् Bn.I; सभ Sc
7 ॰युक्तम्] ॰युक्तिं Sc
8 प्राचीं] प्राचो Np
8 ॰मृदुगति॰] ॰मृदुरति॰ Pb
9 ७३.१] ७३ Bn.I, Pb; ७२ Np; २ Sc
10 विषुववल॰] विषुवल॰ (व) *add. suprascr.* Pm
12 उड्डीयन्ते] उडीयन्ते Bn.II, Rr
12 ॰दुडवः] ॰दुडयः Pm
13 स्तच्चक्रं] स्तचक्रं Bn.II
13 ७३.२] ७१ Bn.II, Br, Pm, Rr

6–9 कुर्वन्मध्ये॰…॰स्वभुक्त्या Reading in MSS Bn.I, Np, Pb, and Sc
10–13 कुर्वन्मध्ये॰…॰भगोलः Reading in MSS Bn.II, Br, Pm, and Rr

तस्यैवान्तर्व्रजति पुरतो येन मार्गेण भानुः
सम्भेद्यद्विर्विषुववलयं सौम्ययाम्यप्रवृत्तः ।
क्रान्तेर्वृत्तं भवनवलयं राशिवृत्तं च तत्स्या-
देतस्यार्धे शशियमदिशोर्मेषजूकादिके स्तः ॥ ७४ ॥

क्रान्तेर्वृत्तादुभयत इह स्तो ध्रुवौ यौ कदम्बौ
तौ तद्योगं दधति सकला राशयो द्वादशैते ।
तुल्यार्कांशैर्भवनवलयेऽजाद्य षड्बाणसूत्रै-
र्भक्ते भ्राम्येद्ध्रुवमभित इह प्रत्य⸢हं स्वः कदम्बः ॥ ७५ ॥ f. 166v Sc

सार्धत्रयोविंशति ⸢२३।३० चापभागैः सौम्येऽथयाम्ये खलु राशिचक्रम् । f. 47r Pm
यस्मिन्सुदूरेऽथ कदम्बयुग्मध्रुवद्वयोपेतमिहायनाख्यम् ॥ ७६ ॥

शङ्खा⸢वर्तवदिन्दुपूर्वकखगः प्राच्यां च येनाध्वना f. 53r Rr
चक्रे भ्राम्यति राशिमण्डलमथो भित्त्वा द्विवारं मुहुः ।
सौम्यावाग्गतवांस्तथा प्रवहतो ⸢नित्यं प्रतीच्यां चल- f. 73v Bn.II
न्किंचित्किंचिदुपेत्य योगमपरं तद्बाणवृत्तं जगुः ॥ ७७ ॥

---

1 ०वान्तर्व्रजति] ०वान्तव्रजति Bn.II, Rr; ०वान्त⟦ा⟧र्व्रजति *corr. ras.* Pm
1 मार्गेण] ⟦मो⟧-मा-र्गेण *corr. ras.* Br
2 ०द्विर्विषुव०] ०द्वि⌐षुव० -वि- *add. in marg. dext. in linea confrac.* Pm
3 ०र्वृत्तं] ०र्व्रतं Bn.II
3 भवनवलयं] भवनवनवलयं० Bn.I
3 राशिवृत्तं] गशिवृत्तं Br
4 ०दिशो०] ०दिशी० Bn.I, Br, Pb, Pm; ०देशे Bn.II; ०दिशि० Rr
4 ०र्मेषजूका०] ०मेषयूका० Bn.I; ०मेषजूका० Br, Pb, Pm, Rr
4 ७४] ७२ Bn.II, Br, Pm, Rr; ७३ Np; ३ Sc
5 ०र्वृत्ता०] ०र्भृत्ता० Np
5 स्तो] स्तौ Rr
5 ध्रुवौ] ध्रु‸-वौ२- *add. in marg. sins.* Sc
6 तद्योगं] तद्योंगं Bn.II; तद्योगे Np; त⟦द्यौ⟧-द्यो-गे *corr. intra lineam* Sc
6 राशयो] राशकलाराशयो Br; राजयो Pm
7 ०कांशैर्भ०] ०कांशेभ० Pb
8 ०द्ध्रुव०] ०द्ध्रुव० Bn.II, Rr
8 ०मभित] ०मभि Sc
8 इह] एह Pb
8 स्वः] स्वाः Sc
8 ७५] ७३ Bn.II, Br, Pm, Rr; ७४ Np; ४ Sc
9 २३।३०] *om.* Np, Sc
9 सौम्ये०] सोम्ये० Bn.II; ऽसौम्ये० Pm
10 यस्मिन्सुदूरेऽथ] यस्मिन्तुदूरेथ Br
10 ०युग्म०] ०युग्मे० Np
10 ०द्वयो०] ०द्वरो० Bn.II
10 ७६] ७४ Bn.II, Br, Pm, Rr; ७५ Np; ५ Sc
11 ०वदिन्दु०] ०वदिंदु० *ut videtur* Pm
11 ०खगः] ०खगाः Np; खग⟦ा⟧ः *corr. ras.* Sc
12 भ्राम्यति] भ्रास्यति Bn.I; म्यति (भ्रा suprascr.) *add. suprascr.* Pb
13 ०वाग्गतवांस्तथा] ०वाग्गधवातथा Np; ०वागात्तवांस्तस्था Pm; वागथवास्त‸ *corr. desunt* Sc
13 प्रवहतो] वहतो Rr
14 ०त्किंचिदुपरि] ०त्कीचिदुपेत्य (ि suprascr.) *corr. suprascr.* Sc
14 ७७] ७५ Bn.II, Br, Pm, Rr; ७६ Np; ६ Sc

प्रवहपवनवेगव्याहताधोर्ध्वगोलो हरिदिशमपि गच्छन्नुच्चतामेति खेटः ।
तदनु तदवरोहो याति यावत्समत्वं तदुदितविपरीतं नीचगत्वे खगस्य ॥ ७८ ॥

[a] f. 167r Sc [b] f. 6v np खगकदम्ब⌈[a]युगोपरि ⌈[b]यद्गतं तदिह खेटशिलीमुखसूत्रकम् ।
अथ खगध्रुवयुग्मगतं च यद्वलयमेतदपक्रमसूत्रकम् ॥ ७९ ॥

f. 4r Pb अस्मिन्भचक्रे ग⌈जवेदसंख्या भमूर्त्तयो या यवनैः प्रदिष्टाः ।
f. 72v Bn.I भानां सहस्रं नयनाक्षियुक्तं १०२२ यथागमे ताः सुबुधैः प्रदेयाः ⌈॥ ८० ॥

तासां भचक्रतः सौम्या मूर्तयश्चैकविंशतिः ।
याम्याः पञ्चदशप्रोक्ता मध्ये द्वादशराशयः ॥ ८१ ॥

कदम्बमभितो येन खेटो गच्छति चाध्वना ।
शराग्रमण्डलं तत्स्याद्यत्स्याद्द्युज्याख्यमण्डलम् ॥ ८२ ॥

---

1 ०वेगव्याहता०] ०वेगादाहतो० Np, Sc
1 ०धोर्ध्व०] ०धोर्द्ध० Pm
1 ०गोलो] ०गोला Pb
1 ०मपि] ०मभि Np
1 ०तामेति खेटः] ०तामे ⌟ खेटः -ति- *add. in marg. sins. in linea confrac.* Pb
2 तदवरोहो] तववरोहो Np; तदवहोरो (२ १) *corr. transp.* Pm
2 यावत्समत्वं] यावत्समत्यं Pm; या‸समत्वं *corr. desunt* Sc
2 नीच०] नीय० Bn.I
2 ७८] ७६ Bn.II, Br, Pm, Rr; ७७ Np; ७ Sc
3 ०कदम्ब०] ०दंकेम्ब० Np
3 ०शिली०] ०शली० Bn.II, Pm, Rr
3 ०मुखसूत्रकं] ०मुसूत्रगं Bn.I
4 खग०] गख० Bn.II, Rr
4 च यद्वलय०] यचद्वलय० Bn.I; चयद्ग(व)⟦यै⟧-य-ले *corr. ras., et add. suprascr.* Pm; (य)चद्वलय० *add. suprascr.* Sc
4 ०मेतदपक्रम०] ०मेतक्रम० Bn.I; मेतद(क्रम)प० *add. suprascr., idem* -क्रम- *in marg. dext.* Pm
4 ०सूत्रकम्] सूत्र(कं) *add. suprascr.* Sc
4 ७९] ७७ Bn.II, Br, Pm, Rr; ७८ Np; ९ Sc
5 अस्मिन्भ०] अस्मिनभ० Bn.II
5 ०संख्या] ०संस्था Bn.II, Br, Pm, Rr
5–6 ०र्त्तयो या०...०ताः सु०] *desunt* Bn.I
5 भमूर्त्तयो] भसूर्त्तयो Pm
5 यवनैः] वनकैः Np, Sc
6 नयना०] नना० Pm
6 १०२२] *om.* Np
6 यथागमे०] यथांगमे० Br, Pm, Rr, Sc
6 प्रदेयाः] प्रदेयम् Bn.I, Bn.II, Br, Pb, Pm, Rr, Sc
6 ८०] ७८ Bn.II, Br, Pm, Rr; ७९ Np; ९ Sc
8 पञ्चदशप्रोक्ता] पञ्चदशःप्रोक्ता Bn.II, Pm, Rr
8 द्वादश०] द्वादशा० Bn.I
8 ८१] ७९ Bn.II, Br, Pm, Rr; ८० Np; ९१० Sc
9 चाध्वना] चाधुना Sc
10 तत्स्याद्य०] तत्स्या(द्)यथाद्यु० *add. suprascr.* Sc
10 ०त्स्याद्द्यु०] ०स्माद्द्यु० Bn.I; ०त्स्याद्यु० Br, Pm, Rr; ०थाद्यु० Np
10 ८२] ८० Bn.II, Br, Pm, Rr; ८१ Np; ११ Sc

यावन्ति वृत्तानि भवन्ति गोले तावन्ति सर्वाणि च कल्पितानि ।
यथान्ध⌈कारे कृतकौतुकानामालातचक्रभ्रमयः शिशूनाम् ॥ ८३ ॥ f. 167v Sc

अत्रैव कक्षा द्युसदां निबन्ध्या उपर्युपर्येव विधोः क्रमेण ।
विक्षेपवृत्तेषु निजेषु तद्वन्मेषादयो राशय एव वेद्याः ॥ ८४ ॥

अथ खगोलरचना ।

भूगर्भक्षितिपृष्ठसूत्रसदृशो यो ना कुगर्भे स्थित-
स्तं यत्खं खलु सर्वतो विजयतेऽनन्तो खगोलोऽस्ति सः ।
वृत्तं तत्र भवेद्यथा परिकरो यस्य ध्रुवौ सर्वदा
खाधःस्वस्तिकसंज्ञकौ बुधजनास्तद्भूजसंज्ञं विदुः ॥ ८५ ॥

प्राक्खस्वस्तिकपश्चिमोपरिगतं यन्मण्डलं यद्ध्रुवौ
याम्योदक्ककुभौ वदन्ति गणका वृत्तं स⌈माख्यं च तत् । f. 41r Br
याम्योद⌈[a]ङ्गरमस्तकोपरि गतं यस्य ध्रुवौ ⌈[b]सर्वदा [a] f. 168r Sc [b] f. 74r Bn.II
प्राक्पश्चादिदमेव सर्वमुनयो मध्याह्नवृत्तं जगुः ॥ ८६ ॥

⌈अथेष्टदिक्तद्विपरीतकाष्ठाखस्वस्तिकाख्योपरि निःसृतं यत् । f. 7r Np
तदेव दृङ्मण्डलनामधेयं कोणेषु कोणाभिधमामनन्ति ॥ ८७ ॥

---

2 ॰मालात॰] ॰मालाभ॰ Np, Sc
2 ८३] ८२ Bn.I, Np; ८१ Bn.II, Br, Pm, Rr; १२ Sc
3 उपर्युपर्येव] उपयुपर्य्येव Bn.I; दुपर्य्युपर्येव Sc
4 विक्षेप॰] ⟦कि⟧विक्षेप॰ *corr. ras.* Pm
4 ८४] ८३ Bn.I, Np; ८२ Bn.II, Br, Pm, Rr; १३ Sc
5 अथ खगोलरचना] अथ खगोलः Bn.II, Br, Pm, Rr; अथ खगोलर[च]ना *add. suprascr.* Sc
6 कुगर्भे] कुगर्भो Bn.II, Rr
7 यत्खं] यखं Sc
9 खाधः॰] ख⌟धः॰ -ा- *add. in marg. dext. in linea confrac.* Bn.II
9 ॰संज्ञकौ] ॰संज्ञको Bn.II; ॰संज्ञितौ Np, Sc
9 ॰जन्नस्तद्भू॰] ॰जनास्त[त्]भू॰ *add. suprascr.* Pm
9 विदुः] जगुः Np, Sc
9 ८५] ८४ Bn.I, Np; ८३ Bn.II, Br, Pm, Rr; १ Sc
10 यद्ध्रुवौ] यद्धुवौ Bn.I, Rr; यद्धवौ Bn.II
11 याम्योद॰] म्योद॰ Pm
11 गणका] [ग]कणका *corr. suprascr.* Pb
12 गतं] गतौ Np, Sc
12 ध्रुवौ] ध्रवौ Bn.II
13 प्राक्पश्चादि॰] प्राक्श्चदि॰ Bn.I
13 ॰दमेव] ॰दमे[व] *add. subscr.* Pm
13 ८६] ८५ Bn.I, Np; ८४ Bn.II, Br, Pm, Rr; २ Sc
14 अथेष्ट॰] अथे⟦ा⟧ष्ट॰ *corr. ras.* Bn.II
14 ॰खस्वस्ति॰] ॰स्वस्ति॰ Bn.I
15 दृङ्मण्डल॰] दृडमण्डल॰ Bn.II; दृग्मण्डल॰ Pb
15 ॰मधेयं] ॰मधेयं☓ -कोणेषु२- *add. in marg. dext.* Rr
15 ८७] ८६ Bn.I, Np; ८५ Bn.II, Br, Pm, Rr; ३ Sc

कदम्बखस्वस्तिकगामि यच्च तदेव दृक्क्षेपकमण्डलं हि ।
पूर्वापराशाध्रुवयुग्ममाप्तमुन्मण्डलं नाम बुधैर्निरुक्तम् ॥ ८८ ॥

f. 53v Rr द्युज्या⌈मण्डलसदृशं द्युज्यावृत्तं स्थिराख्यमत्रापि ।
दक्षिणसौम्यखगोपरि यातं समसूत्रसंज्ञितं वृत्तम् ॥ ८९ ॥

f. 73r Bn.I एवं पूर्वापरखगमाप्तं म⌈ध्याभिधं सूत्रम् ।
f. 47v Pm ⌈पूर्वां वापरकाष्ठां केन्द्रं कृत्वा यदा भतो वृत्तम् ।
f. 168v Sc ⌈द्युज्यामण्डलसदृशं ज्ञेयं पूर्वाभिधं तच्च ॥ ९० ॥

अथ खस्वस्तिकमेव स्पष्टं केन्द्रं प्रकल्प्य यद्वृत्तम् ।
कुर्यादाभात्तत्स्याद्दृग्ज्यावृत्तं सदोर्ध्वमधरं वा ॥ ९१ ॥

द्वादशभिः समसूत्रैर्द्वादशभावाख्यभूजानि ।
क्षितिजादथो धनाद्या भावा ज्ञेया यथा लग्नम् ॥ ९२ ॥

अथो कुजं गच्छति योऽत्र भांशास्तदेव लग्नं क्रियपूर्वकं स्यात् ।
प्रत्यक्कुजे चास्ततनुस्तथैव मध्याह्न वृत्तोपरि मध्यलग्नम् ॥ ९३ ॥

---

1 ०खस्वस्तिक०] ०कस्वस्तिक० Br, Bn.II, Pm, Rr; ०स्वस्वस्तिक० Sc
1 दृक्क्षेप०] दृक्षेप० Pm
2 पूर्वापराशा०] पूर्वापराख्य० Sc
2 मुन्मण्डलं] मन्मण्डलं Bn.II
2 ८८] ८७ Bn.I, Np; ८६ Bn.II, Br, Pm, Rr; ४ Sc
3 द्युज्या०] युज्या० Sc
4 यातं] व्यातं Bn.II
4 वृत्तम्] व्रतं Bn.II
4 ८९] ८८ Bn.I, Np; ८७ Bn.II, Br, Pm, Rr; ५ Sc
5 सूत्रम्] सूत्रं ५ Sc
6 यदा भतो] य अ⌟नक्षत्रात्तो -ा- *add. in marg. dext. in linea confrac.* Bn.II
6 वृत्तम्] वृत्तं ८८ Bn.II, Br, Pm, Rr; वृत्तं ॥ ९० Pb
7 द्युज्यामण्डलसदृशं] मण्डलसदृशं Np; द्युज्यासंज्ञज्ञेयं Sc
7 ज्ञेयं] ज्ञूयं Br; ज्ञेयं ज्ञेयं Np
7 ०भिधं तच्च] ०भिधनन्तच्च Np
7 ९०] ८९ Bn.I, Np; *om.* Bn.II, Br, Pb, Pm, Rr; ६ Sc
8 खस्वस्तिकमेव] खस्वस्तिक *finis imperfecta* Pb
8 यद्वृत्तम्] यद्वृत्तं ८९ Bn.II, Br, Pm, Rr
9 ०त्स्याद्दृग्ज्या०] ०त्स्याद्दग्ज्या० Bn.II, Pm, Rr
9 सदोर्ध्व०] सदोर्द्ध० Bn.II, Pm, Rr
9 ९१] ९० Bn.I, Np; *om.* Bn.II, Br, Pm, Rr; ७ Sc
10 समसूत्रैर्द्वादश०] समसूमसूत्रैद्धदिश० Bn.I
10 ०भूजानि] भूजानि ९० Bn.II, Br, Pm, Rr
11 ०जादथो] ०जादधो Bn.I; ०जादयो Bn.II, Br, Rr; ०ज्यादयो Pm; ०जादतो Rr
11 धनाद्या] धना (द्या) *add. suprascr.* Sc
11 ९२] ९१ Bn.I, Np; *om.* Bn.II, Br, Pm, Rr, Sc
12 भांश०] भास० Bn.II, Rr
13 प्रत्यक्कुजे] प्रत्यक्कुते Bn.I
13 ९३] ९२ Bn.I, Np; ९१ Bn.II, Br, Pm, Rr; *om.* Sc

अधः स्थितं तत्र तु तुर्यसंज्ञं द्वयं द्वयं चान्तरगं हि तेषाम् ।
खलग्नयोर्लग्नचतुर्थयोश्च सुखस्त्रियोः स्त्रीवियतोश्च मध्ये ॥ ९४ ॥

मध्याह्नभूजवलयान्तरसंस्थितानां ना⌈ड्याख्यचक्रसमवृत्तभमण्डलानाम् । f. 169r Sc
त्रींस्त्रीँल्लँवान्निजनिजाङ्किगतान्समानान्भक्त्वा ततो निजनिजध्रुवसम्मुखानि ॥ ९५ ॥

षड्वद्विलिख्य वलयानि च सुस्थि⌈राणि f. 7v Np
भावाभिधानि कुजतोऽधरसंस्थितानि ।
द्रव्यादितः ⌈क्रमतयोपरिगानि रिष्फा- f. 74v Bn.II
द्व्यस्तानि तानि सकलानि गृहाणि विन्द्यात् ॥ ९६ ॥

यस्मिन्यस्मिन्वलये राशिलवो लगति तत्र तत्रैव ।
स सभावः स्फुटसंज्ञस्त्रिविधमिदं भावचक्रं स्यात् ॥ ९७ ॥

उदक्परक्रान्तिरिनस्य येषु लम्बेन तुल्या विषयेषु तेषु ।
कर्कस्थितो भानुरुदेत्युदीच्यामवाचिनक्रः समु⌈पैति चास्तम् ॥ ९८ ॥ f. 169v Sc

---

1 अधः] अथ Bn.II, Rr
1 तत्र तु तुर्य॰] तत्रंनुतुर्य॰ Pm
1 *alterum* द्वयं] द्व्यं⟦द्वयं⟧ *corr. ras.* Bn.II; ⟦यं⟧द्वयं *corr. ras.* Pm
1 तेषाम्] तेषां ९० Sc
2 खलग्न॰] [-1-] लग्न॰ *intra lineam, fort.* ⟨ख⟩ Bn.I
2 ९४] ९३ Bn.I, Np; ९२ Bn.II, Br, Pm, Rr; *om.* Sc
3 नाड्याख्य॰] नाड्याख्यि॰ (ख्य *suprascr.*) *idem suprascr.* Rr
4 त्रींस्त्रीँल्लँवान्नि॰] त्रींस्त्रींल्लयान्नि॰ Pm; त्रीन्त्रीन्लन्नि॰ (वा) *add. suprascr.* Sc
4 ॰जाङ्कि॰] ॰जाह्नि॰ Bn.I; ॰जांहि॰ Bn.II, Pm; ॰जांह्नि॰ Br, Rr
4 ॰न्समा॰] ॰न्सुमा॰ Bn.II, Rr
4 ॰क्त्वा] ॰त्का Bn.I, Bn.II, Br, Pm, Rr; ॰क्ता Sc
4 *alterum* ॰निज॰] *om.* Pm
4 ॰सम्मुखानि] ॰सूंमु[-1-]नि *intra lineam, fort.* ⟨खा⟩ Bn.I; ॰संमुख्यनि Bn.II; ॰संमुखावि Br
4 ९५] ९४ Bn.I, Np; ३० Bn.II, Br, Rr; ९३ Pm; ११ Sc
5 षड्वद्विलि॰] शश्वद्विलि॰ Sc
5 सुस्थिराणि] सुस्थि⟦ा⟧नि (ता) *corr. ras., et add. suprascr.* Pm
6 भावाभि॰…॰स्थितनि] *desunt* Pm
6 कुजतो] कुरुते Bn.II, Br, Rr
7 रिष्फा॰] रिःफा॰ Sc
8 गृहाणि] [-1-] हाणि *intra lineam, fort.* ⟨गृ⟩ Pm
8 ९६] ९५ Bn.I, Np; ३१ Bn.II, Br, Pm, Rr; १३ Sc
10 स सभावः स्फुट॰] समभावः स्फुंत॰ Bn.II
10 ९७] ९६ Bn.I, Np; ३२ Bn.II, Br, Pm, Rr; १३ Sc
11 लम्बेन] लवेषु Sc
11 तेषु] येषु Sc
12 भानुरुदे॰] भानु-रु-दे॰ *add. intra lineam* Sc
12 ९८] ९७ Bn.I, Np; ३३ Bn.II, Br, Pm, Rr; १४ Sc

क्षणं भचक्रं कुजवत्स्थितं हि समुद्गतं स्याद्युगपन्मृगादिम् ।
तथैव कर्कादिकमस्तमेति ततःपरं कर्कमुखं भषङ्क्रम् ॥ ९९ ॥

f. 73v Bn.I क्रमेण लङ्को⌈दयनाडिकाभिर्द्विसंगुणाभिः समुदेति भूजात् ।
नक्रादिकं चास्तमुपैति तद्वत्ततो घटीषष्टिमितं दिनं स्यात् ॥ १०० ॥

रात्रिप्रमाणं क्षणमेव नास्ति द्युःक्षीयतेऽतोऽपि निशैषते च ।
f. 41v Br तुला⌈ङ्गतेऽर्के द्युनिशे समेस्तस्ततोऽपचीयेत दिनप्रमाणम् ॥ १०१ ॥

वर्धेत रात्रिर्धनुरन्तमाप्ते भानौ घटीशुन्यमितं दिनं स्यात् ।
f. 54r Rr रात्रिः परा षष्टि⌈घटीप्रमाणा स्यातां ततो रात्रिदिने विलोमे ॥ १०२ ॥

f. 170r Sc यावद्भिरंशैर्नि⌈जलम्बतोऽधिका खेटस्फुटक्रान्तिरुदक्स्थिता भवेत् ।
तावत्सदा दृश्यतनुः स खेचरः सदास्तगो दक्षिणदिक्स्फुटापमे ॥ १०३ ॥

---

1 ॰वत्स्थितं] ॰संस्थितं Bn.I; ॰वातिस्थितं Bn.II; ॰वात्स्थितं Pm, Rr
1 स्याद्युगपन्मृगादिम्] स्याद्युगन्मर्णादि Pm
2 ॰दिकमस्त॰] ॰दिक^v^स्त॰ -म- *add. in marg. dext.* Pm
2 ततःपरं] तःपरं Bn.I
2 भषङ्क्रम्] भखङ्कं Bn.II
2 ९९] ९८ Bn.I, Np; ३४ Bn.II, Br, Pm, Rr; १५ Sc
3 समुदेति] समुदोति Br
4 नक्रादिकं] नक्रादिकं⟦ा⟧ *corr. ras.* Bn.I
4 स्यात्] स्या Bn.I
4 १००] ९९ Bn.I, Np; ३५ Bn.II, Br, Pm, Rr; *om.* Sc
5 ॰ऽतोऽपि] ॰नापि Np; ॰नोपि Sc
5 निशैषते] निरौद्यते Bn.I; निशैधते Np, Sc
6 द्युनिशे समेस्त] द्युनिसे शमेस्त Bn.II
6 समेस्त] समस्त Np, Sc
6 दिनप्रमाणम्] दिनप्रमा^v^ -णं३- *add. in marg. sup.* Bn.I
6 १०१] १०० Bn.I, Np; ३६ Bn.II, Br, Pm, Rr; *om.* Sc
7 रात्रि॰] राप्रि॰ Bn.I
7 ॰माप्ते] ॰म⟦वि⟧ *corr. ras.*, -प्ते- *add. in marg. dext.* Pm
7 ॰शुन्य॰] ॰शून्य॰ Pm
7 स्यात्] स्यात् १७ Sc
8 रात्रिः] त्रिराः॰ Bn.I
8 ॰घटी॰] ॰घर्यटी॰ Br
8 ॰प्रमाणा] ॰प्रमणा Br; ॰प्रमाण Np
8 रात्रि॰] गत्रि॰ Br
8 विलोमे] विले⌟मे -ा- *add. in marg. dext. in linea confrac.* Bn.II
8 १०२] १०१ Bn.I, Np; ३७ Bn.II, Br, Pm, Rr; १८ Sc
9 ॰रंशै॰] ॰रेशै॰ Bn.II, Rr
9 ॰रुदक्स्थिता] ॰रुक्स्थिता Bn.II, Pm, Rr; ॰रक्स्थिता द *add. suprascr.* Sc
10 तावत्सदा] यावत् सदा Bn.II, Rr
10 १०३] १०२ Bn.I, Np; ३८ Bn.II, Br, Pm, Rr; १९ Sc

अतोऽक्षभागैर्विशरांशखागै ६९|४८ युग्मेन्दुभस्थे द्युमणौ दिनं स्यात् ।
धनुर्मृगस्थे च निशा समस्ता परेषु भेषु द्युनिशोर्व्यवस्था ॥ १०४ ॥

सार्धाष्टशैलैः ७८|३० पलभागकैश्च दिनं वृषाद्ये भचतुष्टयेऽर्के ।
⌈निशा पुनर्वृश्चिकतश्चतुष्के खाङ्का ९० क्षभागैर्द्युसमार्धकेन ॥ १०५ ॥ f. 8r Np

रात्रिस्तथैवास्ति सुमेरुवासिनां मेषादि षट्केऽर्कविलोकनं सदा ।
तुलादिगेऽर्के ⌈[a]रजनीमृगा⌈[b]दिगे निशार्धमेषां द्युदलं च कर्कगे ॥ १०६ ॥ [a] f. 75r Bn.II [b] f. 74r Bn.I

⌈यथा यथा क्रान्तिरुदग्विवर्धने तथा तथार्कस्य समुन्नतांशकाः । f. 170v Sc f. 48r Pm
अतो यदा कर्कटगो दिवाकरः परोन्नतत्वं समुपैत्यहर्दले ॥ १०७ ॥

---

1 अतोऽक्ष॰] अतोक्षि॰ Np
1 ॰खागै] ॰खांगै Bn.II, Br, Pm
1 युग्मेन्दु॰] द्युग्मे॰ Bn.I; द्यग्मे॰ Br
1–2 द्युमणौ दिनं स्यात् ॥ धनुर्मृगस्थे] (द्युमणौ दिनं स्यात् । धनुर्मृगस्थे वनिदलमतः कृष्णशुक्लाष्टमीतोह्नोरात्रेश्चप्रवेशो) (सितसितसितदलयोर्धस्त्ररात्रंप्रसिद्धं ॥ १०९ ॥ एवं समस्तरचना युक्तोज्ञेयः खगोलाख्यः ॥ अथ भू)(पृष्ठगतस्य प्रायश्चैवं भवेत्खगोलोयः । ११० । ज्ञेयः सगोलविज्ञैर्दृग्गोलः केन्द्रभिन्नत्वात् । अथ निज)(मण्डलचापं यल्लक्षणनामकं वदाम्ये तत् ॥ १११ ॥ किषुववृत्तभवृत्तयुतिस्तु या क्रियतुलाधरयोर्मु)(खम) द्युमणौ दिनं स्यात् । धनुर्मृगस्थेशा *secl., ins. per errorum ex* verses 110b, 111a-b, 112a-b, 113a, *alterum* द्युमणौ दिनं स्यात् । धनुर्मृगस्थे *idem* Bn.I
2 ॰र्मृगस्थे च निशा] ॰र्मृगस्थेशा Bn.I
2 समस्ता] प्रदिष्टा Np; सस्ता Sc
2 द्युनिशो॰] द्युनिशौ॰ Bn.II; षुनिशो॰ Sc
2 १०४] १०३ Bn.I, Np; ३९ Bn.II, Br, Pm, Rr; २० Sc
3 ७८|३०] *om.* Bn.I, Bn.II, Br, Pm, Rr, Sc
3 पल॰] पुल॰ Bn.I
3 वृषाद्ये] वृषाखे Bn.I
3 भचतुष्ट॰] भतुष्ट॰ (च) *add. suprascr.* Pm
4 ॰तश्चतुष्के] ॰तश्चतुर्थे Np
4 ९०] *om.* Np
4 क्षभागै॰] क्षभौगै॰ Bn.I; क्षभारौ Br
4 ॰द्युसमार्ध॰] ॰द्येसमार्द्ध॰ Bn.II; ॰द्युसमार्द्धृ॰ Br; ॰द्युसमर्द्ध॰ Np, Sc
4 १०५] १०४ Bn.I, Np; ४० Bn.II, Br, Pm, Rr; *om.* Sc
5 मेषादि षट्के॰] मेषषट्के॰ Br; मेषट्के॰ Rr
5 ॰ऽर्कवि॰] ॰केवि॰ Bn.I
6 ॰मृगादिगे] ॰ष्टगादिगे Br
6 निशार्ध॰] निषाद्धि॰ Bn.II; निषार्ध॰ Pm; निषार्द्ध॰ Rr
6 १०६] १०५ Bn.I, Bn.II, Br, Np, Pm, Rr; २१ Sc
7 ॰रुदग्विव॰] ॰रुग्विव॰ Bn.I
7 ॰वर्धने] ॰वर्धते Np; ॰वर्द्धते Sc
7 समुन्नतां॰] समु⟦ग्तां⟧-तां-॰ *corr. ras.* Pm
7 ॰न्नतांशकाः] ॰न्नतांशकां Bn.II; ॰न्नतांशाः Np, Sc
8 कर्कटगो] कर्कगतो Np, Sc
8 ॰पैत्यहर्दले] ॰पैतिहर्दले Bn.II, Pm, Rr; ॰पैत्यर्हले Np
8 १०७] १०६ Bn.I, Bn.II, Br, Np, Pm, Rr; *om.* Sc

विपर्ययो वाडववह्निवासिनां सुरद्विषां रात्रिदिनव्यवस्थयोः ।
अतः सुराणाममरद्विषां तथा दिवानिशं भास्करवर्षसम्मितम् ॥ १०८ ॥

फलाय संहिताविदो निशार्धतो दिनं जगुः ।
दिनार्धतश्च यामिनीं ततोऽयनाद्यहर्निशम् ॥ १०९ ॥

इन्दोर्गोलोपरिष्टाद्वसति पितृगणोऽमान्तकालेऽर्कतोऽध-
स्तिष्ठन्मूर्धोपरिष्टात्कलयति तरणिं तेन तेषां दिनार्धम् ।
दर्शोऽथो पूर्णमासीरजनिदलमतः कृष्णशुक्लाष्टमीतो-
f. 171r Sc ऽह्नोरात्रेश्च प्रवेशोऽसितसि|तदलयोर्धस्त्ररात्रं प्रसिद्धम् ॥ ११० ॥

एवं समस्तरचनायुक्तो ज्ञेयः खगोलाख्यः ।
अथ भूपृष्ठगतस्य प्रायश्चैवं भवेत्खगोलो यः ॥ १११ ॥

ज्ञेयः स गोलविज्ञैर्दृग्गोलः केन्द्रभिन्नत्वात् ।
अथ निजमण्डलचापं यल्लक्षणनामकं वदाम्येतत् ॥ ११२ ॥

विषुववृत्तभवृत्तयुतिस्तु या क्रियतुलाधरयोर्मुखमस्ति तत् ।
अयनवृत्तभवृत्तयुतिश्च यायनमुखं खलु दक्षिणसौम्यदिक् ॥ ११३ ॥

---

1 विपर्ययो] विपर्यये Bn.II, Rr
1 ॰व्यवस्थयोः] ॰व्य[व]स्थयोः *add. suprascr.* Sc
2 ॰ममर॰] ॰मसुर॰ Np; ॰चसुर॰ Sc
2 भास्करवर्षसम्मितम्] भास्कर^संमितं -वर्ष- *add. in marg. sins.* Pm
2 १०८] १०७ Bn.I, Pm, Rr; ७ Bn.II, Np; १७ Br; *om.* Sc
3 जगुः] जगुः २३ Sc
4 ततोऽयनाद्य॰] तप्तोयनाड्य॰ Bn.II; ततोऽयनाड्य॰ Rr; ततोयनाघ॰ Sc
4 १०९] १०८ Bn.I, Bn.II, Br, Pm, Rr; ८ Np; *om.* Sc
5 ॰गणोऽमान्त॰] ॰गणो[ऽ]मांत॰ *add. suprascr.* Rr
5 ॰कालेऽर्कतोऽध] ॰काले । र्कतोध Br
7 पूर्णमासी॰] पूर्णिमासी॰ Sc
7 ॰रजनि॰] ॰रजनी॰ Bn.II, Np, Rr
8 ॰ऽसितसितदल॰] ॰ऽशितशितदल॰ Bn.II; ॰ऽसितदल॰ Br
8 ११०] १०९ Bn.I; ९ Bn.II, Np, Pm, Rr; १९ Br; २५ Sc
9 ॰युक्तो] ॰युक्ती Bn.II
10 ॰श्चैवं] ॰श्चैचं Pm
10 यः] द्यः Np
10 १११] ११० Bn.I, Np; *om.* Bn.II, Br, Pm, Rr; २६ Sc
11 स गोल॰] खगोल॰ Bn.II
11 ॰र्दृग्गोलः] ॰र्दग्गोलः Bn.II, Rr; र्दग्गोला Sc
11 ॰भिन्नत्वात्] ॰भिन्नत्वात् ११० Bn.II, Br, Pm, Rr; ॰भेदत्वात् Sc
12 ॰मण्डल॰] ॰मण्ड⟦ड⟧ल॰ *corr. ras.* Bn.II; ॰गंडल॰ Br
12 ॰नामकं] ॰मान Np, Sc
12 वदाम्येतत्] वदामेतत् Bn.II
12 ११२] *om.* Bn.II, Br, Pm, Rr; १११ Bn.I; ११ Np; २७ Sc
13 विषुव॰] किषुव॰ Bn.I
13 ॰स्तु] ॰स्त्व Bn.II, Pm, Rr
14 अयनवृत्तभवृत्तयुतिस्च] अधनवृत्तयुतिश्च Bn.I; अयनवृत्तयुतिश्च Pm; अयन॰…॰युतिस्तु *desunt* Sc
14 ॰युतिस्च] ॰युतिस्तु Np
14 ॰सौम्यदिक्] ॰सौम्यदिकु Bn.I
14 ११३] ११२ Bn.I; १११ Bn.II, Br, Rr; १२ Np; ११० Pm; २८ Sc

विषुववृत्त⌈भमण्डलमध्यगं परमिहायनमण्डलजं धनुः । f. 8v Np
यदिदमस्ति स एव परापमो ह्यपमसूत्रगचापमिहापमः ॥ ११४ ॥

विषुवमण्डल⌈[a]खेचरबिम्बयोरपमसूत्रग⌈[b]तं तु शरासनम् । [a] f. 171v Sc [b] f. 54v Rr
स्फुटतरापम एष खगस्य वै भवति याम्यदिगुत्तरतोऽथवा ॥ ११५ ॥

विषुववृत्तभमण्डलमध्यगं विशि⌈खसूत्रधनुर्यदिहास्ति तत् । f. 74v Bn.I
अपम एव पराभिध उच्यते खगभचक्रकयोर्विवरं शरः ॥ ११६ ॥

विषुवमण्डलखेचरमध्यगं विशिखसूत्रधनुर्यदिहास्ति सः ।
स्फुटतरा⌈[a]पमकाङ्क्ष उदीरितो ग⌈[b]णितगोलविचारविचक्षणैः ॥ ११७ ॥ [a] f. 75v Bn.II [b] f. 42r Br

यावद्भिरंशैर्ध्रुव उन्नतः स्यात्तावद्भिरन्यध्रुवकः कुजाधः ।
तावद्भिरेवात्मशिरोनभस्तो नतं नरैर्वैषुववृत्तमूह्यम् ॥ ११८ ॥

तत्रोन्न⌈तांशाः खलु लम्बभागा नतांशकाश्चाक्षलवाः प्रसिद्धाः । f. 172r Sc
मध्याह्नवृत्ते वियदाख्यलग्नखस्वस्तिकान्तः खलु मध्यभागाः ॥ ११९ ॥

---

2 यदि॰] जदि॰ Bn.I
2 ॰मस्ति] ॰मस्ते Pm
2 ह्यपम॰] स्वपम॰ Pm
2 ॰चापमिहा॰] ॰चामपिहा॰ Bn.I; ॰वा‸मिहा॰ *corr. desunt* Sc
2 ११४] ११३ Bn.I, Np; १२ Bn.II, Pm, Rr; ११२ Br; २९ Sc
3 तु शरासनम्] तु शसशनं Bn.II; तनुरासनं Sc
4 एष] एव Bn.I
4 खगस्य] षगस्य Bn.I
4 ११५] ११४ Bn.I, Np; १३ Bn.II, Pm, Rr; ११३ Br; ३० Sc
5–6 विषुववृत्तभमण्डल॰…॰विवरं शरः] *desunt* Bn.II, Br, Pm, Rr
5 विशिखसूत्र॰] विशिषसूत्र॰ Bn.I
6 पराभिध] पराभि Bn.I
6 उच्यते] मुच्यते Sc
6 ११६] ११५ Bn.I, Np; ३१ Sc
7 ॰मण्डलखेचर॰] ॰वृत्तभमण्डल॰ Bn.II, Br, Pm, Rr
7 ॰सूत्र॰] ॰षत्तु॰ Bn.I
8 गणितगोलवि॰] गणिलवि॰ Bn.I; गणितगौखल॰ Bn.II
8 ११७] ११६ Bn.I; १४ Bn.II, Pm, Rr; ११४ Br; १६ Np; ३२ Sc
9 यावद्भिरं॰] याद्भिरं (व) *add. suprascr.* Pm
9 ॰रन्यध्रुवकः] ॰रन्यध्रुवकः (द्) *add. suprascr.* Sc
10 ॰वात्मशिरो] ॰वात्तशिरो Bn.I, Np; ॰वात्रशिरो Sc
10 ॰नभस्तो] ॰नभस्तौ Sc
10 नतं] नरं Sc
10 नरैर्वैषुव॰] नरैर्वेषव॰ Bn.I; नरैर्वैषव॰ Bn.II, Br, Pm, Rr; नवैवैषुव॰ Sc
10 ॰मूह्यम्] ॰मुह्यं Bn.II
10 ११८] ११७ Bn.I; १५ Bn.II, Rr; ११५ Br, Pm; १७ Np; ३३ Sc
11 तत्रोन्न॰] नतोन्न॰ Sc
12 वियदाख्य॰] वि^ख्य॰ -दा२- *add. in marg. sins.* Sc
12 खलु] स्वलु Bn.I
12 ११९] ११८ Bn.I; १६ Bn.II; ११६ Br, Pm, Rr; १८ Np; *om.* Sc

एवं क्ष्माजग्रहबिम्बमध्ये परोन्नतांशा विबुधैर्विचिन्त्याः ।
भूजाभिधे प्राक्खचरोदयान्तरग्रांशकाः स्युः खलु पश्चिमेऽपि ॥ १२० ॥

पूर्वस्वदृङ्मण्डलमध्यभागा दिगंशकाः पश्चिमतोऽपि तद्वत् ।
दृङ्मण्डले भूजखगान्तराले स्युरुन्नतांशा अधरांशका वा ॥ १२१ ॥

तत्रैव खस्वस्तिकखेटमध्ये नतांशका एवमधो भवन्ति ।
f. 172v Sc स्वोन्मण्डलस्वीयधराजमध्ये द्युज्याख्यवृत्तस्य धनुश्चरं स्या|त् ॥ १२२ ॥

f. 9r Np कुजोपरि|ष्टाद्दिनरात्रवृत्ते यद्दृश्यचापं दिनचापमेतत् ।
f. 48v Pm कुजादधस्ताद्दिनरात्रवृत्ते यत्स्याददृश्यं रजनी|धनुस्तत् ॥ १२३ ॥

दृश्यध्रुवासन्नमहो महत्स्याददृश्यपार्श्वे लघु वासरं हि ।
दिनाद्विलोमं रजनीप्रमाणं महत्तया स्वल्पतया च नित्यम् ॥ १२४ ॥

f. 75r Bn.I पूर्वापरोन्मण्डलमध्यगामि द्युज्याख्यवृत्तस्य धनुर्भवे|द्यत् ।
सदैव तच्चक्रदलांश १८० तुल्यं चराढ्यहीनं स्वदिनप्रमाणम् ॥ १२५ ॥

---

1 ॰बिम्बमध्ये] ॰विंबनध्ये Br
1 ॰विचिन्त्याः] ॰विंचिंत्याः Br
2 ॰रग्रांशकाः] ॰रग्राभिधाः Np, Sc
2 स्युः] सुः Bn.II
2 १२०] ११९ Bn.I; ११७ Bn.II, Br, Pm, Rr; १९ Np; ३५ Sc
3 पूर्वस्व॰] पूर्वेग्रांशकात्स्व॰ Np; पूर्वेस्व॰ Sc
4 ॰गान्तराले] ॰गांतरा⟦क⟧ले *corr. ras.* Sc
4 अधरांशका] अधरांशका (पश्चिमतोपि त)(द्वत् ॥ दृण्मण्डले भूजखगांतराले स्युरुन्नतांशा अध) *secl., ins. per errorum ex* verse 121 Bn.I; अधारांशका Bn.II
4 १२१] १२० Bn.I; १८ Bn.II, Pm, Rr; ११८ Br; २० Np; ३६ Sc
5 खस्वस्ति॰] षस्वस्ति॰ Bn.II
6 ॰जमध्ये] ॰जमध्य Bn.I
6 द्युज्याख्य॰] द्युज्याख्यु॰ Pm
6 धनुश्चरं] धनुश्च⟦इ⟧रं *corr. ras.* Pm
6 १२२] १२१ Bn.I, Np; १९ Bn.II, Rr; ११९ Br, Pm; ३७ Sc
7 ॰रात्रवृत्ते] ॰रात्रवृत्त Bn.I; ॰रात्रित्ते (वृ) *add. suprascr.* Sc
7 यद्दृश्य॰] यदृश्य॰ Pm
7 दिनचाप॰] ⟦दिनचाप⟧दिनचाप॰ *corr. ras.* Sc
8 ॰स्ताद्दिन॰] ॰स्तादिन॰ Bn.II
8 ॰रात्रवृत्ते] ॰रात्रि c वृत्ते *ut videtur, ign. litt.* Bn.II
8 यत्स्याद॰] यत्स्य⌐द॰ -ा- *add. in marg. dext. in linea confrac.* Bn.II
8 १२३] १२२ Bn.I, Np; १२० Bn.II, Br, Pm, Rr; ३८ Sc
9 महत्स्या॰] मरहत्स्या॰ Bn.I
9 ॰दृश्यपार्श्वे] ॰अदृश्यध्रुवश्यपार्श्वे Bn.II
9 हि] हि २१ Bn.II, Rr; हि १२१ Br; हि १२० Pm
10 महत्तया] महत्तथा Bn.II, Br, Rr
10 १२४] १२३ Bn.I, Np; *om.* Bn.II, Br, Pm, Rr; ३९ Sc
11 ॰वृत्तस्य] ॰कृत्तस्य Pm
11 ॰भवेद्यत्] ॰भवेद्यत् १२२ Bn.II, Br, Pm, Rr
12 तच्चक्रदलांश १८०] तच्चक्र ३६० दलांश १८० Np; तदं^v^लांश *add. desunt* Sc
12 चराढ्य॰] राढ्य॰ Bn.I; चराह्य॰ Pm
12 स्वदिन॰] स्व⟦दी⟧-दि-न॰ *corr. ras.* Pm
12 १२५] १२४ Bn.I, Np; *om.* Bn.II, Br, Pm, Rr; ४० Sc

स्वोन्मण्डलं व्यक्षकुजं सदैव ततो निरक्षे द्युनिशोः समत्वम् ॥ १२६ ॥

यथा यथापक्रममण्डलस्य स्याद्दण्डवच्चापमृजु क्षमाजे ।
तथा तथोद्गच्छ⌈ति भूयसैव कालेन तिर्यक्पुनरल्पकेन ॥ १२७ ॥ f. 173r Sc

⌈निरक्षदेशात्खलु सौम्यभागे नक्रादयः षड्क्षितिजं तिरोगाः । f. 76r Bn.II
व्रजन्ति कर्कादय एव भूजमृजुत्वमाप्ताः परतोऽन्यथा ते ॥ १२८ ॥

दृक्क्षेपवृत्ते त्रिभहीनलग्नखस्वस्तिकान्तर्धनुरस्ति यच्च ।
दृक्क्षेपचापं गदितं तदेव खनन्द ९० शुद्धं खलु दृग्गतिः स्यात् ॥ १२९ ॥

दृक्क्षेपवृत्ते स्वकदम्बभूजम⌈ध्ये च दृक्क्षेपधनुर्भवेद्वा । f. 55r Rr
कदम्बखस्वस्तिकमध्यचापं सादृग्गतिर्वा गणकैर्निरुक्ताः ॥ १३० ॥

कोणमण्डलगतोन्नतांशकाः कोणशङ्कुलवका बुधैः स्मृता ।
एवमेव सममण्डलोन्नता भागकाः सम⌈नरोत्थिता मताः ॥ १३१ ॥ 173v Sc

---

1 स्वोन्मण्डलं व्यक्ष॰] स्वोन्मण्ड लंख्यंच॰ *add. suprascr.* Sc
1 सदैव] तदैव Br, Pm, Rr
1 १२६] *om.* Bn.I, Np, Sc; १२३ Bn.II, Br, Pm, Rr
2 स्याद्दण्डवच्चा॰] स्याद्वंनुवच्चा॰ Bn.I
2 ॰क्षमाजे] ॰क्षमाजे ॥ १२५ ॥ Bn.I, Np; ॰क्षमाजे ४१ Sc
3 तिर्यक्पुनर॰] तिर्यक्पनर॰ Pm
3 १२७] *om.* Bn.I, Np, Sc; १२४ Bn.II, Br, Pm, Rr
4 सौम्य॰] मौम्य॰ Br
4 षड्क्षितिजं] षट्क्षितिजे Np, Sc
4 तिरोगाः] तिरोगाः ॥ १२६ ॥ Bn.I, Np; ⟦ि⟧तुरोगाः *corr. ras.* Pm; तिरोगाः ४२ Sc
5 व्रजन्ति] वृजंति Pm
5 ॰माप्ताः] ॰मानाः Sc
5 १२८] *om.* Bn.I, Np, Sc; १२५ १५ Bn.II, Rr; १२५ Br, Pm
6 दृक्क्षेप॰] दृक्षे॰ Pm
6 ॰वृत्ते] ॰वृत्तं Bn.I
6 त्रिभहीन॰] नविहीन॰ Np, Sc
6 ॰कान्तर्ध॰] ॰कान्तेध॰ Np, Sc
6 ॰नुरस्ति] ॰नरस्ति Bn.II
6 यच्च] यच्च ॥ १२७ ॥ Bn.I, Np; यच्च ४३ Sc
7 ९०] *om.* Bn.I, Bn.II, Br, Pm, Rr, Sc
7 खलु] खल Bn.II, Pm, Rr
7 दृग्गतिः] दृगातिः Sc
7 १२९] *om.* Bn.I, Np, Sc; १२६ Bn.II, Br, Pm, Rr
8 दृक्क्षेप॰] दृक्षे॰ Pm
8 स्वकदम्ब॰] स्वदम्ब॰ Np
8 दृक्क्षेप॰] दृक्षेप॰ *corr. suprascr.* Sc
8 ॰र्भवेद्वा] ॰र्भवेद्धा ॥ १२८ ॥ Bn.I
9 ॰दृग्गतिर्वा] ॰दृग्गतिवा Bn.II
9 ॰र्निरुक्ताः] ॰र्निरुक्त Bn.II
9 १३०] *om.* Bn.I; १२७ Bn.II, Br, Pm, Rr; १२८ Np; ४४ Sc
10 ॰लवका] ॰लवन्नाः Sc
10 स्मृता] स्मृताः ॥ २९ ॥ Bn.I
11 सममण्डलोन्नता] समंडलान्नता Bn.II; सममंडलान्नता Rr; सममण्डलोन्मताः Sc
11 भागकाः] भंगकाः *add. suprascr.* Sc
11 मताः] मता २८ *add. suprascr.* Pm
11 १३१] *om.* Bn.I, Bn.II, Pm, Sc; १२८ Br; १२९ Np; २८ Rr

उन्मण्डलं यदि रविः समुपैति सम्य-
ग्दृङ्मण्डलोन्नतलवोन्मितयस्तदानीम् ।
उद्वृत्तशङ्कुलवका गणकैर्विचिन्त्या
दृग्ज्यावृतावपि भुजांशमितिः स्वदिक्स्यात् ॥ १३२ ॥

f. 9v Np ⌈एवं गोलविचक्षणो गणितविच्छिल्पागमज्ञः सुधीः
f. 42v Br स्वेच्छा⌈कल्पनया महान्ति बहुशः स्वल्पानि वृत्तानि वा ।
गोलस्थानि विचिन्त्य कार्यसमये तत्तद्धनुःसिञ्जिनी-
र्ज्ञात्वा कोटिभुजोत्थिताः स विशिखास्त्रैराशिकं साधयेत् ॥ १३३ ॥

इत्यादि गोलरचना सुमहाप्रवीणैर्गोला मृदा किमुत धातुभिरुत्तमाख्यैः ।
[a] f. 75v Bn.I [b] f. 174r Sc काष्ठेन ⌈[a]वा शरलवंशश⌈[b]लाकिकाभिः कार्या विषाणफलकुञ्जरदन्तपूर्वैः ॥ १३४ ॥

इत्येतस्यामिन्द्रपुर्यां वसन्सन्नित्यानन्दो देवदत्तस्य पुत्रः ।
सारोद्धारे सर्वसिद्धान्तराजे गोलाध्यायं प्रापयत्तत्र पूर्तिम् ॥ १३५ ॥

---

1 समुपैति] समुपै[v] -ति- *add. in marg. inf.* Pm
2 ॰न्नतलवो॰] ॰न्मितलवो॰ Bn.I, Np, Sc
2 ॰स्तदानीम्] ॰स्तदानीम् ॥ १३० ॥ Bn.I
4 १३२] *om.* Bn.I; २९ Bn.II, Pm, Rr; १२९ Br; १३० Np; ४⟦९⟧-६- *corr. in ras.* Sc
5 ॰विच्छिल्पा] ॰विछिल्पा॰ Bn.I, Bn.II, Br, Pm, Rr
6 महान्ति] महानि Bn.II, Rr; महान्नि Br, Pm; नपंति Sc
6 वृत्तानि वा] वृत्तनि वा ॥ १३१ ॥ Bn.I
7 सिञ्जिनी॰] सिनी॰ Bn.I
8 ॰भुजो॰] ॰भुजौ॰ Bn.I
8 ॰विशिखा॰] ॰विशिषा॰ Bn.I
8 साधयेत्] सधयेत् Bn.II, Rr
8 १३३] १३२ Bn.I; १३० Bn.II, Br, Pm, Rr; १३१ Np; ४७ Sc
9 इत्यादि] इतिआदि Pm
9 ॰र्गोला] ॰र्गोलां Pm
9 किमुत] किमत Bn.II
10 शरलवंश] शरलवांश Bn.II, Pm, Rr; सरलवंश Np, Sc
10 १३४] १३३ Bn.I; १३१ Bn.II, Br, Pm, Rr; १३२ Np; ४८ Sc
11 ॰पुर्यां] ॰पूर्यां Bn.II, Rr
12 पूर्तिम्] पूर्त्तिम् ॥ शुभम् ॥ Np
12 १३५] *om.* Bn.I, Np, Sc; १३२ Bn.II, Br, Pm, Rr

4 Edited Sanskrit Text and Its English Translation

DOI: 10.4324/9780429506680-4

अथ गोलाध्यायो व्याख्यायते ।

तं वन्दे गणनायकं सुरगणा यत्पादपद्मद्वयं
सेवन्तेऽभिमतार्थसाधनविधौ विघ्नौघविध्वंसकृत् ।
यं सांख्याः पुरुषं प्रधानमपि वा शैवाः शिवं मेनिरे
नित्यं ब्रह्म चिदात्म सर्वमपरं मिथ्येति वेदान्तिनः ॥ १ ॥

यत्किंचिद्गणितं ग्रहस्य गणको जानन्विना वासनां
पृष्टः सन्नपरेण गोलविदुषा प्रश्नप्रपञ्चोक्तिभिः ।
किं ब्रूते प्रति तं तदुत्तरमयं गोलागमाज्ञो यत-
स्तस्माद्गोलविचारचारुरचनां वक्ष्ये सतां प्रीतये ॥ २ ॥

रात्रौ सद्मविचित्रचित्ररचनं दीपं विना किं यथा
त्रैलोक्यं सकलार्थपूर्णमपि किं भास्वत्प्रकाशं विना ।
वैदग्ध्येन विना प्रगल्भतरुणी सौन्दर्ययुक्तापि किं
तद्वद्गोलविवर्जितो गणितवित्प्रश्नोत्तरे किं विभुः ॥ ३ ॥

पाटीकुट्टकबीजगोलनिपुणो वेदाङ्गपारङ्गमः
काव्यालंकृतिशिल्पशास्त्रकुशलो यो न्यायशास्त्रादिवित् ।
सिद्धान्तेऽत्र समस्तवस्तुसहिते तस्याधिकारो भवे-
च्चेदेवं न यथा कथंचिदभिधामात्रं प्रसिद्धिं नयेत् ॥ ४ ॥

कीदृग्भूगोलसंस्था कथय कथमहो चक्रसंज्ञो भगोलः
कीदृग्भूयः खगोलः स्थिर इह सततं कीदृशो दृष्टिगोलः ।
तत्तत्स्थं वस्तु कीदृक्सकलमपि यथा भूमिगोलेऽब्धिशैल-
द्वीपाद्यं तद्यथा वा विषुववृतिमुखं चक्रसंज्ञे भगोले ॥ ५ ॥

Now, the chapter on spheres is described.

1 I praise him, the leader of the multitude whose two lotus feet the divinities worship according to the process propitious to [their] desires, the destroyer of the flood of obstacles, [the one] who the Sāṃkhyas considered Puruṣa or even Pradhāna, the Śaivas [considered] Śiva, [and] the Vedāntins [considered] the eternal Brahman [who is] pure thought [and] the All [with] everything else false.

2 A mathematician, knowing whatever little computation of a planet without [its] justification [and] having been interrogated with various questions [about the visible world] by another scholar [who was] versed in the study of spheres, what could he reply as an answer [to those questions] being ignorant in the science on spheres [himself]? Therefore, I shall proclaim [this] beautiful composition on the investigation of spheres for the spleasure of all beings.

3 What good is the variegated and colourful arrangement of houses at night without light? What [good] are the three worlds, even if full of all opulence, without the light of the Sun? Without intelligence, what [good] is a confident, young maiden even if [she were to be] endowed with beauty? [And] likewise, deprived of [the knowledge of] spheres, what can a knower of computations be capable of in question and answer?

4 The one who is adept in arithmetic, indeterminate equations, algebra, and [the science of] spheres; [who is] learned in the auxiliary branches of the Vedas; skilful in the science of poetics, rhetoric, and the creative and mechanical arts; [the one] who knows the science of logic, etc., his authority in the canonical texts, along with their entire content, should be [considered] in this matter. If indeed [it is] not thus, somehow he [i.e., the unskilled person] will only bring out the express sense of the words [and nothing else].

5 Pray tell, of what kind is the established order [of things] on the Earth-sphere? Why is the sphere of asterisms called a wheel? Furthermore, of what kind is the oblique celestial sphere [that is] fixed in this world? Of what kind is the visible celestial sphere? Of what kind is the thing, [considered] in its entirety, [present] in each of them, just as there are the oceans, mountains, islands, etc. on the Earth-sphere, or just as [there is] the principal perimeter of the equinoxes [i.e., the celestial equator] in the wheel-like sphere of asterisms?

पूर्वापरं दक्षिणसौम्यवृत्तं भूजाभिधं कोणगमण्डलं च ।
उन्मण्डलाद्यं च खगोलसंस्थं तथैव दृग्गोलगतं यथा वा ॥ ६ ॥

अथ भूगोलभगोलखगोलदृग्गोलास्तत्तत्स्थवस्तूनि च प्रोच्यन्ते ।

निराधारोऽजस्रं भवति निजशक्त्या स्थिरतरः
क्षितेर्गोलः क्ष्वम्बुज्वलनपवनाकाशनिचितः ।
यथाऽयोगोलोऽन्तर्वियति परितश्चुंबकमये
गृहे ग्रामारामाचलजलधिदेवासुरनृभृत् ॥ ७ ॥

उत्तुङ्गं शैलशृङ्गं खगुण ३० परिमितैर्योजनैर्दृश्यते चे-
दर्वागेवाध्वमध्ये कथय च कियती शैलशृङ्गोन्नतिः सा ।
किं वा शृङ्गोच्छ्रयो मे भवति सविदितो दृश्यते योजनैः कै-
रेवं गोलाकृतिश्चेद्भवति वसुमती त्वन्मते मां प्रचक्ष्व ॥ ८ ॥

तदुत्तरस्य गणितम् ।

नृपादशैलान्तरमार्गयोजनै ३० र्भचक्रभागा ३६० गुणिता हृताः पुनः ।
वसुंधरायाः परिणाहयोजनै ६००० स्तदाप्तचापं विवरांशका मताः ॥ ९ ॥

तत्कोटिज्योना त्रिभज्या ६० विधेया शेषं भूव्यासार्द्धमानेन निघ्नम् ।
तत्कोटिज्यामानभक्तं च लब्धं शृङ्गौन्नत्यं योजनैः प्रस्फुटं स्यात् ॥ १० ॥

भूव्यासार्धं शैलशृङ्गोच्छ्र्याढ्यं हारः कल्पोऽथोन्नतिस्त्रिज्यकाघ्नी ।
हारेणाप्ता तद्विहीना त्रिभज्या तत्कोटिज्या कल्पनीयात्र तज्ज्ञैः ॥ ११ ॥

6 Just as the prime vertical, the meridian, the horizon, the vertical circle passing through the intercardinal directions, the equinoctial colure, etc. [are] situated in the oblique celestial sphere, in a similar manner, [those great circles are also] situated in the visible celestial sphere.

Now, the Earth-sphere, the sphere of asterisms, the oblique celestial sphere, the visible celestial sphere, and the things in each of them are declared.

7 The Earth-sphere, eternally unsupported and immovable due to its own [inherent] energy, is composed of earth, water, fire, air, and ether. Just as a ball of iron remains in the air [when] surrounded by a house made of lodestone, [similarly the Earth-sphere remains unsupported in space, while itself] supporting villages, gardens, mountains, oceans, gods, demons, and people.

8 If a lofty mountain summit is indeed seen in the middle of the road at a distance of thirty measured *yojana*s from a certain point, state how great is that elevation of the mountain's summit. Or perhaps, the elevation of the mountain's summit is [known] to me, [and] knowing that, with how many *yojana*s is [the mountain] seen from this point? Thus, if the Earth is of the shape of a sphere, tell me your opinion [in this matter].

The calculation of the answer to those [questions].

9 The 360 degrees of a circle multiplied by the 30 *yojana*s of the distance between the foot of a man and the mountain is then divided by the 6000 *yojana*s of the circumference of the Earth. The arc of that quotient is considered as the degrees of difference.

10 The *sinus totus* 60 should be taken [and] diminished by the Cosine of that [arc]. The remainder [from the subtraction] is multiplied by the measure of the radius of the Earth, and divided by the measure of the Cosine of that [arc]. The quotient is the elevation of the mountain's summit [measured] in *yojana*s.

11 The proper divisor is the radius of the Earth increased by the elevation of the mountain's summit. Then, the elevation [of the mountain peak] is multiplied by the *sinus totus* [and] divided by the divisor. The *sinus totus* is [then] diminished by that [amount, and] that [remainder] should be regarded as the Cosine by those knowledgeable in this matter.

तत्कोटिज्यावर्गहीनात्त्रिमौर्व्या वर्गान्मूलं तस्य चापस्य भागाः ।
चक्रांशाप्ता ३६० भूपरिध्युन्मिति ६००० घ्ना ज्ञेयः सोऽध्वा शैलपुंपादमध्ये ॥ १२ ॥

अथो कुमध्ये यदि वास्ति लङ्का प्राच्यां तदा स्याद्यमकोटिरेव ।
अवाचि वेद्यो वडवानलाख्यो मेरुः सदोदक्पररोमकं च ॥ १३ ॥

चेद्भूमिमध्ये यमकोटिपूः स्यात्प्रच्यां तदा सिद्धपुरी विचिन्त्या ।
प्रत्यक्चलं काथ यथा कुमध्ये सिद्धाभिधाना नगरी प्रकल्प्या ॥ १४ ॥

प्राच्यां तदा रोमकपत्तनं हि पश्चात्स्थिता स्याद्यमकोटिरेषा ।
चेद्रोमकं नाम कुमध्यसंस्थं प्राच्यां हि लङ्का परतश्च सिद्धा ॥ १५ ॥

एतासु सर्वासु नृणां पुरीषु मेरुः सदोदग्वडवानलोऽवाक् ।
मूर्ध्नि स्थितं भ्राम्यति कालचक्रं ध्रुवद्वयं तिष्ठति भूजलग्नं ॥ १६ ॥

परस्परं षड्धरणीस्थलानि भूपादभागैः ९० सततं स्थितानि ।
एतानि भूमेः परिणाहतुर्यभागेन १५०० भूयः परिकल्पितानि ॥ १७ ॥

मेरुर्यदा तिष्ठति भूमिमध्ये प्राच्यां यमाख्यापररोमकं हि ।
सिद्धोत्तरा दक्षिणगा च लङ्का सौम्यो ध्रुवः खेऽङ्घ्रितले च याम्यः ॥ १८ ॥

12 The square root from the square of the *sinus totus* diminished by the square of that Cosine [is found]. The degrees of its arc are divided by 360 degrees of a circle [and] multiplied by the 6000 [*yojana*s of the] circumference of the Earth. That [amount] should be known as the distance going between the foot of a man and the mountain.

13 Now, if [the city of] Laṅkā is in the middle of the Earth, then the very [city of] Yamakoṭi is to the east [of Laṅkā. The place] called Vaḍavānala should be understood to be to the south [of Laṅkā], Mt Meru is always northwards [of Laṅkā], and [the city of] Romaka is on the other side [i.e., to the west of Laṅkā].

14 If the city of Yamakoṭi is in the middle of the Earth, then [the city of] Siddhapurī is to be considered to the east [of Yamakoṭi]. Now, what [city is found] moving westwards insofar as the city named Siddhā [i.e., Siddhapurī] is [thought] to be positioned in the middle of the Earth?

15 In that case, on account of the city of Romaka [being] to the east [of Siddhapurī], this [city of] Yamakoṭi is situated westwards [of Siddhapurī]. If [the city] named Romaka is located in the middle of the Earth, indeed [the city of] Laṅkā is to the east [of Romaka] and [the city of] Siddhā [i.e., Siddhapurī] is on the opposite side [i.e., to the west of Romaka].

16 In all of these [equatorial] cities of men, Mt Meru [is] always northwards [and] Vaḍavānala [is always] southwards. [Furthermore, in these cities] the celestial equator, situated overhead [i.e., at the zenith], revolves [westwards, while] the two celestial poles remain fixed on the horizon.

17 The six terrestrial places are always situated with quarter of the Earth in degrees (90°) from one another. Moreover, all of them are imagined [to be mutually separated] by the fourth part of the circumference of the Earth (1500 *yojana*s).

18 When Mt Meru stands in the middle of the Earth, [the city] called Yama [i.e., Yamakoṭi] is to the east because [the city of] Romaka is on the other side [i.e., to the west of Mt Meru], and [the city of] Siddhā [i.e., Siddhapurī] is [to the] north, [with the city of] Laṅkā being southern. The north celestial pole is [directly overhead] in the sky [i.e., at the zenith] and the south celestial pole is [below] the soles of one's feet [i.e., at the nadir].

कूमध्यगश्चेद्वडवानलाख्यः प्राक्पश्चिमे स्तो यमरोमकाख्ये ।
लङ्कोत्तरस्था यमदिक्च सिद्धा याम्यो ध्रुवः खेऽङ्घ्रितले च सौम्यः ॥ १९ ॥

भूजासक्तमिवाभितः सुरगणा मेरुस्थिता वैषवं
वृत्तं गच्छदसव्यतोऽष्टककुभां मार्गेण पश्यन्ति हि ।
सव्यं व्युत्क्रमतो दिशां वडवगा दैत्याः सदा प्राङ्मुखा-
स्तद्वृत्तेन धराखिला विजयते गोलाकृतिः सर्वतः ॥ २० ॥

यत्र क्षितौ तिष्ठति मानवोऽसौ तस्मात्कुपादान्तरितं समन्तात् ।
वृत्तं तिरश्चीनमवैहि भूजं खं मूर्ध्नि भूम्यर्धगतं ३००० पदाधः ॥ २१ ॥

तथा हि लङ्का क्षितिपृष्ठमध्ये तद्भूजगा एव सुमेरुपूर्वाः ।
कुबेरदिक्तोऽङ्घ्रितलस्थिता च सिद्धापुरी तद्वदतोऽपरत्र ॥ २२ ॥

रोमकाभिधपुराच्च पश्चिमे खालदातपुरमस्ति वारिणि ।
नेत्रलोचन २२ लवैरतः पुरो गङ्गदुज्दमिह भार्धभागकैः १८० ॥ २३ ॥

लङ्कारोमकखालदातनगरं संस्पृश्य यन्मण्डलं
यायाद्भूपरिधिः ६००० स एव गणकैरुक्तोऽत्र मध्याभिधः ।
तत्रस्थो विषयो निरक्ष इह ये लोका वसन्ति ध्रुवौ
ते पश्यन्त्यवनीजगौ च विषुवद्वृत्तं स्वमूर्ध्नि भ्रमत् ॥ २४ ॥

19 If [the city] called Vaḍavānala is in the middle of the Earth, to the east and west [of Vaḍavānala] are [the cities] called Yama [i.e., Yamakoṭi] and Romaka [respectively, the city of] Laṅkā is situated to the north, and [the city of] Siddhā [i.e., Siddhapurī] is [in] the southern direction. The south celestial pole is [directly overhead] in the sky [i.e., at the zenith] and the north celestial pole is [below] the soles of one's feet [i.e., at the nadir].

20 The multitude of gods situated on Mt Meru see the celestial equator that is moving counterclockwise [towards the left] following the path of the eight directions as if [it were] fastened to the horizon on all sides. [And] on account of the reversal of the directions, the demons situated at *Vaḍavā* [i.e., Vaḍavānala, see the celestial equator moving] clockwise [towards the right; both the gods and the demons] always facing eastwards. Likewise, for that reason, the Earth extends everywhere in the shape of a sphere.

21 Wherever a man stands on [the surface of] the Earth, from that [place] separated by a quarter of [the circumference of] the Earth on all sides, [one should] understand the horizontal circle [to be] the horizon. [At that place,] the sky is [understood to be] overhead [i.e., at the zenith and] the distance of half [the circumference of] the Earth (3000 *yojana*s) is [understood to be] below the feet [i.e., antipodal].

22 Accordingly, as [the city of] Laṅkā is in the middle of the surface of Earth, [the places] beginning with Mt Sumeru are on its horizon from the northern direction, and hence, likewise, the antipodal city [called] Siddhā [i.e., Siddhapurī] is in the opposite place.

23 The city of Khāladāta is in the ocean to the west [of Laṅkā] and [separated] from the city called Romaka by 22°, [and] eastwards from this place [i.e., east of the city of Khāladāta, the city of] Gaṅgadujda is here [in the ocean, separated] by 180°.

24 Whatever circle, having touched the cities of Laṅkā, Romaka, and Khāladāta, extends [forth], indeed that earth circumference (6000 *yojana*s) is then called by the name middle[-circle, i.e., the terrestrial equator] by mathematicians. A country situated here has no latitude, and the people who live here see the two celestial poles on the horizon [with] the celestial equator revolving at their zenith.

आगङ्गदुज्दादथ खालदातं यावत्पुनर्मेरुमुदक्स्थितं च ।
एतावती भूमिरिह त्रिकोणा पादोन्मिता तिष्ठति वारिमुक्ता ॥ २५ ॥

षष्ट्या घटीनां द्युनिशव्यवस्था यावद्भवेत्तावदिहापि लोकाः ।
तिष्ठन्ति नीहारमयाद्रिभीत्या तत्रापि वस्तुं विभवो न सन्ति ॥ २६ ॥

पश्यन्तु काश्मीरमुखप्रदेशान्हिमागमे दुर्गतरान्नराणाम् ।
सदैव यत्र प्रपतेद्धिमानी शक्नोति को गन्तुमिहोत्तरेषु ॥ २७ ॥

लङ्काकाञ्चीश्वेतशैलोज्ञयिन्यो रोहीताख्यं सांनिहित्यं सुमेरुः ।
रेखा भूमेर्मध्यनाम्नी निरुक्ता तस्यां किंचिन्नास्ति देशान्तरं हि ॥ २८ ॥

निरक्षदेशाच्चलितः पुमान्यथा कुबेरकाष्ठां च तथोन्नतं ध्रुवम् ।
विलोकयेन्नम्रमपीह वैषवं नभस्तलाद्याम्यविभागसंस्थितम् ॥ २९ ॥

यथा यथा गच्छति दक्षिणां पुमांस्तथा तथा पश्यति सध्रुवं नतम् ।
अतो ध्रुवो यत्र शिरःस्थितो भवेन्निरक्षतो भूचरणान्तरे तु सः ॥ ३० ॥

यत्र यत्र वसति क्षमोपरि प्राणभृद्धरणिगर्भकर्षितः ।
तत्र तत्र किल मन्यते स्वयं भूपरिष्ठमिव चेतसा स्वकम् ॥ ३१ ॥

भूचतुर्थलवकान्तरे यथा तिर्यगेव खलु मन्यते स्थलम् ।
भूदलान्तरमधःस्थितं स्वतोऽन्यत्र तद्वदपि कल्पनान्यथा ॥ ३२ ॥

25 Now, the amount of distance [there is] from [the city of] Gaṅgadujda to [the city of] Khāladāta, and again to the northward-situated Mt Meru, of such an amount, this triangular [region of] the Earth measuring a quarter [of the Earth] remains free of water.

26 As far as [the northern region where] the duration of the day or the night is sixty *ghaṭī*s, people still remain in this place. Beyond that, [however,] with the fear of the snow-covered mountains, people are not able to live [there].

27 At the beginning of winter, [people] must see the regions towards Kāśmīra being [highly] inaccessible for men. Who is then able to go [to regions] where it always snows, [further afar] to the north of here?

28 [The localities of] Laṅkā, Kāñcī, Śvetaśaila, and Ujjayinī; [the one] called Rohita; Saṃnihitya; and Mt Sumeru. The line [joining these localities] is declared as the [prime] meridian of the Earth because there is no longitudinal difference [for localities] situated on it.

29 Moreover, like a man having been moved from the equatorial region to the northern region, so too is the polestar elevated [above the horizon in the northern region]. Here, [one] should observe the [circle] belonging to the equinox [i.e., the celestial equator] bent downwards from the zenith [and] directed towards the southern part [with equal relative displacement].

30 The more a man goes southwards, the more he sees the polestar being depressed. Hence, wherever the polestar is situated [directly] overhead [i.e., at the zenith], that [place] is then at a distance of a quarter of the Earth [i.e., 90°] from the terrestrial equator.

31 Wherever one lives on the surface of the Earth, drawn towards the womb of [this] life-supporting Earth [i.e., towards the interior of the Earth], indeed there he considers himself as standing above [the surface of] the Earth with his own mind.

32 Just as one considers the land at a distance of a quarter of the Earth in degrees [i.e., at a distance of 90° from one's position] being just horizontal, [he should also consider the land] at a distance of half the Earth from his own position as being situated below, and likewise elsewhere, imagining [it] differently.

निरक्षतः सौम्यविभागसंस्थे यत्र प्रदेशे परमद्युमानम् ।
व्यष्टांशदन्तोन्मितनाडिकाभि ३१|५२|३० स्तत्राद्यखण्डस्य भवेत्प्रवेशः ॥ ३३ ॥

सपादनाडी १|१५ सहिते च तस्मिन्द्वितीयखण्डस्य तथा प्रवेशः ।
एवं तृतीयादिकखण्डकानां ज्ञेयाः प्रवेशाः किल सप्त यावत् ॥ ३४ ॥

अथ पुराणमतेन द्वीपसमुद्रपर्वतखण्डादीनां संस्थोच्यते ।

परिधेस्तु शतांशको यतः समवद्धाति ततः समन्ततः ।
पृथिवीं च पुराणपण्डिताः सकलामेव समां समूचिरे ॥ ३५ ॥

भूपृष्ठस्पृक्दृष्टिसूत्रं विलग्नं व्योम्ना तिर्यग्यत्र तद्दृष्टिभूजम् ।
चक्राकारं सर्वतो दृश्यते यल्लोकालोकं पर्वतं तं वदन्ति ॥ ३६ ॥

दृश्यादृश्यगिरिं केचिन्नभःकक्षां जगुर्बुधाः ।
तेन किं खचरा नित्यमुदयास्तौ प्रकुर्वते ॥ ३७ ॥

क्षितिजमेव भवेदुदयाचलश्चलति येन कुतोऽपि न नित्यशः ।
स्वविषयेऽल्पतरं हि यदन्तरं पतति तच्चलनेऽपि खगोदये ॥ ३८ ॥

सुवर्णशैलं विपुलोन्नतं रविस्तदीय कट्यां परितः सदा भ्रमन् ।
कथं न कुर्याद्द्युनिशोर्व्यवस्थितिं तदेकपार्श्वे वसतां सुपर्वणाम् ॥ ३९ ॥

33 In the region situated to the north of the equator where the measure of the longest day is 32 *nāḍikā*s minus an eighth part [of a *nāḍikā* i.e., 31;52,30 *nāḍikā*s], there lies the beginning of the first clime.

34 And with that [measure of day-length 31;52,30 *nāḍikā*s] increased by a *nāḍī* with a quarter [i.e., increased by 1;15 *nāḍikā*s] is likewise the beginning of the second clime. In this way, the beginnings of the third and other climes are to be known [with successive increments in day-lengths of 1;15 *nāḍikā*s] up to seven.

Now, in the opinion of the Purāṇas, the established order of the islands, oceans, mountains, continents, etc. is declared.

35 Since the one-hundredth part of the circumference [of the Earth] appears [locally] flat, therefore the scholars of the Purāṇas declared the entire Earth to be completely flat everywhere.

36 Where [one's] line of sight touching the surface of the Earth is connected with the sky going obliquely, there the visible horizon is seen in the shape of a circle from all sides. [The scholars of the Purāṇas] declare that [visible horizon] as the Lokāloka mountain.

37 Some wise men declared the periphery of the sky as the Lokāloka mountain. With that said, how do the planets regularly accomplish rising and setting?

38 The eastern mountain [i.e., the Udayācala mountain] should be the very horizon because it never moves anywhere [i.e., it is always stationary. Yet] in one's own country, whatever [longitudinal] difference occurs, even a small amount, that [difference] is in the very variation in the rising of a planet [i.e., the difference in terrestrial longitudes causes a change in rising times of a planet in different places].

39 The golden mountain [i.e., Mt Meru] is [a mountain with] large elevation. The Sun is always wandering about its hip on all sides [i.e., revolving around the girth of Mt Meru]. How could this not cause the [regular] separation of day and night for the gods who [only] dwell on one of its sides [namely, the eastern face of Mt Meru]?

क्षाराम्बुधिः क्षितिकटीस्थितमेखलेव व्यासोऽस्य योजनशतं मुनिभिः प्रदिष्टः ।
तस्मादुदीचिकुदले सकलेऽस्ति जम्बूद्वीपं परत्र कुदले तु पराणि षट्स्युः॥४०॥

क्षाराब्धिः पयसां सिन्धुस्ततो दधिघृताम्बुधी ।
इक्षुमद्यार्णवौ स्वादुसिन्धुः सप्ताब्धयोऽत्र वै ॥ ४१ ॥

तेषां मध्ये क्षारसिन्धुर्महीयान्व्यक्षो देशः प्रोच्यते सोऽयमेव ।
तस्मादल्पाल्पप्रमाणाः परे स्युः स्वल्पोऽत्यन्तं सप्तमः स्वादुसिन्धुः ॥ ४२ ॥

स्वादूदकान्तर्वडवानलोऽस्ति वसन्ति दैत्या इह मेरुमूले ।
पाताललाका धरणीपुटस्थाः प्रकाशिताः सर्पमणिप्रकाशैः ॥ ४३ ॥

द्वयोर्द्वयोरन्तरगं समुद्रयोरुदाहृतं द्वीपमुदारबुद्धिभिः ।
महद्यदाद्यं परमल्पमल्पकं ततोऽस्ति षड्द्वीपमिलादलं परम् ॥ ४४ ॥

शाकं च शाल्मलं कौशं क्रौञ्चं गोमेदपुष्करे ।
सप्तमं जाम्बवं द्वीपं भूगोलार्धे व्यवस्थितम् ॥ ४५ ॥

लङ्कापुरीतोऽपि कुबेरदिक्स्थो हिमाचलो नाम नगाधिराजः ।
प्राक्पश्चिमाम्भोनिधिसीमदैर्घ्यस्ततोऽप्युदीच्यामिह हेमकूटः ॥ ४६ ॥

40 The saline ocean is like the girdle situated at the hip of the Earth. Its width is decreed by the sages as one hundred *yojana*s. Therefore, [the island-continent of] Jambūdvīpa is in the entire northern hemisphere of the Earth, whereas the other six [island-continents] are in the hemisphere of the Earth elsewhere [i.e., in the southern hemisphere].

41 The ocean of salt water, the ocean of milk, after that, the oceans of yogurt and clarified butter, the oceans of sugarcane [juice] and wine, [and finally] the ocean of sweet water are indeed the seven oceans here.

42 Among them, the saline ocean is the greatest. The equatorial region is declared to be this very [saline ocean]. From that [place], the progressively smaller extents [of the other oceans] are farther away [with] the seventh ocean of sweet water being exceedingly small.

43 Vaḍavānala is in the middle [of the ocean] of sweet water [and] the demons live in this place at the foot of Mt Meru. The nether worlds are situated in the concavity of the Earth [and] are illuminated with the brightness of the serpent-stone [worn by the race of serpent-demons or *nāga*s who live in the *nāgaloka*, the lowest subterranean region].

44 Between each pair of oceans, an island-continent is declared by those with great intelligence. The greatest [island-continent] is that which is the first [i.e., Jambūdvīpa, and] from there, the progressively smaller six island-continents [that make] the other half of the Earth [i.e., the southern hemisphere] are...

45 ...Śāka, Śalmala, Kauśa, Krauñca, Gomeda, [and] Puṣkara. The seventh [island-continent] is the island-continent bearing the Jambū tree [i.e., the Jambūdvīpa] situated [entirely] in one-half of the sphere of the Earth [i.e., in the northern hemisphere].

46 The king of mountains called Himācala, being situated in the northern direction from the city of Laṅkā, extends between the eastern and western boundaries of the ocean. And then, [further] to the north of that place [i.e., beyond the Himācala mountain] is the Hemakūṭa [mountain].

तद्वत्परं निषधशैल उदीचिभागे सिद्धात्पुराद्गिरिरुदक्प्रभवेत्पुरावत् ।
शृङ्गी च शुक्ल इति नीलगिरिः क्रमेण तेषां द्वयोर्विवरदेशगतं च वर्षम् ॥ ४७ ॥

भारतं किन्नरं तस्माद्धरिवर्शं यथोत्तरम् ।
लङ्कातोऽथ पुरात्सिद्धात्कुरुवर्षं तथोत्तरम् ॥ ४८ ॥

हिरण्मयं रम्यकनामखण्डकं सुमाल्यवद्गन्धगिरी क्रमादथ ।
यमादिकोट्यन्तपुराच्च रोमकादुदक्ककुप्स्थौ च पुराणसन्मतौ ॥ ४९ ॥

निषधपर्वतनीलनगावधी भवत उत्तरदक्षिणगामिनौ ।
जलधिमाल्यवदन्तरवर्ति यत्तदिह भद्रतुरङ्गमखण्डकम् ॥ ५० ॥

गन्धमादनपयोधिमध्यगं यज्ञगुस्तदिति केतुमालकम् ।
नीलगन्धनिषधाद्रिमाल्यवन्मध्यसंस्थितमिलावृतं बुधाः ॥ ५१ ॥

इलावृतान्तर्निवसन्ति निर्जराः सुचारुचामीकरभूतलान्विते ।
इहैव मेरुः किल मध्यसंस्थितः सुरालयः काञ्चनरत्नवान्गिरिः ॥ ५२ ॥

श्रीमद्विरिञ्चिजननाय वसुंधराब्जे पौराणिका जगुरमुं किल कर्णिकेति ।
यस्माद्धरा भवति नाभिसरोरुहं हि पुंसो विराज इतरत्र षडङ्गसंस्था ॥ ५३ ॥

47 Likewise, the Niṣadha mountain is [further] beyond that [i.e., beyond the Hemakūṭa mountain]. In the northern part [of Jambūdvīpa], from the city of Siddhā [i.e., from Siddhapurī], the mountains appear northwards towards the east in the order of Śṛṅgī, Śukla, and Nīlagiri. And the [individual] *varṣa*s (physiographic regions) are situated in the intermediate space between two of them [i.e., between the pairs of mountains].

48 Just as the *varṣa*s of Bhārata, Kinnara, [and] from that, Hari are north of [the city of] Laṅkā, so also [are the following *varṣa*s] north of the city of Siddhā [i.e., Siddhapurī]: Kuru *varṣa*...

49 ...Hiraṇmaya, [and] the region named Ramyaka. Now, the Sumālyavat [and] Gandha mountains are situated in the region towards the north from the boundary of the city of Yamakoṭi and [the city of] Romaka respectively, according to the noble opinion of the Purāṇas.

50 [With] the boundaries being the Niṣadha mountain and the Nīla mountain towards the north and south, what [*varṣa*] is situated between the ocean and the Mālyavat [mountain], that here is the whole Bhadraturaṅga [*varṣa*].

51 What wise men declared as lying between the ocean and the Gandhamādana [mountain], that is the Ketumālaka [*varṣa*; and also] the [central] Ilāvṛta [*varṣa*] being situated between the Nīla, Gandha [i.e., Gandhamādana], Niṣadha [and] Mālyavat [mountains].

52 The immortals live in Ilāvṛta [*varṣa*] endowed with beautiful golden earth. [Mt] Meru, situated right in the middle [of Ilāvṛta], is the abode of the gods [and] a mountain abounding in gold and precious stones.

53 Those well versed in the Purāṇas declared that very mountain as the pericarp in the earth lotus for the birth of Brahmā, as the Earth forms the lotus emerging from the navel of the supreme spirit [i.e., Garbhodakaśāyī Viṣṇu]; elsewhere, [the other *loka*s or realms] are arranged as the six limbs [of Viśvarūpa Viṣṇu].

विष्कंभागा मन्दरो गन्धशैलो विस्तीर्णाद्रिः पार्श्वभूभृत्क्रमेण ।
तेषु प्रोचुः केतुवृक्षान्मुनीन्द्रा नीपं जम्बूं सद्वटं पिप्पलं च ॥ ५४ ॥

यज्जम्बूफलतोऽगलद्रसमयी जम्बूनदीप्रस्रुता
या या तद्रसमिश्रिता मृदभवत्तत्तत्सुवर्णं स्फुटम् ।
तस्मादुत्तरतः समागतनदीवाहेषु जाम्बूनदं
किंचित्किंचिदवाप्यतेऽपि सिकतामध्ये नरैरुत्तमम् ॥ ५५ ॥

विष्कम्भशैलेषु वनानि चैत्ररथं तथा नन्दननामधेयम् ।
धृत्याख्यवैभ्राजकसंज्ञिते च सरांस्यथैतेषु यथा क्रमेण ॥ ५६ ॥

अरुणमानसके च महाह्रदं सितजलं पुनरेषु विलासिभिः ।
सुरतकेलिसमाकुलमालसाः स्वपतिभिश्च रमन्त्यमराङ्गनाः ॥ ५७ ॥

शृङ्गत्रयं कनकरत्नमयं सुमेरौ तस्मिन्विरञ्चिहरिशम्भुपुराणि सन्ति ।
इन्द्राग्निकालनिर्ऋताम्बुपवायुसोमेशानां पुराणि तदधोऽपि भवन्ति चाष्टौ ॥ ५८ ॥

स्वर्णाद्रौ पतिता हि विष्णुपदतो जाता चतुर्धा ततो
विष्कम्भाचलमस्तकोपरिसरः प्राप्ता चतस्रो दिशः ।
गच्छन्ती गिरिगह्वरादिषु रटत्कल्लोललोलाम्बुभृ-
त्सीताख्यालकनन्दिका सुरसरिच्चक्षुः सुभद्रा क्रमात् ॥ ५९ ॥

54 The buttress mountains [to Mt Meru] are the Mandara [mountain], the Gandha mountain, Vistīrṇa mountain, and the Pārśva mountain in regular order [i.e., starting from the eastern side of Mt Meru]. The chiefs of the sages declared the signature trees on these [mountains] being Nīpa (*kadamba* or burflower), Jambū (black plum), Vaṭa (Indian fig or Banyan), and Pippala (sacred fig) [respectively].

55 The Jambū river, formed of the juice that is oozing from the Jambū (black plum) fruit, flowed forth [and] whatever earth mixed with that juice, all of that became pure gold. Therefore, in the flows of the confluent rivers from the north, just the smallest amount of the best *Jambūnada* [gold] is obtained by men in the sand [of that river].

56 The [four] forests in the [four] buttress mountains [of Mt Meru] are Caitraratha, and likewise, [those] named Nandana, Dhṛti and Vaibhrājaka. Moreover, the [four] lakes in these [forests] in regular order [i.e., starting from the eastern Caitraratha forest] are...

57 ...Aruṇa, Mānasa, Mahāhrada, and Sitajala. The celestial nymphs, their minds filled with dalliance, sport with their playful lords repeatedly in [the waters of] these lakes.

58 The three mountain summits on [Mt] Sumeru are made of gold and precious stones. In them are [located] the cities of Brahmā, Viṣṇu, and Śiva. Just beneath that [i.e., beneath these three cities] are the eight cites of [the eight regents of the directions] Indra, Agni, Kāla, Nirṛta, Ambupa, Vāyu, Soma, and Īśāna.

59 [The one] born from the foot of Viṣṇu [i.e., the river Gaṅgā], having just descended upon the golden mountain [Mt Meru], became four-fold from that place [i.e., separated into four streams and] reached the lakes beyond the summits of the [four] buttress mountains [of Mt Meru], going through the mountains, caves, etc. in the four directions [and] carrying roaring waves of rolling water. [These five rivers are] Sītā, Alakanandikā, Surasarit [i.e., Gaṅgā], Cakṣu, and Subhadrā in regular order [i.e., starting from the east].

शम्भोर्मस्तकभूषणाय विमला मुक्तावलीवत्स्थिता
या सोपानपरम्परेव विदिता स्वर्गोन्मुखानां नृणाम् ।
या विष्णोश्चरणाम्बुजस्य विलसल्लोकत्रये विस्तृता
सा गङ्गा हृदयस्थिता भवतु मे पापौघविध्वंसिनी ॥ ६० ॥

ऐन्द्रं खण्डमथो कशेरुशकलं स्यात्ताम्रपर्णं ततो
विज्ञेयं च गभस्तिमादनमतः खण्डं कुमारीति च ।
नागं सौम्यमथापि वारुणमतः प्रात्यं च गन्धर्वकं
वर्षेऽस्मिन्नपि भारते नव सदा खण्डानि सन्ति ध्रुवम् ॥ ६१ ॥

विप्रक्षत्रियवैश्यादि व्यवस्था या पृथक्पृथक् ।
कुमारीखण्डमध्यस्था दृश्यते नेतरत्र सा ॥ ६२ ॥

सप्तैव सन्ति कुलभूमिधरा महेन्द्रः शुक्तिस्तथा मलय ऋक्षकपारियात्रौ ।
सह्यः सविन्ध्य इह भारतवर्षमध्ये प्रोक्ताः पुराणविबुधैर्जगति प्रसिद्धाः ॥ ६३ ॥

यद्वीपषट्कं खलु दक्षिणस्थं व्यक्षात्स भूर्लोक इह प्रसिद्धः ।
सौम्यो भुवर्लोक इतश्च मेरुः स्वर्लोक उक्तश्च पुराणविद्भिः ॥ ६४ ॥

ततः परं व्योम्नि महो जनोऽतस्तपश्च सत्यं किल सप्त लोकाः ।
केचिज्जगुर्लोकचतुर्दशत्वमधः स्थिताः सप्त तथोर्ध्वसंस्थाः ॥ ६५ ॥

लङ्कापुरे रविरुदेति यदा तदानीं मध्याह्नसंज्ञसमयो यमकोटिपुर्याम् ।
सिद्धाभिधेऽस्तसमयोऽथ सदा प्रतीच्यां स्याद्रोमके स्फुटतरं रजनीदलं हि॥६६॥

60 The stainless one [who is] established like the pearl necklace of the head ornament of Śiva, [she] who is known as the staircase for men who desire heaven, [she] who expanded through the glistening three worlds of the lotus-like feet of Viṣṇu, may that [river] Gaṅgā, the destroyer of the flood of evil, [always] abide in my heart.

61 The Aindra region, then the Kaśeru region, and beyond that the Tāmraparṇa [region] should be known; from there the Gabhastimādana [region] and the Kumārī region, and then the [regions of] Nāga and Saumya, from this [the region of] Vāruṇa and towards the end, the Gandharvaka [region]. Indeed in this Bhārata *varṣa*, there are always [these] nine regions assuredly.

62 What arrangement of Brāhmaṇa, *Kṣatriya*, *Vaiśya*, etc., separate from one another, is [practised] in the region of Kumārī [in Bhārata *varṣa*], that [arrangement] is not seen anywhere else.

63 There are seven principal mountain ranges Mahendra, Śukti, then Malaya, Ṛkṣaka, and Pāriyātra, [and] Sahya along with Vindhya [situated] here in Bhārata *varṣa* [that are] proclaimed in the world by the learned men of the Purāṇas as well known.

64 What indeed consists of an aggregate of six island-continents [and is] situated south of the equator, that is the well-known Bhūrloka here. And from this place, Bhuvarloka is said to be the northern and Svarloka is [Mt] Meru by those who know the Purāṇas.

65 Beyond that in the heaven is Mahar, from this Jana, Tapas, and Satya. [These are] seven worlds so stated. Some [scholars] have proclaimed the existence of fourteen worlds, [with] seven situated below and likewise [seven] situated above.

66 When the Sun rises in the city of Laṅkā, at that very moment, the time is called midday in the city of Yamakoṭi, sunset in [the city called] Siddhā [i.e., Siddhapurī], and then in the city of Romaka always to the west [of Laṅkā], it should be precisely midnight.

यत्रोदयो भवति भानुमतश्च तत्र पूर्वा दिशाऽस्तमुपयाति पुनः स यत्र ।
प्रत्यक्तयोस्तिमिकृता खलु दक्षिणोदक्तस्मात्समस्तविषये शशिदिक्सुमेरुः॥६७॥

लङ्कानगर्या अथवोज्जयिन्याः श्रीगर्गराटात्कुरुजाङ्गलाद्वा ।
सुमेरुतो वा कुचतुर्यभागे प्राचीप्रतीच्योर्यमरोमके स्तः ॥ ६८ ॥

ताभ्यां न पश्चादथवा पुरस्ताल्लङ्का विना यत्ककुभो विचित्राः ।
गोलाकृतिं तेन वदन्ति धात्रीं ज्योतिर्विदो दर्पणसंनिभां न ॥ ६९ ॥

भूगोलखण्डस्य शरप्रमाणं निघ्नं धरायाः परिणाहमित्या ।
तद्भूमिखण्डोपरिगं फलाख्यं क्षेत्रोद्भवं योजनकैः स्फुटं हि ॥ ७० ॥

सुमेरुमभितो निजं पुरमवाप्य वृत्तं च य-
त्स्फुटः कुपरिधिः स्मृतः स तु भचक्र ३६० भागाङ्कितः ।
तथा खरस ६० नाडिकाङ्कित इहाथवा योजनैः
स्फुटाख्यपरिणाहजैर्भवति खेटदेशान्तरम् ॥ ७१ ॥

इति भूगोलरचना ।

अथ भगोलरचना ।

भूमेर्वायोरुपरि परितो व्योमकक्षावधिस्थः
पिण्डीभूतः प्रवहपवनो गोलकाकारमूर्तिः ।
अन्तः स्वच्छः स्फटिकघटवन्नीरवत्सावकाशः
शश्वद्भ्राम्यन्नपरककुभं दक्षिणोदग्ध्रुवाभ्याम् ॥ ७२ ॥

67 And where the rising of the Sun occurs, there is the eastern direction. And again, where it obtains setting [that is] the western [direction]. The north-south [direction] is indeed determined with [the figure of] the fish for those two [places]. Therefore, in the entire world, [Mt] Sumeru is [always] to the north [of one's location].

68 From the city of Laṅkā or from [the cities of] Ujjayinī, Śrī Gargarāṭ, Kurujāṅgala, or [Mt] Sumeru, at a distance of a fourth part of the Earth [i.e., at a distance of 90°] towards the east and west [respectively] are the [cities of] Yama[*koṭi*] and Romaka.

69 [However], because the directions are strange, [there are] no [equatorial cities] westwards or eastwards from those two [antipodal equatorial cities, i.e., from Yamakoṭi and Romaka] with the exception of Laṅkā. For that reason, the astronomers declare the Earth as spherical in form [and] not resembling a [flat] mirror.

70 The measure of the height of [any] clime on the sphere of the Earth is multiplied by the measure of the circumference of the Earth. That is the true result [i.e., the surface area] of the geometric figure produced on the surface of [that] clime of the Earth [and measured] with *yojana*s.

71 And what circle is all around [Mt] Sumeru, having extended to one's own city, that is then declared as the corrected [local] circumference of the Earth [i.e., a parallel of latitude] marked with 360 degrees, or, in this case, likewise marked with 60 *nāḍikā*s. The longitudinal correction of a planet [for a particular time and place on Earth] is [computed] with [this] corrected circumference [measured] in *yojana*s.

Thus [ends the section on] the composition of the sphere of Earth.

Now, the composition of the sphere of asterisms.

72 [The sphere of asterism is] situated up to the periphery of the heavens, [extending] all around, above the [outermost] winds of the Earth [i.e., above the Earth's atmosphere], joined [together with] the Pravaha wind, [and] embodied in the shape of a sphere. [Its] interior is transparent like a crystalline pitcher, like water, with [expansive] space, perpetually revolving in the western direction about the north and south celestial poles.

Reading 1 in Bn.I, Np, Pb, and Sc

कुर्वन्मध्ये विषुववलयं पार्श्वजातद्युवृत्तं
तस्यान्तःस्थं श्लथमिव सभं तुङ्गपातादियुक्तम् ।
प्राचीं गच्छन्मृदुगतियथा ज्ञेयमेतद्धचक्रं
तस्यान्तःस्थाः सकलखचराः प्राचि यान्ति स्वभुक्त्या ॥ ७३.१ ॥

Reading 2 in Bn.II, Br, Pm, and Rr

कुर्वन्मध्ये विषुववलयं पार्श्वजातद्युवृत्तं
यस्मिन्नन्तर्जल इव उषास्तारकाः प्राक्तरन्ति ।
उड्डीयन्ते युगपदुडवः खे यथा खेचरेन्द्रा-
स्तच्चक्रं वा भवलयमिति प्रोच्यते वा भगोलः ॥ ७३.२ ॥

तस्यैवान्तर्व्रजति पुरतो येन मार्गेण भानुः
सम्भेद्यद्विर्विषुववलयं सौम्ययाम्यप्रवृत्तः ।
क्रन्तेर्वृत्तं भवनवलयं राशिवृत्तं च तत्स्या-
देतस्यार्धे शशियमदिशोर्मेषजूकादिके स्तः ॥ ७४ ॥

क्रान्तेर्वृत्तादुभयत इह स्तो ध्रुवौ यौ कदम्बौ
तौ तद्योगे दधति सकला राशयो द्वादशैते ।
तुल्यार्कांशैर्भवनवलयेऽजाद्य षड्बाणसूत्रै-
र्भक्ते भ्राम्येद्ध्रुवमभित इह प्रत्यहं स्वः कदम्बः ॥ ७५ ॥

सार्धत्रयोविंशति २३|३० चापभागैः सौम्येऽथयाम्ये खलु राशिचक्रम् ।
यस्मिन्सुदूरेऽथ कदम्बयुग्मध्रुवद्वयोपेतमिहायनाख्यम् ॥ ७६ ॥

शङ्खावर्तवदिन्दुपूर्वकखगः प्राच्यां च येनाध्वना
चक्रे भ्राम्यति राशिमण्डलमथो भित्वा द्विवारं मुहुः ।
सौम्यावाग्गतवांस्तथा प्रवहतो नित्यं प्रतीच्यां चल-
न्किंचित्किंचिदुपेत्य योगमपरं तद्बाणवृत्तं जगुः ॥ ७७ ॥

73.1 Making the celestial equator in the middle, the diurnal circle produced on its side is situated in its interior [i.e., inside the sphere of asterisms], unfastened in some measure, parallel, and possessed of elevation, depression, etc. Accordingly, [what is] moving eastwards with a slow speed should be understood as this wheel of asterisms. All the planets situated in its interior [i.e., inside the wheel of asterisms] move eastwards with their own daily motion.

73.2 Making the celestial equator in the middle, the diurnal circle produced on its side [is the one] inside of which, like fish in water, the stars swim from the east. The stars fly in the sky side by side just as the best of planets do so, and that wheel is either called the ring of asterisms or the sphere of asterisms.

74 Indeed, within it [i.e., within the sphere of asterisms], the path by which the Sun travels eastwards, having been brought into contact with the celestial equator twice [in its] northern and southern progression, that [path] is [called] the circle of declination, the circle of lunar mansions, or the zodiacal circle. The two halves of this [i.e., the two halves of the zodiacal circle] in the north and south directions are [the six zodiacal signs] beginning with Aries and Libra [respectively].

75 The two fixed [poles] that are here on both sides of the circle of declination, those are the two ecliptic poles. All these twelve zodiacal signs produce their union in the circle of lunar mansions [i.e., they constitute the ecliptic] having been divided into twelve equal parts with six direction [circles] of ecliptic latitude and [beginning] from the zodiacal sign Aries. Its own ecliptic pole revolves here around the celestial pole [i.e., around the polestar] everyday.

76 With 23;30 degrees of arc to the north, and indeed [also] to the south, [of the celestial equator] is the zodiacal circle. Now, in it, at its furthest distance, having [an amount equal to the distance between] the pair of ecliptic poles and the pair of celestial poles [i.e., the point of the ecliptic intersecting the solstitial colure] is known here as the solstitial [point].

77 Like the convolution of a conch-shell, the course by which a planet, beginning with the Moon, revolves in a circle towards the east, then, having passed through the zodiacal circle momentarily on two occasions going north [and] south, is likewise always moving towards the west [impelled] by the Pravaha wind [and] approaching the next conjunction little by little, [wise men have] declared that [course as] the latitudinal circle [i.e., the planet's orbit].

प्रवहपवनवेगव्याहताधोर्ध्वगोलो हरिदिशमपि गच्छन्नुच्चतामेति खेटः ।
तदनु तदवरोहो याति यावत्समत्वं तदुदितविपरीतं नीचगत्वे खगस्य ॥ ७८ ॥

खगकदम्बयुगोपरि यद्गतं तदिह खेटशिलीमुखसूत्रकम् ।
अथ खगध्रुवयुग्मगतं च यद्वलयमेतदपक्रमसूत्रकम् ॥ ७९ ॥

अस्मिन्भचक्रे गजवेदसंख्याभमूर्त्तयो या यवनैः प्रदिष्टाः ।
भानां सहस्रं नयनाक्षियुक्तं १०२२ यथागमे ताः सुबुधैः प्रदेयाः ॥ ८० ॥

तासां भचक्रतः सौम्या मूर्तयश्चैकविंशति ।
याम्याः पञ्चदशप्रोक्ताः मध्ये द्वादशराशयः ॥ ८१ ॥

कदम्बमभितो येन खेटो गच्छति चाध्वना ।
शराग्रमण्डलं तत्स्याद्यत्स्याद्द्युज्याख्यमण्डलम् ॥ ८२ ॥

यावन्ति वृत्तानि भवन्ति गोले तावन्ति सर्वाणि च कल्पितानि ।
यथान्धकारे कृतकौतुकानामालातचक्रभ्रमयः शिशूनाम् ॥ ८३ ॥

अत्रैव कक्षा द्युसदां निबन्ध्या उपर्युपर्येव विधोः क्रमेण ।
विक्षेपवृत्तेषु निजेषु तद्वन्मेषादयो राशय एव वेद्याः ॥ ८४ ॥

अथ खगोलरचना ।

78 The upward and downward circle [i.e., the orbit of the planet stands] impeded by the [constant westward] circulation of the Pravaha wind. Going towards the direction of the Sun [i.e., eastwards], the planet goes towards the apex of its orbit [i.e., the planet's orbital apogee]. Thereafter, its descent advances until it reaches equality [with the amount of] ascension [in the] reversed [direction] at the lowest point of the orbit [i.e., the orbital perigee] of the planet.

79 What [great circle] goes over the planet and the pair of ecliptic poles, that here is the direction [circle] of ecliptic latitude of the planet. Now, what [great] circle is connected with the planet and the pair of celestial poles, this is the direction [circle] of declination.

80 In this sphere of asterisms, the 48 total constellations [and] likewise the 1022 of the stars [in them], which have been pointed out by the foreigners in [their] tradition, all of that should be taught by wise men.

81 Among all of these [48 constellations], 21 figures [i.e., constellations] are northern with respect to the circle of asterisms, 15 are said to be southern, and 12 zodiacal signs are in the middle.

82 And the course by which a planet goes all around the ecliptic pole, [and] what [resembles] the [diurnal] circle called the day-radius [of the planet], that is the [parallel] circle of ecliptic latitude.

83 However many circles are produced on the sphere, all of those [circles] are indeed imaginary like the whirling circles of firebrand [made] for playful children in the darkness.

84 The orbits of the planets should be fastened right here [in the *bhagola*], indeed higher and higher above the Moon, in regular order, in their own orbital circles. Likewise, even the zodiacal signs beginning with Aries should be understood [to be fastened in *bhagola*].

Now, the composition of the oblique celestial sphere.

भूगर्भक्षितिपृष्ठसूत्रसदृशो यो ना कुगर्भे स्थित-
स्तं यत्खं खलु सर्वतो विजयतेऽनन्तो खगोलोऽस्ति सः ।
वृत्तं तत्र भवेद्यथा परिकरो यस्य ध्रुवौ सर्वदा
खाधःस्वस्तिकसंज्ञकौ बुधजनास्तद्भूजसंज्ञं विदुः ॥ ८५ ॥

प्राक्खस्वस्तिकपश्चिमोपरिगतं यन्मण्डलं यद्ध्रुवौ
याम्योदक्ककुभौ वदन्ति गणका वृत्तं समाख्यं च तत् ।
याम्योदङ्नरमस्तकोपरि गतं यस्य ध्रुवौ सर्वदा
प्राक्पश्चादिदमेव सर्वमुनयो मध्याह्नवृत्तं जगुः ॥ ८६ ॥

अथेष्टदिक्तद्विपरीतकाष्ठाखस्वस्तिकाख्योपरि निःसृतं यत् ।
तदेव दृङ्मण्डलनामधेयं कोणेषु कोणाभिधमामनन्ति ॥ ८७ ॥

कदम्बखस्वस्तिकगामि यच्च तदेव दृक्क्षेपकमण्डलं हि ।
पूर्वापराशाध्रुवयुग्ममाप्तमुन्मण्डलं नाम बुधैर्निरुक्तम् ॥ ८८ ॥

द्युज्यामण्डलसदृशं द्युज्यावृत्तं स्थिराख्यमत्रापि ।
दक्षिणसौम्यखगोपरि यातं समसूत्रसंज्ञितं वृत्तम् ॥ ८९ ॥

एवं पूर्वापरखगमाप्तं मध्याभिधं सूत्रम् ।
पूर्वां वापरकाष्ठां केन्द्रं कृत्वा यदा भतो वृत्तम् ।
द्युज्यामण्डलसदृशं ज्ञेयं पूर्वाभिधं तच्च ॥ ९० ॥

85 A man situated at the centre of the Earth [is oriented in a direction] resembling [the direction of the] line joining the Earth's centre to the Earth's surface. The space that extends in every direction [from that geocentric observer] is the eternal oblique celestial sphere. In that place, there is a [great] circle, like a girdle, whose two poles are always called the zenith and nadir. Wise men consider that [great circle] the [celestial] horizon by name.

86 What [great] circle, having gone over the east, the zenith, and the west, [and] whose two poles are the southern and northern directions, mathematicians declare that as the prime vertical by name. Having gone over the south, north, and the zenith, [the great circle] whose two poles are always to the east and the west, all the sages have indeed declared this [great circle] as the meridian.

87 Now, what [great circle] has gone forth over the direction of the desired [object], its reversed direction, and the zenith, [wise men] call that very [circle] the vertical circle of altitude by name; [when it is] in the [intercardinal] corners, [they call it] the intercardinal circle by name.

88 And what [great circle] extends to the zenith and ecliptic pole, that is indeed the secondary circle to the ecliptic. [The great circle] reaching the eastern and western regions and the pair of celestial poles is declared as the equinoctial colure [or six o'clock hour circle] by name.

89 The circle of day-radius, resembling the diurnal circle [in the *bhagola*], is also known as a stationary [circle] here [i.e., in the *khagola*]. The [great] circle situated over the south, the north, and the planet is called the secondary circle to the prime vertical.

90 In the same way, [the great circle] that has reached the east, the west, and the celestial body is called the secondary circle to the meridian. Having made the centre of a circle [as] the eastern or the western direction, whenever a [small] circle [is made] with respect to a celestial object similar to the diurnal circle [in the *khagola*], that [small circle] should be understood by the previous name [i.e., by the word *madhya*, implying the parallel of meridian].

अथ खस्वस्तिकमेव स्पष्टं केन्द्रं प्रकल्प्य यद्वृत्तम् ।
कुर्यादाभात्तत्स्याद्दृग्ज्यावृत्तं सदोर्ध्वमधरं वा ॥ ९१ ॥

द्वादशभिः समसूत्रैर्द्वादशभावाख्यभूजानि ।
क्षितिजादथो धनाद्या भावा ज्ञेया यथा लग्नम् ॥ ९२ ॥

अथो कुजं गच्छति योऽत्र भांशस्तदेव लग्नं क्रियपूर्वकं स्यात् ।
प्रत्यक्कुजे चास्ततनुस्तथैव मध्याह्न वृत्तोपरि मध्यलग्नम् ॥ ९३ ॥

अधः स्थितं तत्र तु तुर्यसंज्ञं द्वयं द्वयं चान्तरगं हि तेषाम् ।
खलग्नयोर्लग्नचतुर्थयोश्च सुखस्त्रियोः स्त्रीवियतोश्च मध्ये ॥ ९४ ॥

मध्याह्नभूजवलयान्तरसंस्थितानां नाड्याख्यचक्रसमवृत्तभमण्डलानाम् ।
त्रींस्त्रींल्लँवान्निजनिजाङ्घ्रिगतान्समानान्भक्त्वा ततो निजनिजध्रुवसम्मुखानि ॥ ९५ ॥

षड्वद्विलिख्यवलयानि च सुस्थिराणि
भावाभिधानि कुजतोऽधरसंस्थितानि ।
द्रव्यादितः क्रमतयोपरिगानि रिष्फा-
द्व्यस्तानि तानि सकलानि गृहाणि विन्द्यात् ॥ ९६ ॥

91 Now, having fixed the true centre [of a circle as] the very zenith, the [small] circle which one makes with respect to a celestial object, that is the parallel of altitude. [It is] always upwards or downwards [with respect to the horizon].

92 [Going] from the [celestial] horizon, the sides [i.e., arcs of the ecliptic] called the twelve astrological houses [are created] with twelve secondary circles to the prime vertical. Now, the houses should be understood as being first [*dhanādya*, lit. immediately preceding the second house of wealth] according to [the house in which] the Ascendant [falls].

93 Now, what part of an asterism [i.e., the degrees of zodiacal signs] approaches the [eastern] horizon in this case, that is indeed the Ascendant [i.e., the first astrological house] beginning with [0° of] Aries; and [what part of an asterism is] at the western horizon, [that is] the setting point [i.e., the Descendant or the seventh astrological house]; and likewise, [what is situated] over the meridian, [that is] the Medium Coeli [i.e., the tenth astrological house];...

94 ...and [the part of an asterism] situated below [at the nadir] is known as the fourth [astrological house or the Imum Coeli]. And situated between them [i.e., between the four angular houses] are pairs of two [intermediate astrological houses, namely, the pairs] lying in the middle of [the houses of] the sky and the Ascendant [i.e., between the Medium Coeli and the Ascendant]; the Ascendant and the fourth [i.e., between the Ascendant and the Imum Coeli]; happiness and women [i.e., between the Imum Coeli and the Descendant]; and women and the sky [i.e., between the Descendant and the Medium Coeli respectively].

95 [The arcs] of the celestial equator, prime vertical, and ecliptic that are situated between the circles of the [celestial] horizon and the [local] meridian, having divided [these arcs of the] three [great circles] into three equal parts each in their own respective quadrants, are, consequently, oriented towards their own respective fixed poles.

96 And having marked [these twelve arcs as] six by six [with respect to the celestial horizon], the [twelve] fixed zones called astrological houses are placed downwards with respect to the [celestial] horizon, beginning with the first house [*dravyādi*, lit. immediately preceding the second house of wealth and] going over the triads [of houses] up to the twelfth house. One should understand all those [astrological] houses as distinct.

यस्मिन्यस्मिन्वलये राशिलवो लगति तत्र तत्रैव ।
स सभावः स्फुटसंज्ञस्त्रिविधमिदं भावचक्रं स्यात् ॥ ९७ ॥

उदक्परक्रान्तिरिनस्य येषु लम्बेन तुल्या विषयेषु तेषु ।
कर्कस्थितो भानुरुदेत्युदीच्यामवाचिनक्रः समुपैति चास्तम् ॥ ९८ ॥

क्षणं भचक्रं कुजवत्स्थितं हि समुद्गतं स्याद्युगपन्मृगादिम् ।
तथैव कर्कादिकमस्तमेति ततःपरं कर्कमुखं भषङ्कम् ॥ ९९ ॥

क्रमेण लङ्कोदयनाडिकाभिर्द्विसंगुणाभिः समुदेति भूजात् ।
नक्रादिकं चास्तमुपैति तद्वत्ततो घटीषष्टिमितं दिनं स्यात् ॥ १०० ॥

रात्रिप्रमाणं क्षणमेव नास्ति द्युःक्षीयतेऽतोऽपि निशैषते च ।
तुलाङ्गतेऽर्के द्युनिशे समेस्तस्ततोऽपचीयेत दिनप्रमाणम् ॥ १०१ ॥

वर्धेत रात्रिर्धनुरन्तमाप्ते भानौ घटीशुन्यमितं दिनं स्यात् ।
रात्रिः परा षष्टिघटीप्रमाणा स्यातां ततो रात्रिदिने विलोमे ॥ १०२ ॥

यावद्भिरंशैर्निजलम्बतोऽधिका खेटस्फुटक्रान्तिरुदक्स्थिता भवेत् ।
तावत्सदा दृश्यतनुः स खेचरः सदास्तगो दक्षिणदिक्स्फुटायमे ॥ १०३ ॥

अतोऽक्षभागैर्विशरांशखागै ६९|४८ र्युग्मेन्दुभस्थे द्युमणौ दिनं स्यात् ।
धनुर्मृगस्थे च निशा समस्ता पुरेषु भेषु द्युनिशोर्व्यवस्था ॥ १०४ ॥

97 In whichever respective zone [i.e., in whichever one of the twelve houses], the degree of a zodiacal sign comes in contact, in those exact places, it is correctly identified as that [respective] astrological house. [In this way,] this circle of astrological houses is three-fold.

98 In all the places where the maximum northern declination of the Sun is equal to the co-latitude, in those places, the Sun, [being] situated in Cancer, rises to the north and Capricorn attains setting to the south.

99 The ecliptic, remaining just like the horizon for an instant, is risen up simultaneously with the [six zodiacal signs] beginning with Capricorn [and] likewise, the [six zodiacal signs] beginning with Cancer set. Thereafter, the six zodiacal signs beginning with Cancer...

100 ...in regular order, rise together from the horizon with *nāḍikā*s of right ascension multiplied by two, and the [six zodiacal signs] beginning with Capricorn likewise attain setting. Therefore, the [solstitial] day [at the Arctic Circle] is equal to sixty *ghaṭīkā*s.

101 There is no length of the night even for an instant [on the solstitial day]. And thereafter, [the length of] the day diminishes and [the length of] the night grows. When the Sun is in Libra, the day and night are equal. Thereafter, the length of the day is decreased,...

102 ...[and the length of] the night increases. The Sun, having reached the end of Sagittarius, [the length of] the day is equal to zero *ghaṭī*s [and] the night is the maximum 60 *ghaṭī*s in length. Thereafter, both [the lengths of] the night and day are reversed.

103 As long as the degrees with which the true declination of a planet is situated to the north [and] in excess of one's own co-latitude, to such an extent that planet always has a visible form [i.e., it is always visible; it] always remains set when the true declination is in the southern direction [exceeding one's own co-latitude].

104 Therefore, [at places] with 69;48 degrees of terrestrial latitude, when the Sun is in the lunar mansion *indubha* [i.e., the *nakṣatra Mṛgaśirṣa*] of Gemini there is [just] day and [when it] is between Sagittarius and Capricorn, [it is] entirely night. [When the Sun is] in other zodiacal signs, there is a proportion of [both] day and night.

सार्धाष्टशैलैः ७८|३० पलभागकैश्च दिनं वृशाद्ये भचतुष्टयेऽर्के ।
निशा पुनर्वृश्चिकतश्चतुष्के खाङ्का ९० क्षभागैर्द्युसमार्धकेन ॥ १०५ ॥

रात्रिस्तथैवास्ति सुमेरुवासिनां मेषादि षट्केऽर्कविलोकनं सदा ।
तुलादिगेऽर्के रजनीमृगादिगे निशार्धमेषां द्युदलं च कर्कगे ॥ १०६ ॥

यथा यथा क्रान्तिरुदग्विवर्धने तथा तथार्कस्य समुन्नतांशकाः ।
अतो यदा कर्कटगो दिवाकरः परोन्नतत्वं समुपैत्यहर्दले ॥ १०७ ॥

विपर्ययो वाडववह्निवासिनां सुरद्विषां रात्रिदिनव्यवस्थयोः ।
अतः सुराणाममरद्विषां तथा दिवानिशं भास्करवर्षसम्मितम् ॥ १०८ ॥

फलाय संहिताविदो निशार्धतो दिनं जगुः ।
दिनार्धतश्च यामिनीं ततोऽयनाद्यहर्निशम् ॥ १०९ ॥

इन्दोर्गोलोपरिष्टाद्वसति पितृगणोऽमान्तकालेऽर्कतोऽध-
स्तिष्ठन्मूर्धोपरिष्टात्कलयति तरणिं तेन तेषां दिनार्धम् ।
दर्शोथो पूर्णमासीरजनिदलमतः कृष्णशुक्लाष्टमीतो-
ऽह्नोरात्रेश्च प्रवेशोऽसितसितदलयोर्धस्त्ररात्रं प्रसिद्धम् ॥ ११० ॥

105 And, [at places] with 78;30 degrees of terrestrial latitude, there is [just] day when the Sun is in the set of four zodiacal signs beginning with Taurus; it is [just] night again [when the Sun is] in the set of four [zodiacal signs] with respect to Scorpio. [At the place] with 90 degrees of terrestrial latitude, with equal half as the day...

106 ...is likewise the night for [the gods] residing on Mt Sumeru; always observing the Sun [i.e., day] when [it is] in the set of six [zodiacal signs] beginning with Aries; night when the Sun is in the [set of six zodiacal signs] beginning with Libra; midnight [when it is] in the beginning of Capricorn; and their midday when [the Sun is] in Cancer.

107 The more the declination [of the Sun] is on the rise to the north, the more are the degrees of increased elevation of the Sun. Therefore, when the Sun is at the [start of] Cancer, it attains maximum elevation [in the sky] at midday.

108 There is a reversal in the proportion of night and day for the enemies of the gods who live in [the regions] of submarine fires [i.e., for the demon inhabitants of the south pole]. Therefore, for the gods, and likewise for the enemies of the gods, [the period of] a day and night is equal to a solar year.

109 For the sake of [its] fruit [i.e., for its computational advantage], men learned in the doctrine of astrology have declared the day [beginning] from the middle of the night and the night [beginning] from middle of the day. Consequently, [the period of] a day and night [at the poles] begins with the [northward or southward] movement [of the Sun from its solstitial positions on the ecliptic].

110 The forefathers live above the sphere of the Moon, [and hence] at the time of the new Moon [while] being situated below the Sun, [they] observe the Sun above at the highest summit [in the sky, i.e., at their zenith]. Hence, the day of the new moon is their midday [and] likewise, the full moon is [their] midnight. Therefore, the start of the day and of the night is from the eighth lunar day [*tithi*] of the waning phase of the Moon and the waxing phase of the Moon [respectively]. [The period of] a day and night [for the lunar manes] is [thus] arranged in [the] dark and light halves [of a synodic lunar month].

एवं समस्तरचनायुक्तो ज्ञेयः खगोलाख्यः ।
अथ भूपृष्ठगतस्य प्रायश्चैवं भवेत्खगोलो यः ॥ १११ ॥

ज्ञेयः स गोलविज्ञैर्दृग्गोलः केन्द्रभिन्नत्वात् ।
अथ निजमण्डलचापं यल्लक्षणनामकं वदाम्येतत् ॥ ११२ ॥

विषुववृत्तभवृत्तयुतिस्तु या क्रियतुलाधरयोर्मुखमस्ति तत् ।
अयनवृत्तभवृत्तयुतिश्च यायनमुखं खलु दक्षिणसौम्यदिक् ॥ ११३ ॥

विषुववृत्तभमण्डलमध्यगं परमिहायनमण्डलजं धनुः ।
यदिदमस्ति स एव परापमो ह्यपमसूत्रगचापमिहापमः ॥ ११४ ॥

विषुवमण्डलखेचरबिम्बयोरपमसूत्रगतं तु शरासनम् ।
स्फुटतरापम एष खगस्य वै भवति याम्यदिगुत्तरतोऽथवा ॥ ११५ ॥

विषुववृत्तभमण्डलमध्यगं विशिखसूत्रधनुर्यदिहास्ति तत् ।
अपम एव पराभिध उच्यते खगभचक्रकयोर्विवरं शरः ॥ ११६ ॥

विषुवमण्डलखेचरमध्यगं विशिखसूत्रधनुर्यदिहास्ति सः ।
स्फुटतरापमकाङ्क उदीरितो गणितगोलविचारविचक्षणैः ॥ ११७ ॥

यावद्भिरंशैर्ध्रुव उन्नतः स्यात्तावद्भिरन्यध्रुवकः कुजाधः ।
तावद्भिरेवात्मशिरोनभस्तो नतं नरैर्वैषुववृत्तमूह्यम् ॥ ११८ ॥

111 In this way, endowed with its entire composition, [the sphere] called the oblique celestial sphere should be known. Now, what oblique celestial sphere is for the most part connected to the surface of the Earth [i.e., centred on the surface of the Earth]...

112 ...that is to be known as the visible celestial sphere by men learned in the science of spheres on account of the difference in centres. Now, what is the characteristic name [of the] individual arc of a [celestial] circle, I state this here.

113 What is the union of the celestial equator and the ecliptic, that is the beginning of the two zodiacal signs Aries and Libra, and what is the union of the solstitial colure and the ecliptic, that is indeed the beginning of the movement of the Sun towards the south and the north.

114 What arc is produced on the solstitial colure here [at its] greatest [distance] between the celestial equator and the ecliptic, that is indeed the greatest declination [of the Sun]. An arc of the direction [circle] of declination is the declination in this case.

115 But [what] arc of the direction [circle] of declination is between the celestial equator and the disk of a planet, this is certainly the true declination of the planet towards the southern direction or northwards, or else...

116 ...what arc of the direction [circle] of [ecliptic] latitude is here between the celestial equator and the ecliptic, that very [arc of] declination is called the other by name [i.e., the arc of second declination]; the [arc of the direction circle of ecliptic latitude corresponding to the] difference between the planet and the ecliptic is the latitude [of the planet].

117 What arc of the direction [circle] of [ecliptic] latitude is here between the celestial equator and the planet, that [arc] is declared as the curve of true declination by those who are wise in the investigations of computations and spheres.

118 By as many degrees as the celestial pole is elevated [above the local horizon in the sky], with that amount [of degrees] the other celestial pole is below the horizon. Indeed with that same amount, the celestial equator should be regarded by men to be depressed from one's own head in the sky [i.e., from one's local zenith].

तत्रोन्नतांशाः खलु लम्बभागा नतांशकाश्चाक्षलवाः प्रसिद्धाः ।
मध्याह्नवृत्ते वियदाख्यलग्नखस्वस्तिकान्तः खलु मध्यभागाः ॥ ११९ ॥

एवं क्षमाजग्रहबिम्बमध्ये परोन्नतांशा विबुधैर्विचिन्त्याः ।
भूजाभिधे प्राक्खचरोदयान्तरग्रांशकाः स्युः खलु पश्चिमेऽपि ॥ १२० ॥

पूर्वस्वदृङ्मण्डलमध्यभागा दिगंशकाः पश्चिमतोऽपि तद्वत् ।
दृङ्मण्डले भूजखगान्तराले स्युरुन्नतांशा अधरांशका वा ॥ १२१ ॥

तत्रैव खस्वस्तिकखेटमध्ये नतांशका एवमधो भवन्ति ।
स्वोन्मण्डलस्वीयधराजमध्ये द्युज्याख्यवृत्तस्य धनुश्चरं स्यात् ॥ १२२ ॥

कुजोपरिष्टाद्दिनरात्रवृत्ते यद्दृश्यचापं दिनचापमेतत् ।
कुजादधस्ताद्दिनरात्रवृत्ते यत्स्याददृश्यं रजनीधनुस्तत् ॥ १२३ ॥

दृश्यध्रुवासन्नमहो महत्स्याददृश्यपार्श्वे लघु वासरं हि ।
दिनाद्विलोमं रजनीप्रमाणं महत्तया स्वल्पतया च नित्यम् ॥ १२४ ॥

119 Certainly, at that place, the degrees of elevation [of the celestial equator from the horizon] are the degrees of co-latitude and the degrees of depression [of the celestial equator from the zenith] are well known [as the] degrees of latitude. On the meridian, [the distance] between the meridian ecliptic point and the zenith is then the degrees [of depression] of the meridian-ecliptic point.

120 In this way, the degrees of maximum elevation [of a planet, measured along the local meridian,] are considered [to lie] between the horizon and the disk of the planet by wise men. The degrees of rising [or ortive] amplitude [of the planet] are [considered to lie] between the [due] east and the rising point of the planet on the [local] horizon, and, then, also in the west.

121 The degrees of azimuth [of a planet] are the degrees between the [due] east and [the planet's] vertical circle of altitude, [and] likewise, also with respect to the [due] west. The degrees of elevation or degrees below [i.e., the degrees of depression of a planet] are on the vertical circle of altitude between the horizon and the planet.

122 At that very place, between the zenith and the planet, the degrees of zenith distance [of a planet] are thus [situated] below [the zenith]. The arc of the circle of day-radius [of the planet] between one's own equinoctial colure [i.e., the six o'clock hour circle] and one's own [local] horizon is the [total] ascensional difference.

123 What visible arc is on the diurnal circle above the horizon, this is the arc of day. Below the horizon on the diurnal circle, what is invisible, that is the arc of night.

124 [The length of] the day, being in the vicinity of the visible polestar [i.e., the Sun being northwards on the ecliptic, seen from the northern hemisphere], is large just as [the length of] the day is small [closer] towards the invisible side [i.e., the Sun being southwards on the ecliptic, seen from the northern hemisphere]. The length of the night is always reversed from [that of] the day with largeness [towards the south] and with smallness [towards the north].

पूर्वापरोन्मण्डलमध्यगामि द्युज्याख्यवृत्तस्य धनुर्भवेद्यत् ।
सदैव तच्चक्रदलांश १८० तुल्यं चराढ्यहीनं स्वदिनप्रमाणम् ॥ १२५ ॥

स्वोन्मण्डलं व्यक्षकुजं सदैव ततो निरक्षे द्युनिशोः समत्वम् ॥ १२६ ॥

यथा यथापक्रममण्डलस्य स्याद्दण्डवच्चापमृजु क्षमाजे ।
तथा तथोद्गच्छति भूयसैव कालेन तिर्यक्पुनरल्पकेन ॥ १२७ ॥

निरक्षदेशात्खलु सौम्यभागे नक्रादयः षड्क्षितिजं तिरोगाः ।
व्रजन्ति कर्कादय एव भूजमृजुत्वमाप्ताः परतोऽन्यथा ते ॥ १२८ ॥

दृक्षेपवृत्ते त्रिभहीनलग्नखस्वस्तिकान्तर्धनुरास्ति यच्च ।
दृक्षेपचापं गदितं तदेव खनन्द ९० शुद्धं खलु दृग्गतिः स्यात् ॥ १२९ ॥

दृक्क्षेपवृत्ते स्वकदम्बभूजमध्ये च दृक्षेपधनुर्भवेद्वा ।
कदम्बखस्वस्तिकमध्यचापं साद्दृग्गतिर्वा गणकैर्निरुक्ताः ॥ १३० ॥

कोणमण्डलगतोन्नतांशकाः कोणशङ्कुलवका बुधैः स्मृताः ।
एवमेव सममण्डलोन्नता भागकाः समनरोत्थिता मताः ॥ १३१ ॥

125 What arc of the circle called the day-radius is between the eastern and western [points of intersection with the] equinoctial colure, that [arc] is always equal to degrees of a semicircle (180°). The measure of one's own day-length is increased or decreased with the ascensional difference.

126 The equatorial horizon is always [coincident with] its own local equinoctial colure [and] consequently, there is equality of day and night at the terrestrial equator.

127 The extent to which an arc of the ecliptic, like a straight line, is perpendicular to the horizon, to such an extent [it] rises with longer amounts of time; however, [going] obliquely, [it rises] with shorter [time].

128 Certainly, in the region north of the terrestrial equator, the six [zodiacal signs] beginning with Capricorn cross the horizon moving obliquely exactly as [the six zodiacal signs] beginning with Cancer [cross] the horizon [having] obtained straightness. On the other side [i.e., to the south of the equator], they are the other way around [i.e., they are reversed in their angles of ascension].

129 And what arc between the nonagesimal point and the zenith is on the secondary vertical circle to the ecliptic, that is indeed called the arc of the zenith distance of the nonagesimal point. [That arc] subtracted from ninety degrees produces the arc of the altitude of the nonagesimal point [i.e., the complement of the zenith distance of the nonagesimal point].

130 Or, [what is the arc] on the secondary vertical circle to the ecliptic and also between its ecliptic pole and the horizon, [that] is [equivalent to] the arc of the zenith distance of the nonagesimal point. Or, [what] arc is between the ecliptic pole and the zenith, that is [also] declared by the astronomers as the altitude of the nonagesimal point.

131 The degrees of elevation [of a celestial object measured] on the vertical circle passing through the intercardinal directions are declared by the wise as the degrees of [its] intercardinal altitude. Similarly, the degrees [of the celestial object] elevated on the prime vertical are regarded as [the degrees of the] prime vertical altitude produced [on it].

उन्मण्डलं यदि रविः समुपैति सम्य-
ग्दृङ्मण्डलोन्नतलवोन्मितयस्तदानीम् ।
उद्धृत्तशङ्कुलवका गणकैर्विचिन्त्या
दृग्ज्यावृतावपि भुजांशमितिः स्वदिक्स्यात् ॥ १३२ ॥

एवं गोलविचक्षणो गणितविच्छिल्पागमज्ञः सुधीः
स्वेच्छाकल्पनया महान्ति बहुशः स्वल्पानि वृत्तानि वा ।
गोलस्थानि विचिन्त्य कार्यसमये तत्तद्धनुःसिञ्जिनी-
र्ज्ञात्वा कोटिभुजोत्थिताः स विशिखास्त्रैराशिकं साधयेत् ॥ १३३ ॥

इत्यादि गोलरचना सुमहाप्रवीणैर्गोला मृदा किमुत धातुभिरुत्तमाख्यैः ।
काष्ठेन वा शरलवंशशलाकिकाभिः कार्या विषाणफलकुञ्जरदन्तपूर्वैः ॥ १३४ ॥

इत्येतस्यामिन्द्रपुर्यां वसन्सन्नित्यानन्दो देवदत्तस्य पुत्रः ।
सारोद्धारे सर्वसिद्धान्तराजे गोलाध्यायं प्रापयत्तत्र पूर्तिम् ॥ १३५ ॥

132 When the Sun reaches the equinoctial colure properly, at that time, the measure of [its] elevation in degrees on the vertical circle of altitude [passing through the Sun] is considered the equinoctial colure altitude by astronomers. On the very circle [i.e., on the vertical circle of altitude], the Sine of the zenith distance [of the Sun] is [an amount] measuring the degrees of the perpendicular towards one's own direction [i.e., the horizontal distance of the Sun from one's zenithal direction].

133 In this way, the one who is wise in [the science of] spheres, the knower of computations, the knower of the science of form [i.e., the creative and mechanical arts], a learned man, having considered the many kinds of great and small circles situated in the sphere by imagining [them] according to his own will, at the time of performing [calculations], having known the Sines of the respective arcs, with the sagittas [i.e., the versed Sines] produced by the *bhujā* and the *koṭi* [i.e., the complement of *bhujā*], he should demonstrate the rule of three.

134 Beginning in this manner, the fabrication of the armillary sphere should be done by those who are exceptionally skilful with balls of clay, or with metals known to be the best, with wood, or with straight pieces of bamboo, or with old pieces of animal horns, or ivory.

135 In this manner, the wise Nityānanda, the son of Devadatta, residing in this city of Indrapurī, caused the chapter on spheres to attain completion there in the quintessential 'King of All Treatises' *Sarvasiddhāntarāja*.

5 Critical Notes and Technical Analyses

Introducing the chapter on spheres

Nityānanda begins the chapter (*adhyāya*) on spheres (*gola*) with an incipit statement in prose using the inceptive particle *atha* 'now'. In Sanskrit texts, the word *atha* is often used at the beginning of a chapter (or section) as an auspicious allure. It has been considered equivalent to the *śabda-brahman* 'the primal sound identified with Brahman' in philosophical discussions, and with regard to divine aetiology, its use at the beginning of an utterance has also been compared to that of *Oṃ* (the *praṇava*) in the Upanishads (see, e.g., Minkowski 2008, footnote 20 on p. 13). On a more practical level, *atha* simply indicates the start of a new chapter (or a new section) in the organisation of a text.

Verse 1

A *maṅgalācaraṇa* is a form of an encomium where one's chosen deities (or patrons) are praised in a panegyric poem seeking benediction (*āśāsya*) and propitiousness (*kalyāṇamaya-daśā*). In literary works, a *maṅgalācaraṇa* is composed at the beginning of a treatise (*śāstrārambha*) as a divine supplication (*prārthanā*) to remove any obstacles that may hinder the very act of composition and ensure a successful completion of the work.[1]

Nityānanda starts his *golādhyāya* by paying homage to Gaṇeśa—an invocatory practice followed by most Hindus even in the present day. Gaṇeśa is extolled as the 'leader of men' (*gaṇa-nāyaka*), the one whose 'two feet are like lotuses' (*padma-pada-dvaya*)—understood metaphorically as the one who is the sublime foundation of all beings—and, more felicitously, as the 'destroyer of the flood of all obstacles' (*vighnaugha-vidhvaṃsa-kṛt*). These epithets follow the traditional Hindu custom of describing the gods by both their physical and divine attributes. In the

DOI: 10.4324/9780429506680-5

second hemistich of the verse, Nityānanda offers his reverence to the absolute and ultimate reality (or God), variously regarded as the Puruṣa or Pradhāna by the Sāṃkhyas , Śiva by the Śaivas, or Brahman by the Vedāntins.[2]

In the tradition of Sanskrit astronomy (*jyotiḥśāstra*), most authors like Āryabhaṭa I (b. 473), Brahmagupta (b. 598), Lalla (ca. late eight or early ninth century), Vaṭeśvara (fl. 899/904), Śrīpati (ca. mid-eleventh century), Bhāskara II (b. 1114), and Jñānarāja (fl. 1503), to name just a few, also began their work with a *maṅgala* verse. Often, the object of their devotion, i.e., the individual deity (*īśvara* or *bhagavat*) an author chose to supplicate, alluded to their sectarian belief. Some authors like Varāhamihira (ca. 550) were more secular in their *maṅgalācaraṇa*, choosing to offer their obeisance to the ancient seers (*muni*s) who revealed the astronomy of the gods to men out of benevolence; see *Pañcasiddhāntikā* (I.1). Others like Lalla venerated the *navagraha*s, the deified forms of the luminaries and the planets, as well as different sectarian gods; see *Śiṣyadhīvṛddhidatantra* (I.1).

In his *Sarvasiddhāntarāja*, Nityānanda composes separate *maṅgala* verses for the chapters on computations (*gaṇitādhyāya*) and on spheres (*golādhyāya*). In the *gaṇitādhyāya* (I.1.1), his *maṅgalācaraṇa* (in the *śālinī* meter) puns on his own name (*nitya-ānanda* 'eternal happiness') to eulogise his chosen deity Mukunda 'the giver of liberation', an epithet of Kṛṣṇa:

> *Oṃ namaḥ śrīgaṇādhipataye vighnahartre* |
>
> *nityānandaṃ nityam avyaktam īśaṃ*
> *saccidrūpaṃ nirguṇaṃ nirvikāraṃ* |
> *ekaṃ vyāpibrahmavedāntavedyaṃ*
> *yat taṃ vande viśvarūpaṃ mukundaṃ* ||
>
> Obeisance to Lord Gaṇapati, the remover of all obstacles.
>
> I praise Lord Mukunda in his universal form, he who is perpetual bliss, eternal, unmanifested, the supreme lord, the form of pure existence and thought, without attributes, without change, the One [who] is regarded in the Vedānta as the [all] pervasive Brahman.

Verses 2–3

In these two verses, Nityānanda stresses the importance of the science of spheres (*gola-āgama*) for any mathematician (*gaṇaka* or *gaṇitavid*, lit. the one who knows how to compute), especially for those who wish to uphold their reputations in an assembly of scholars. Learned assemblies (*sabhā*s) were places where scholars regularly gathered to interrogate one another on their subject knowledge, and among an assembly of mathematicians, the knowledge of spheres was thought to be foundational to the study of mathematical astronomy. Therefore, the horror of being speechless among one's peers was a strong incentive for any novice mathematician to take the study of spheres very seriously.

By Nityānanda's time, the importance of spheres (*gola*) to understand the mathematics (*gaṇita*) of the heavens was an established fact. The knowledge of spheres allowed a mathematician to conceive a proof, demonstration, or justification (variously understood as *upapatti*) for a particular computational algorithm. In fact, Bhāskara II, in his *Siddhāntaśiromaṇi* (1150), states *...upapattiṃ vinā...gaṇako niḥsaṃśayo na svayam* (II.1.2) '...without *upapatti*...the astronomer himself would not be free of doubts'. He adds that these *upapatti*s are clearly perceived in the sphere, just like holding a *āmlaka* berry (Indian gooseberry, *Phyllanthus emblica*) in one's hand, and hence his lecture on spheres (Srinivas 2008, p. 1833).

For Nityānanda, not understanding proper rationales (*vāsanā*) is a sign of a mathematician's nescience (*ajñāta*), and, ultimately, the source of their humiliation. In Siddhāntic literature, *vāsanā* commonly refers to an expository text (commentary) that explains meaning of another text, e.g., Pṛthūdakasvāmin's *Vāsanābhāṣya* (860) on Brahmagupta's *Brāhmasphuṭasiddhānta* or Nṛsiṃha Daivajña's *Vāsanāvārttika* (1621) on Bhāskara II's *Siddhāntaśiromaṇi*. Seen in this light, Nityānanda's 'beautiful composition' (*cāru-racanā*) on the 'investigations of spheres' (*golavicāra*) is then the rational commentary that delivers a mathematician from the ignominy of his nescience.

In their own expositions on spheres, most Siddhāntic authors also play on the motif of the fear of public humiliation to emphasise the study of spheres; see, e.g., Lalla's *Śiṣyadhīvṛddhidatantra* (XIV.2–3), Vaṭeśvara's *Vaṭeśvarasiddhānta* (II.1.3), Śrīpati's *Siddhāntaśekhara* (XV.2–3), or Bhāskara's *Siddhāntaśiromaṇi* (II.1.3–4). Nityānanda goes a step further (in verse 3) by using metaphorical examples (*dṛṣṭānta*), disguised as rhetorical questions, to convince his readers of the importance of truly understanding the science of spheres. He indirectly suggests that his text on spheres is the very light that can guide readers from the dark realms of ignorance into the company of knowledgeable and intelligent men. Just as the pulchritude of a young maiden pales without the beauty of her mind, so too does the reputation of a mathematician, deprived of the knowledge of spheres, wane when questioned about his capability.

Nityānanda's language in verse 3 is replete with several figures of speech (*alaṃkāra*), e.g., at the very beginning of the verse, he employs an alliteration of words (*pādānuprāsa*) in describing the variegated and colourful arrangement (*vicitra-citra-racanā*) of houses (*sadman*). The word *sadman*, ordinarily indicating a seat or an abode, also implies the astrological houses (constellations) or the very heaven and earth itself. In his (rhetorical) statement on a young maiden's countenance, the adjective *pragalbha* can mean someone who is confident, just as easily as it implies someone who is arrogant. Its use creates a sense of intentional ambiguity (*śleṣa*) meant to provoke the reader's curiosity. Perhaps, 'without intelligence' (*vaidagdhya-vinā*), a young maiden 'endowed with great beauty' (*saundarya-yukta*) naturally turns to become arrogant (and hence unattractive).

Beyond these linguistic devices, Nityānanda also uses expressions that are rich in their sociocultural context. For example, when describing the three worlds as *sakala-artha-pūrṇa* 'full of all things one desires', he is referring to *sakala-artha* as the totality of objects men desire: objects that may be spiritual, material, or sensual in nature. In Hinduism, these objects are considered the tripartite goals (*dharma*, *artha*, and *kāma* respectively) of human life (*puruṣārtha*)—a concept with which most of his readers would be very familiar.

Verse 4

Nityānanda gives a list of academic qualifications required for a scholar to be recognised as a mathematician. According to him, a true mathematician must know arithmetic (*pāṭīgaṇita*), the methods of solving linear indeterminate equations (*kuṭṭaka*), algebra (*bījagaṇita*), and theory of spheres (*gola*). Moreover, he should also be qualified in the auxiliary branches of the Vedas (*vedāṅga*), namely, (i) *śikṣā* or the science of phonetics, phonology, and morphophonology; (ii) *chanda* or the science of poetic meter; (iii) *vyākaraṇa* or the science of grammar; (iv) *nirukta* or the science of etymology; (v) *jyotiṣa* or the science of mathematical astronomy and astrology; and (vi) *kalpa* or the science of ceremonies or rituals. Furthermore, the scholar needs to be proficient in the science (*śāstra*) of poetics (*kāvya*), rhetoric (*alaṃkāra*), and the creative and mechanical arts (*śilpa*), and, at the same time, also be skilled in various other sciences like logic (*nyāyaśāstra*).

It is interesting to compare Nityānanda's list of qualifications (*yogyatā*) of a mathematician with those prescribed by Varāhamihira in his *Bṛhat Saṃhitā* (ca. sixth century, II.2–10); see, e.g., Subbarayappa and Sarma (1985, pp. 9–10). While Varāhamihira's list of requirements is specific to the technical domains of mathematics and astronomy, Nityānanda's requirements demand a wider grasp of knowledge across the disciplines (*śāstras*) of mathematics and the humanities. If the *śāstras* are understood as a form of timeless knowledge grounded in traditional epistemology, then its learning not only enables a mathematician to be adept in interpreting secular or scriptural works, but also grants him the social distinction of being a treasurer of the collective common knowledge of his society.

Those mathematicians who successfully master the aforementioned topics are then considered as subject experts in canonical literature (*siddhāntas*), and, accordingly, they are able to provide nuanced explanations (*sūkṣmārtha*) of mathematical concepts beyond their literal or express meanings (*abhidhā*). Without this intellectual training, they would run the risk of being dismissed and derided in the assembly of their peers. In fact, Nityānanda's admonition is also a veiled commendation of his own credentials: his exegesis on spheres is meant to be regarded as substantial (and not merely nominal) since he has himself mastered the aforementioned topics in the highest degree (and is consequently an authority on the subject).

Verse 5

Nityānanda's *golādhyāya* includes discussions on four types of spheres. The first is the Earth-sphere (*bhūgola* or *bhūmigola*) that contains the different islands, oceans, forests, rivers, and mountains that make up the oecumene (the habitable world, from the Greek *οἰκουμένη* 'inhabited'). The second is the sphere of asterisms (*bhagola*). The sphere of asterisms, with the stars, planets, and constellations situated within it, forms the vault of the heavens and its diurnal westward movement causes the rising and setting of the stars. Essentially, the *bhagola* embodies the cyclical eternal movement of the heavens and hence is thought to be the wheel (*cakra*) of time. The celestial equator (*viṣuvavṛtta*, lit. the circle enveloping the equinoxes) is the great circle on this sphere that measures the rising and setting times of celestial objects.

The third is the oblique celestial sphere (*khagola*). The oblique celestial sphere is motionless in the sense that the local zenith and celestial horizon remain fixed to a geocentric observer, while the stars, planets, and constellations of the sphere of asterisms revolve. Others scholars on the topic of spheres have also differentiated between a moving 'starry sphere' (*bhagola*) and a motionless 'celestial sphere' (*khagola*); e.g., Parameśvara (fl. 1431/1450) in his *Goladīpikā* (I.14) states

> *bhramati hy aparābhimukhaṃ pravahākṣepāt sadā bhagolo'yam |*
> *sthira eva khagolaḥ syād digādi siddhyai prakalpito hy eṣa ||*
>
> The Starry sphere revolves constantly westwards, blown by the *pravaha* wind. The Celestial sphere, however, remains stationary; it is constructed in order to reckon directions, etc.
>
> (K. V. Sarma 1957, p. 71)

And finally, the fourth is the visible celestial sphere (*dṛggola* or *dṛṣṭigola*). The visible celestial sphere is the spherical dome of the sky visible to a terrestrial observer. In this case, the horizon corresponds to one's local geographic horizon (unimpeded along the line of sight) and the zenith lies directly overhead. The visible celestial sphere captures the combined effect of viewing the heavens turning westwards (in the rotating sphere of asterisms) against the fixed reference circles (in the oblique celestial sphere). Bhāskara II, in his *Siddhāntaśiromaṇi* (II.6.9), provides a similar description of a visible 'double sphere' (*dṛggola*):

> *bhagolavṛttaiḥ sahita khagolo dṛggolasaṃjño 'pamamaṇḍalādyaiḥ |*
> *dvigolajājātaṃ khalu dṛṣyate 'tra kṣetraṃ hi dṛggolam ato vadanti ||*
>
> When the system of the *khagola*, celestial sphere, is mixed with the ecliptic; and all the other circles forming the *bhagola* (which will be presently shown) it is then called the *ḍrig-gola* [sic] double sphere. As in this the figures formed by the circles of the two spheres *khagola* and *bhagola* are seen, it is therefore called *ḍriggola* [sic] double sphere.
>
> (L. Wilkinson and B. D. Śāstrī 1861, p. 153)

See Figure 1.1 in Section 1.3.4 for a schematic drawing of the three orientations of the celestial sphere forming the *bhagola*, *khagola*, and *dṛggola*.

Stylistically, Nityānanda composes this verse as a series of questions to stimulate a reader's inquisitiveness (*kathaṃkathikatā*), and in what follows in this chapter, he himself provides the answers to these questions (and much more). This form of anthypophora allows Nityānanda to educate the reader about the different topics he intends to discuss in this chapter transitionally. His use of alliterations in this verse is particularly noteworthy for its sonorous ornamentation (*śabda-alaṃkāra*), e.g., the interrogative word *kidṛś* is purposefully repeated several times to create an alliteration on words (*pādānuprāsa*) and the terminal word *golaḥ* is compounded at the end of the first two quarters (*pāda*) to create an alliteration on line-endings (*antyānuprāsa*).

Verse 6

Continuing in the prefatory spirit, Nityānanda simply states the names of some of the great circles commonly seen in the oblique celestial sphere (*khagola*) and the visible celestial sphere (*dṛggola*). These include:

1 the prime vertical (*pūrvāparavṛtta*) passing through the local zenith and extending along the east-west direction;

2 the meridian (*dakṣiṇasaumyavṛtta*) passing through the local zenith and extending along the north-south direction;

3 the horizon (*bhūja*), understood as the larger (geocentric) celestial horizon in the *khagola* and the smaller (topocentric) geographic horizon for a local terrestrial observer in the *dṛggola*;

4 the vertical circles (*koṇagamaṇḍala*) passing through the local zenith and the intercardinal directions; and

5 the equinoctial colure or the six o'clock hour circle (*unmaṇḍala*) passing through the due east and due west points on the horizon (i.e., positions on the horizon marking the equinoctial rising of the Sun) and the north celestial pole.

Introducing the four kinds of spheres, beginning with the Earth-sphere

The four spheres described in the *golādhyāya* are the Earth-sphere (*bhū-gola*), the sphere of asterisms (*bhagola*), the oblique celestial sphere (*kha-gola*), and the visible celestial sphere (*dṛggola*). The rest of the chapter describes (in individual sections) the different constituent elements in each of these spheres, e.g., the geographical features of the Earth-sphere or the great circles on the sphere of asterism .

The first section on the Earth-sphere describes (i) its physical and political geographies based on mythological narratives from the Purāṇas, (ii) its geometrical sphericity, and (iii) the astronomical consequences of the Earth being a sphere.

Verse 7

Nityānanda begins this section by describing the Earth-sphere (*kṣiti-gola* or *bhūgola*) as a celestial object that remains eternally unsupported (*nira-adhara*) in space. The mythemic conception of Earth in the Purāṇas is a flat, circular, disc-like structure supported from below by cosmic creatures of enormous dimensions, variously described as turtles, tortoises, elephants, or even serpents in different stories. By Nityānanda's time, the infinite regress in determining the original support had lead many Siddhāntic authors to abandon the Purāṇic view of a supported Earth in favour of an unsupported Earth-sphere; see, e.g., Lalla's *Śiṣya-dhīvṛddhidatantra* (XX.41) or Bhāskara II's *Siddhāntaśiromaṇi* (II.3.4).

Early Siddhāntic scholars had argued that the very nature of the Earth—composed of the five great elements (*pañca-mahābhūta*) earth (*ku*), water (*ambu*), fire (*jvalana*), air (*pavana*), and ether (*ākāśa*)—is to remain unsupported in line with the nature of its constituent elements. Nityānanda agrees with the elemental composition of the Earth, but adds that the Earth-sphere remains motionless (*sthiratara*) and unsupported (*nira-adhara*) in space due to its own inherent energy (*sva-śakti*). The concept of inherent energy or a self-supporting force can be found in many philosophical schools of thought prevalent at that time.[3]

To explain the idea of an unsupported Earth suspended in space, Nityānanda uses a simile (*upamā*) to compare the Earth-sphere to an iron ball (*ayas-gola*) suspended in a magnetic trap. This construction follows the classical structure of a Sanskrit simile where (i) the subject of comparison (*upameya*) is the unsupported Earth; (ii) the object of comparison (*upamāna*) is the suspended iron ball; (iii) the shared property (*upamā-dharma*) is their inherent energy, i.e., the Earth's self-sustaining energy and the iron ball's interactive energy (in a magnetic confinement); and (iv) the simile-denoting morpheme (*aupamyavācaka*) is the particle *yathā*. This analogy can be found in many Siddhāntic texts from Varāhamihira's *Pañcasiddhāntikā* (XIII.1) to Śrīpati's *Siddhāntaśekhara* (XV.22).

Moreover, Nityānanda adds that the Earth-sphere remains unsupported in the vast expansiveness of space despite bearing the weight of the villages, gardens, mountains, oceans, gods, demons, and people. This apparent antithesis of an object being unsupported and yet bearing the weight of all other things is a contradictory idea (*virodha*); however, its reconciliation (*parihāra*) is found in the foundational belief that only the Earth (among all other celestial objects) has the exclusive nature to support while itself being unsupported.

Many earlier Siddhāntic authors have refuted the opinions of the Jains and Buddhists on the motion of the Earth (and the Sun), and, in their own respective compositions, also declared the Earth to be an unsupported sphere suspended in space by its own inherent energy; see, e.g., Varāhamihira's *Pañcasiddhāntikā* (XIII.1, 2, 8), Lalla's *Śiṣyadhīvṛddhidatantra* (XVII.1, 2, 6 and XX.38, 42, 43), Śrīpati's *Siddhāntaśekhara* (XV.11–13, 14, 15–19a), and Bhāskara II's *Siddhāntaśiromaṇi* (II.3.2, 7–10). Nityānanda has repeated this Siddhāntic opinion here in the *golādhyāya* (verse 7) of his *Sarvasiddhāntarāja*, and again in the *gaṇitādhyāya* (I.3.180) where he composes an almost identical verse (also in the *śikhariṇī* meter) saying

> *svaśaktyaiva prāyaḥ sthiratara ilākendram abhito*
> *mṛdaḥ piṇḍākāro jalaśikhimarudvyomanicitaḥ* |
> *khamadhye bhūgolaḥ sthita iha saritsindhunagarī-*
> *girigrāmāraṇyaiḥ saha jagad idaṃ bhāti sakalam* ||

> The Earth-sphere, with [its] rivers, oceans, cities, mountains, villages and forests, having the shape of a lump of earth, is situated here in space, remaining steady solely by its own power, probably, around the centre of the Earth, on which are piled water, fire, air, and the sky. That is what appears as this world.

Verses 8–12

In verses 8–12, Nityānanda discusses two geodetic methods to determine the elevation (*unnati* or *ucchraya*) and the distance of an arbitrarily high mountain summit (*uttuṅga-śailaśṛṅga*). These didactic interrogations appear interposed in an otherwise narratorial account of the Earth-sphere in what follows. The two questions are mutually invertible where one unknown quantity is determined knowing the second; e.g., the unknown height of the mountain summit in *yojanas*[4] is calculated knowing the corresponding distance to the mountain in *yojanas*. The solutions to these geodetic problems of height and distance measurements require a familiarity with spherical geometry and trigonometry.

Calculating an unknown height corresponding to a known distance

Verses 9 and 10 describe Nityānanda's method to compute the elevation or height h of a mountain's summit (*śailaśṛṅga-unnati*) in *yojanas* when the mountain is at a distance of $d = 30$ *yojanas* from a terrestrial observer (*nṛ-pāda*, lit. the foot of a man). The following steps describe his algorithm (using modern notations):

1 Calculating the degrees of arc (*cāpa*) corresponding to an arc-length of 30 *yojanas*; see Figure 5.1.

Nityānanda uses the equivalence of ratios to estimate this quantity. The circumference of the Earth, taken as 6000 *yojanas*, corresponds to an angular (arc-measure) of 360° (*bhacakra-bhāgas*, lit. the

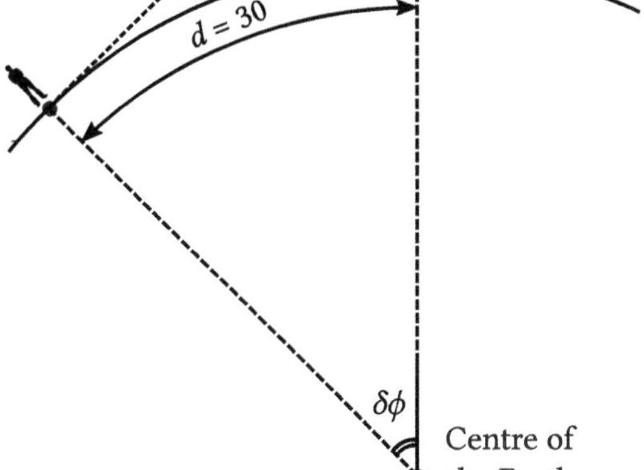

Figure 5.1 The degrees of an arc corresponding to an arc-length of 30 *yojanas* along the circumference of the Earth of 6000 *yojanas*.

degrees of the circle of asterisms). Therefore, an arc-length of 30 *yojana*s corresponds to an arc-measure $\delta\phi$—called the 'degrees of difference' (*vivara-aṃśa*)—following the ratio:

$$\frac{30^{\,yojanas}}{\delta\phi} = \frac{6000^{\,yojanas}}{360^\circ} \Rightarrow \delta\phi = \frac{30^{\,yojanas} \times 360^\circ}{6000^{\,yojanas}} = 1.8^\circ.$$

This calculation follows the rule of three (*trairāśika*) where

$$\underset{\delta\phi\,=\,1.8^\circ}{\underset{(\textit{icchā-phala})}{\text{Desired resultant}}} = \frac{\overbrace{\underset{30\ yojanas}{\text{Resultant amount }(\textit{phala-rāśi})} \times \underset{360^\circ}{\text{Desired amount }(\textit{icchā-rāśi})}}^{\text{Product obtained }(\textit{labdha})}}{\underset{\text{argument }(\textit{pramāṇa})}{6000\ yojanas}}.$$

The method of *trairāśika* lies at the heart of most astronomical computations (see, e.g., S. R. Sarma 2002).

2 Calculating the height h of the mountain summit in *yojana*s using a *sinus totus* (*tribhajyā*) of 60, a radius R_e of the Earth in *yojanas*, and the Cosine of the arc-measure (degrees of difference) obtained in the previous step.

According to Nityānanda, the height h can be computed by

i taking the difference between the *sinus totus* and the Cosine of the arc-measure (degrees of difference), i.e., $60 - \text{Cos}\,\delta\phi$ or $60 - \text{Cos}\,1.8^\circ$;

ii multiplying this difference by the radius of the Earth, i.e., $R_e\,(60 - \text{Cos}\,1.8^\circ)$; and finally

iii dividing the product by the Cosine of the arc-measure (degrees of difference); in other words, $h = \dfrac{R_e\,(60 - \text{Cos}\,1.8^\circ)}{\text{Cos}\,1.8^\circ}$.

Nityānanda offers no geometric verification (*pratyaya-karaṇa*) or proof (*upapatti*) of this algorithm; he merely states the procedural method for obtaining the result. Also, the final answer to this computation is left unstated, perhaps, as a challenge for dedicated readers.

The derivation of this expression relies on a simple geometrical construction and the use of trigonometric ratios. In Figure 5.2, O is the centre of the Earth, T is the top of the mountain, and S represents the foot of the observer. R_e indicates the radius of the Earth,

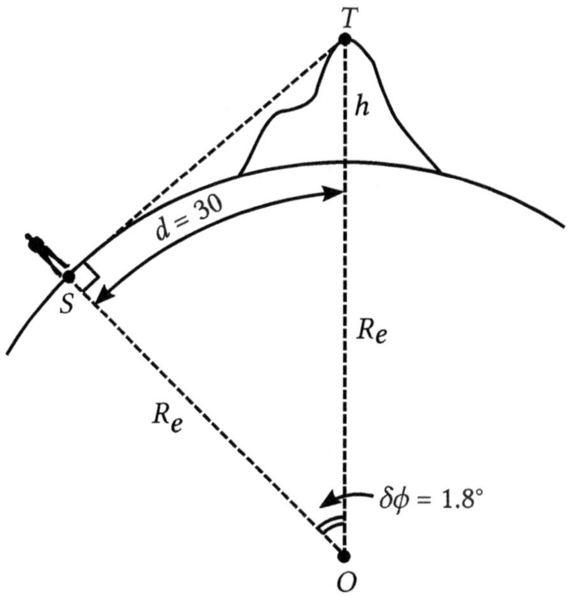

Figure 5.2 The height h *yojanas* of a mountain at a distance of $d = 30$ *yojanas* with a corresponding arc-measure $\delta\phi = 1.8°$ in relation to the radius of the Earth R_e.

h is the height of the mountain summit in *yojanas*, $d = 30$ *yojanas* is the distance from the mountain (along its plumb-line) to S, and $\delta\phi = \measuredangle SOT = 1.8°$. The inequality $h \ll R_e$ ensures that the angle subtended along the line of sight from the mountain summit to the surface of the Earth is a right angle, i.e., OS $\perp$ TS, making $\triangle$OST a right-angled triangle. This gives an expression for the height of the mountain summit as

$$\cos 1.8° = \frac{R_e}{R_e + h} \Rightarrow \text{Cos}\, 1.8° = \frac{\mathscr{R}\, R_e}{R_e + h}$$
$$\Rightarrow h = \frac{R_e\,(\mathscr{R} - \text{Cos}\, 1.8°)}{\text{Cos}\, 1.8°}, \text{ where the } \textit{sinus totus}\ \mathscr{R} \text{ is } 60. \tag{5.1}$$

***On the requirement that** $h \ll R_e$*

The smallness of the ratio of the height of the mountain summit to the radius of the Earth is an important requirement to ensure $\triangle$OST is a right-angled triangle. As Figure 5.3 shows, although the tangent to the circle (here, the surface of the Earth at R) from an exterior point T is orthogonal to the radius of curvature and hence forms the right-angled $\triangle$ORT, the arc-length (or distance) $\widehat{\text{RM}}$ is not *necessarily* the same as

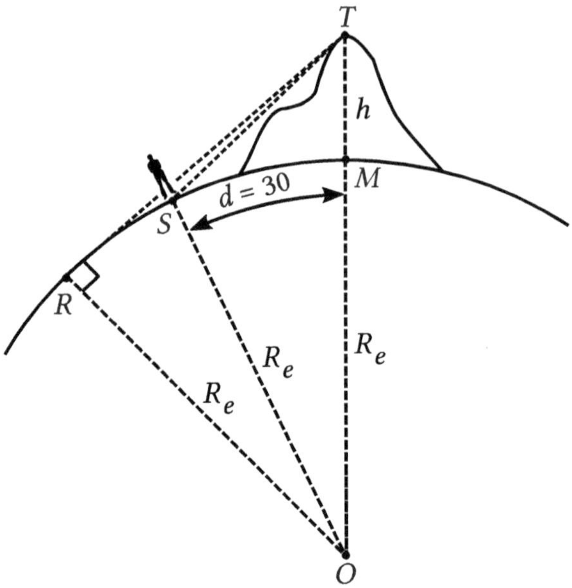

Figure 5.3 The requirement $h \ll R_e$ ensures RT $\sim$ ST which, in turn, makes the right-angled triangle $\triangle$ORT $\sim$ $\triangle$OST.

the distance $\widehat{SM}$ for our observer stationed at S. Nityānanda places the observer at fixed distance of 30 *yojana*s, and hence, the $\triangle$OST need not be a right-angled triangle. The requirement $h \ll R_e$ (or effectively RT $\sim$ ST) ensures the tangent ST is perpendicular to the radial line OS at the observer's position *S*.

Calculating an unknown distance corresponding to a known height

Verses 11 and 12 describe Nityānanda's method to compute the distance *d* to a mountain (*śaila-adhvan*) in *yojana*s from a terrestrial observer's location (*puṃs-pāda*, lit. the foot of a man) when the elevation or height of its summit (*śṛṅga-ucchraya*), say, *h yojana*s is known. This method essentially reverses the method described in verses 9 and 10, and can be understood as follows:

1 Compute the divisor as radius of the Earth R_e increased by the height of the mountain summit h (the known quantity in this case), i.e., $R_e + h$. This corresponds to OT in Figure 5.4.

2 Compute the dividend by multiplying the height of the mountain summit h with the *sinus totus* (*trijyakā*) 60, i.e., $60h$.

3 Compute the quotient of the division, i.e., $60h/(R_e + h)$, and subtracting it from the *sinus totus* 60, equate the result to the Cosine of

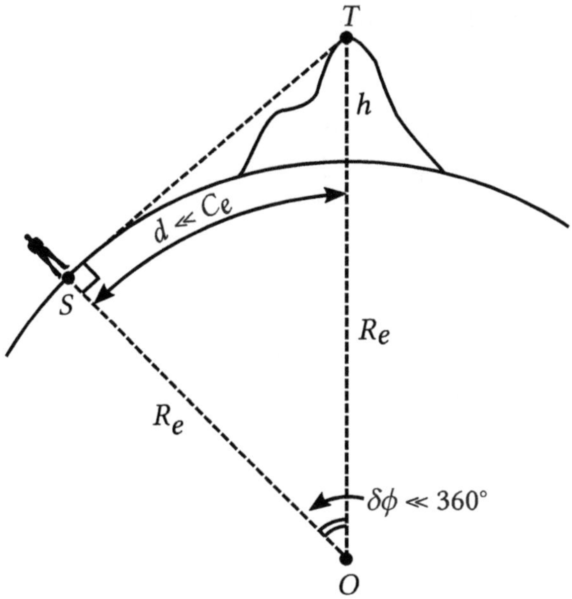

Figure 5.4 The known height h *yojanas* of a mountain at a distance of unknown d *yojanas* with a corresponding arc-measure $\delta\phi \ll 360°$. The height of the mountain $h \ll R_e$ where R_e is the radius of the Earth, ensuring $d \ll C_e$ where C_e is the circumference of the Earth (measured in *yojanas*).

an arc, say, $\delta\phi$. This gives an equation of the form

$$\text{Cos}\,\delta\phi = 60 - \frac{60h}{R_e + h}.$$

4 Compute the square root of the difference of squares of the *sinus totus* and the Cosine of the arc, i.e., $\sqrt{60^2 - \text{Cos}^2\delta\phi}$. Essentially, this gives $\text{Sin}\,\delta\phi$, and with $\delta\phi$ being small, $\text{Sin}\,\delta\phi \sim \delta\phi$, implying the arc-measure $\delta\phi° \approx \sqrt{60^2 - \text{Cos}^2\delta\phi}$.

The disproportionately smaller height of the mountain compared to the radius of the Earth, i.e., $h \ll R_e$, ensures that the point of tangency (point R in Figure 5.3) to the surface of the Earth from the mountain peak forms an arc-length $\widehat{\text{RM}}$ that is the furthest distance from the mountain at which the mountain peak would just be visible along an observer's line of sight. Therefore, for an observer positioned on the surface of the Earth; e.g., at point S, we find $\widehat{\text{SM}} \leq \widehat{\text{RM}}$. Now, as $\widehat{\text{RM}} \ll C_e$ where $C_e = 2\pi R_e$ is the circumference of the Earth, it reasons that $\delta\phi$ is a small measure of arc in degrees.

5 Finally, compute the distance d to the mountain in *yojana*s from an observer's position as $d = \dfrac{\delta\phi^\circ}{360^\circ} \times 6000$.

Here, we find the inverse rule of three (*vyasta-trairāśika*) in effect:

$$\begin{array}{c}\text{Desired resultant}\\ (\textit{icchā-phala})\\ d\ \textit{yojanas}\end{array} = \begin{array}{c}\text{Quotient}\\ \text{obtained}\\ (\textit{labdha})\end{array} \left\{ \frac{\begin{array}{c}\text{Resultant}\\ \text{amount}\\ (\textit{phala-rāśi})\\ \delta\phi^\circ\end{array}}{\begin{array}{c}360^\circ\\ \text{desired}\\ \text{amount}\\ (\textit{icchā-rāśi})\end{array}} \right. \times \begin{array}{c}\text{Argument}\\ (\textit{pramāṇa})\\ 6000\ \textit{yojanas}\end{array} .$$

Nityānanda's geodetic methods to calculate the height of a mountain summit or the distance to it closely resemble al-Bīrūnī's method of computing the radius of the Earth from his *al-Qānūn al-Masᶜūdī* (Masudic Canon); see Appendix A for the mathematical equivalence between al-Bīrūnī's and Nityānanda's expressions.

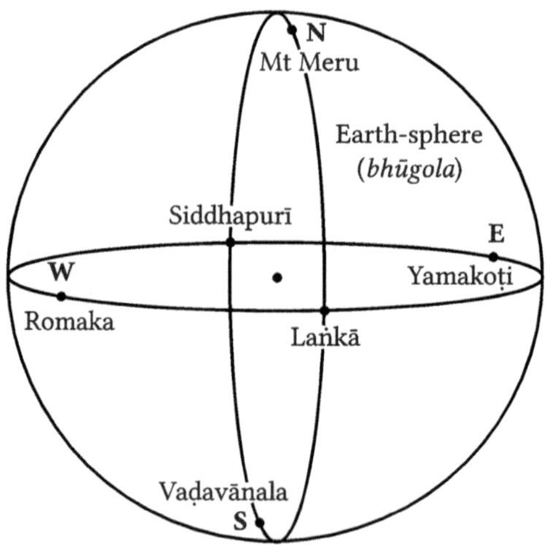

Figure 5.5 The relative positions of the cardinal localities on the Earth-sphere from the central perspective of Laṅkā.

Table 5.1 The relative positions of the six cardinal localities with respect to the central city of Laṅkā

Position	*Orientation (with respect to Laṅkā)*	*Cardinal localities*
Equatorial	East	Yamakoṭi (or Yama)
	West	Romaka
	Antipodal	Siddhapurī (or Siddhā)
Polar	North	Mt Meru (or Mt Sumeru)
	South	Vaḍavānala (or Vaḍavā)

Verses 13–15

In verses 13–15, Nityānanda describes the relative positions of the six cardinal localities on Earth, namely, the four equatorial cities of Laṅkā, Yamakoṭi, Siddhapurī, and Romaka, and the two polar regions of Mt Meru and Vaḍavānala. Figure 5.5 depicts the arrangement of these cardinal places on the Earth-sphere (*bhūgola*). With Laṅkā in the middle, i.e., along a central meridian, the relative positions of the other places are described in Table 5.1.

This paradigmatic geography can be found in almost all Siddhāntic works; e.g., Lalla's *Śiṣyadhīvṛddhidatantra* (XVII.3), Śrīpati's *Siddhānta-śekhara* (XV.30), and Bhāskara II's *Siddhāntaśiromaṇi* (II.3.17) include identical descriptions of the cardinal localities on the Earth-sphere in relation to the central city of Laṅkā.

Verse 16

In the first *pāda* of this verse, Nityānanda identifies equatorial observers as those who are *etāsu sarvāsu nṛṇāṃ purīṣu* 'in all of these cities of men', in reference to the four equatorial cities (described in verses 13–15). It is, however, a little unclear if Siddhapurī can be considered a city where men dwell. While all the four cities, i.e., Laṅkā, Yamakoṭi, Siddhapurī, and Romaka, are conceptualised as equatorial localities (rather than inhabited cities), Siddhapurī, in particular, stands out being antipodal to Laṅkā and, hence, situated in an ocean devoid of landmass.[5] It is, nevertheless, clear that the polar regions (Mt Meru to the north and Vaḍavānala to the south) are identically oriented for any equatorial observer, regardless of whether they are stationed at a real or hypothetical location on the equator of the Earth-sphere. Moreover, to an equatorial observer, the celestial equator, situated directly overhead, appears to revolve westwards daily as it carries the stars and the planets with it. Its

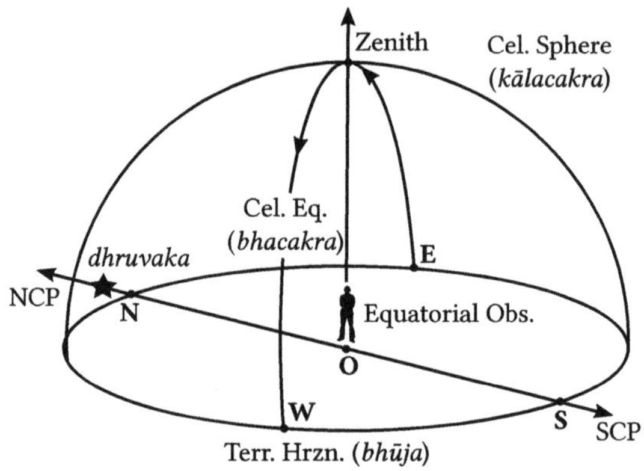

Figure 5.6 The relative positions of the celestial equator (at the zenith) and the celestial poles (along the horizon) with respect to an equatorial observer.

diurnal rotation measures the movement of time and hence its epithet *kāla-cakra* 'the wheel of time'.[6]

As Figure 5.6 shows, the *kāla-cakra* can be understood as the celestial equator that embodies the daily revolutions of the stars and the planets. At equatorial locations, the celestial sphere (*bhagola*) rises and sets vertically with the celestial equator being directly overhead (*mūrdhan-sthita*), i.e., at the zenith. The pair of north and south celestial poles (*dhruva-dvaya*) are considered fixed to the due north and due south points on the terrestrial horizon (*bhūja*) respectively. At the terrestrial equator, the polestar (*dhruvaka*) marks the due north point on the terrestrial horizon.

Earlier Siddhāntic authors have also made similar statements about the configuration of the celestial sphere from an equatorial perspective; see, e.g., Śrīpati's *Siddhāntaśekhara* (XV.54) or Bhāskara II's *Siddhānta-śiromaṇi* (II.3.48).

Verse 17

Having previously located the six cardinal localities on the Earth-sphere in relation to one another (see verses 13–15), Nityānanda now explicitly defines their mutual separation to be 90° (*bhū-pāda-bhāga*, lit. the degrees of a quarter of the Earth). This arc-measure of separation (in degrees) is related to the arc-length of distance (in *yojana*s) on a spherical Earth. An arc-measure of 90° corresponds to an arc-length of a quarter of the circumference of the Earth. If the circumference of the Earth-sphere is taken as 6000 *yojanas* (as stated in verses 9 and 12), the cardinal localities are separated by ¼ of 6000 or 1500 *yojanas*. However, this value appears to be inserted (in Nāgarī numerals) in the verse without any word-numerals (e.g., following the *bhūtasaṁkhyā* system) describing it in the Sanskrit text of the verse.

The geometrical constraint in positioning these six terrestrial localities (*dharaṇī-sthalas*) at a mutual separation of 90° necessitates a spherical Earth. For instance, for all internal angles of a triangle joining any two equatorial cities with a polar region (e.g., Laṅkā, Romaka, and Mt Meru) to be ninety degrees, the surface needs to be spherical. Earlier Siddhāntic authors have also placed these cardinal localities with similar spherical geometrical constraints, suggesting that, perhaps, these places were hypothetical cardinal locations on the Earth-sphere rather than real geographical places; see, e.g., Lalla's *Śiṣyadhīvṛddhidatantra* (XVII.4, 7) or Bhāskara II's *Siddhāntaśiromaṇi* (II.3.18, 52).

Verses 18–20

In verses 18 and 19, Nityānanda describes the relative positions of the four equatorial cities and the two celestial poles for polar observers, namely, for an observer on Mt Meru, i.e., at the north pole of the Earth, and for an observer at Vaḍavānala, i.e., at the south pole of the Earth.

Figure 5.7 depicts the mutually reversed arrangements of the equatorial cities and the celestial poles from the northern and southern polar perspectives. Nityānanda's verses on the relative positions of cardinal localities (in verses 13–15 and 18–19) share a recurring style of composition. In each case, the perspective is centred on a particular cardinal

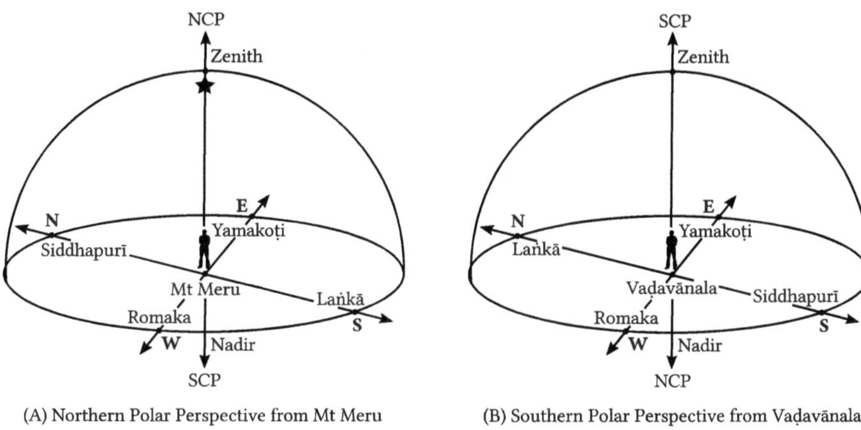

Figure 5.7 The relative positions of the equatorial cities and celestial poles for polar observers from (A) a northern polar perspective and (B) a southern polar perspective.

location—similar to designating a centre (*kendra*) to a circle (*vṛtta*)—and then situating the other localities in relation to this central place along its circular periphery. Typically, for an observer located at Laṅkā (verse 13), Mt Meru (verse 18), and Vaḍavānala (verse 19), the equatorial cities of Yamakoṭi and Romaka always remain due east and due west respectively, whereas the other localities are arranged according to their respective orientations, always facing northwards.

In verse 20, Nityānanda adds some mythohistorical descriptions of the polar regions. The multitude of gods (*suragaṇa*s) who live on Mt Meru, i.e., the observers close to the north pole, as well as the demons (*daitya*s) who live in the submarine regions of Vaḍavānala, i.e., observers close to the south pole, find the celestial equator (*vaiṣava-vṛtta*) to be *bhūjāsaktam ivābhitaḥ* 'as if [it were] fastened to the horizon on all sides'; in effect, the celestial equator is just the horizon for polar observers. Moreover, the observers close to the north pole find the westwards-rotating celestial sphere (strictly speaking, the northern hemisphere of it) moving from the left towards the right (*asavya-diś*) as they look eastwards. Whereas, the observers close to the south pole see the celestial sphere (or more precisely, the southern hemisphere of it) moving from the left to the right (*savya-diś*) as they look eastwards.

Nityānanda identifies the movement of the heavens as the movement of the celestial equator going through the eight cardinal directions (*asta-kakubhā*)—in other words, going all around a central observer. However, the movement of the celestial equator, or any imaginary great circle for that matter, is not a discernible phenomenon; its motion is revealed by looking at the stars and constellations that appear along its perimeter. Bhāskara II, in his *Siddhāntaśiromaṇi* (II.3.5), identifies the difference between the directions of rotation for northern and southern polar observers by referring to movement of the circle of constellations (*ṛkṣa-cakra*). Figure 5.8 depicts the directions of motion of the celestial sphere (with its constituent planets, luminaries, and constellations) from the northern and southern polar perspectives.

The orientation of the gods and the demons—both said to be facing eastwards (*prāṅ-mukha*)—appears a little irrelevant to how they would see the celestial sphere rotate. A northern or southern polar observer would see the celestial sphere moving rightwards or leftwards respectively, in any direction along their horizon. Perhaps, the east-facing orientation of these antipodal gods and demons alludes to the auspiciousness of the eastern direction. It is a common Hindu practice to circumambulate religious objects in a *savya* (or *pradakṣiṇa*) direction that keeps the central object to one's right-hand side (*dakṣiṇa*). Here the word *dakṣiṇa*, ordinarily referring to the southern direction, takes on the meaning of right-hand side when facing eastwards. As Figure 5.8 shows, the gods (northern polar observers) would see the dome of heaven move counterclockwise (*asavya* or *uttara*) about the central polestar, while the

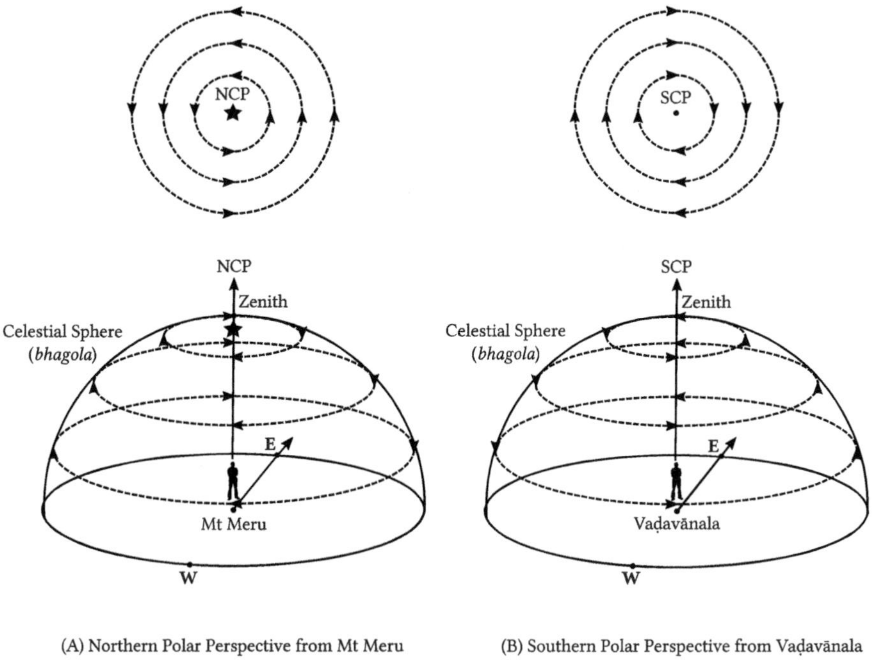

Figure 5.8 The directions of rotation of the celestial sphere from (A) a northern polar perspective and (B) a southern polar perspective. From the perspective of a northern polar observer facing eastwards, the motion of the celestial sphere is rightwards, whereas for a southern polar observer, the celestial sphere appears to move leftwards.

demons (southern polar observers) would observe the dome of heaven move clockwise (*savya* or *dakṣiṇa*).

Nityānanda's discussions (through verses 13–20) on the relative positions of the four equatorial cities, the two polar regions, the celestial equator (or the celestial sphere), and the two celestial poles—from the perspectives of both equatorial and polar (hypothetical) observers—culminate in him concluding that the Earth (*dharā* or *bhūmi*) is therefore established to be spherical in shape (*golākṛti*) everywhere (as an geometrical necessity).

Verse 21

The relative positions of the terrestrial horizon (*bhūja*), the local zenith (*mūrdhan*), and the antipodal location (*pada-adhas*) for an observer (*mānava*) standing on the surface of the Earth (*kṣiti*) are depicted in Figure 5.9.

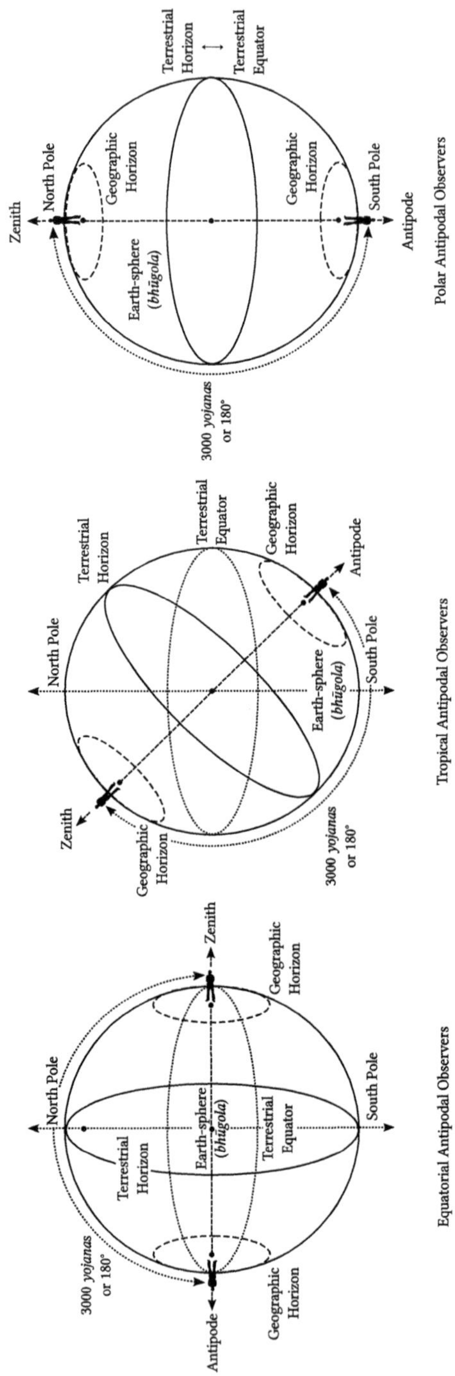

Figure 5.9 The terrestrial horizon, the zenith, and its antipodes for observers at various locations on the surface of the Earth-sphere.

For a terrestrial observer, the (conceived) terrestrial horizon differs from the (visible) geographic horizon. The geographic horizon extends along the uninterrupted line of sight from the observer's position and is tangential to the Earth's surface, while the terrestrial horizon envelopes half the Earth's hemisphere from one's local position, i.e., at a distance of half the circumference of the Earth from the local position. If the circumference of the Earth-sphere is taken as 6000 *yojana*s (e.g., from verse 9), the terrestrial horizon is at a distance of ¼ of 6000 or 1500 *yojana*s from a local observer's position, and the corresponding antipodal place is ½ of 600 or 3000 *yojana*s from their position. The number 3000 appears to be inserted (in Nāgarī numerals) in the verse without any word-numerals (e.g., following the *bhūtasaṁkhyā* system) describing it in the Sanskrit text of the verse. The description of antipodal places on the Earth separated by half the Earth's circumference also finds mention in Lalla's *Śiṣyadhīvṛddhidatantra* (XVII.5) where he describes two antipodal observers as 'a man standing on the bank of a river and his reflection in the water'.

On a grammatical note, the word *avaihi* in the third *pāda* is another form of the word *avehi* according to an older grammar called *mahāvyākaraṇa*. Śrī Madhavācārya (1238–1317), the Dvaita Vedāntic philosopher of the Bhakti movement, describes the existence of other forms of grammar prior to Pāṇini's *vyākaraṇa* and uses the older *mahāvyākaraṇa* form *avaihi* instead of *avehi* in his *Mahābhārata Tātparya Nirṇaya* (VII.38). The word *avaihi* is then the second singular imperative active form of $\sqrt{}$*ave* (class 2 verb) meaning 'consider' or 'know this'. I use the expression '[one should] understand' to render the Sanskrit second-person imperative into English as it retains the deontic (irrealis) mood and reads more naturally.

Verses 22–23

As seen before, Figure 5.5 depicts the localities along the four cardinal directions of their terrestrial horizons for the equatorial cities of Laṅkā and Siddhapurī. When the city of Laṅkā is considered to be in the middle (*kṣiti-pṛṣṭha-madhya*), going clockwise (*savya*) from the northern direction (*kubera-diś*), the cardinal localities are Mt Sumeru (or Mt Meru) (north), Yamakoṭi (east), Vaḍavānala (south), and Romaka (west). Whereas, when the city of Siddhapurī is considered to be in the centre, going clockwise (*savya*) from the northern direction, the cardinal localities are Mt Sumeru (north), Romaka (east), Vaḍavānala (south), and Yamakoṭi (west). To avoid repetition, Nityānanda simply mentions the ordering of these cities in verse 22 as *sumerupūrvāḥ* 'beginning with Mt Sumeru'.

In verse 23, he describes two other equatorial cities that are antipodal to one another, namely, the city of Khāladāta (to the west of Romaka) at a longitudinal difference of 22° (expressed as the *bhūtasaṁkhyā* numerals *netra-locana*), and its antipodal location Gaṅgadujda (*bha-ardha-bhāga* or 180°from Khāladāta). See Section 1.3.9.3 for a discussion on the cities of Khāladāta and Gaṅgadujda.

Verse 24

The terrestrial equator, called the *madhya-maṇḍala*, lit. the middle circle, extends over the cities of Laṅkā, Romaka, and Khāladāta, and constitutes the circumference of Earth measuring 6000 *yojana*s. Again, the number 6000 appears to be inserted (in Nāgarī numerals) in the verse without any word-numerals (e.g., following the *bhūtasaṁkhyā* system) describing it in the Sanskrit text of the verse. Being equatorial cities, the cities of Laṅkā, Romaka, and Khāladāta have no latitude (*nirakṣa*, lit. devoid of latitude). Also, as already discussed in verse 16, for an equatorial observer, the two celestial poles (*dhruva-dvaya*) appear fastened to the horizon (*avanīja*) along the due north and due south directions and the celestial equator (*viṣuvatvṛtta*) is directly overhead (*mūrdhan*) at their zenith. The celestial sphere (represented by its equator) revolves westwards in an upright manner at these equatorial places; see Figure 5.6.

With the exception of Khāladāta as a special equatorial city along the (Islamicate) prime meridian, verse 24 repeats what has already been described in verses 13–16. The repetition is perhaps aimed at creating an equivalence between the perspectives shared by observes at Laṅkā (Indian prime meridian) and Khāladāta (Islamicate prime meridian).

Verse 25

Here, Nityānanda demarcates a triangular (*trikoṇā*) region on the Earth's surface that extends between the polar region of Mt Meru and the two equatorial cities of Gaṅgadujda (east of Laṅkā), and Khāladāta (west of Laṅkā). This triangular region measures a quarter of the surface of Earth (*bhūmi-pāda-unmita*) and is free from water (*vāri-mukta*)—in other words, the oecumene or the habitable world.

Nityānanda refers to the shape (traced by these three localities) on the surface of the Earth-sphere as a triangle occupying a quarter of the Earth's surface; strictly speaking, the geometrical object is a non-degenerate digon or spherical lune (a spherical polygon with two vertices and two sides) with internal right angles and antipodal vertices. Siddhāntic astronomers have often commented on spherical lunes in connection to the surface area of spheres; see, e.g., Bhāskara II's *Siddhāntaśiromaṇi* (II.3.53–61) and his auto-commentary *Vāsanābhāṣya* (passages corresponding to these verses).

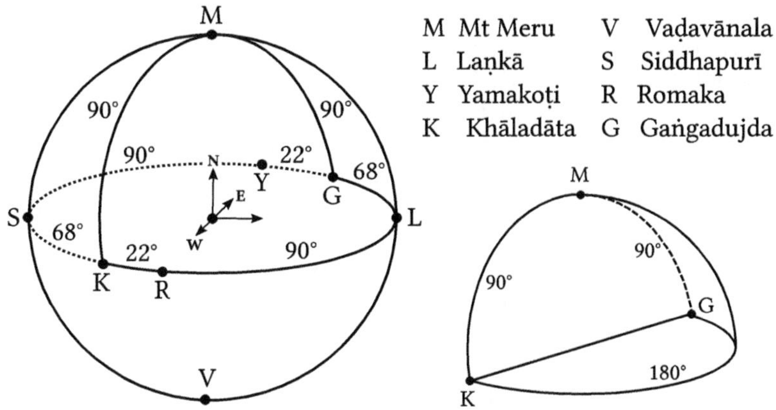

Figure 5.10 The relative arrangement of the terrestrial localities and the habitable world (oecumene) on the Earth-sphere (spherical lune KMG).

In al-Bīrūnī's writings, we find an identical description of the oecumene as a quadrant of the habitable world (*al-rub^c al-ma^{c}mūr*) surrounded by the encompassing ocean that separates the habitable lands from the uninhabited waters (and islands) beyond (Sachau [1910] 2013, p. 196). Nityānanda's description of a quarter of the Earth bound by specific terrestrial localities and devoid of water (i.e., the oecumene) is both novel and unique for a Sanskrit Siddhāntic text.

Indeed, the reference to Khāladāta strongly suggests that he acquired this information from an Islamicate source, presumable, the celestial globes and planar astrolabes that where known in Indian since the late fourteenth century (see, e.g., Mahendra Sūri's *Yantrarāja* in S. R. Sarma 1999) and started to be manufactured in Mughal India from the early sixteenth century (see, e.g., S. R. Sarma 2011, 2018). In describing the parameters inscribed by gazetteers in Islamic astrolabes, S. R. Sarma (2019, p. 44) states that the longitudes of the places were measured from the Fortunate Islands (*al-Jazā$^\circ$ir al-Khālidāt*), a practice followed by the Indo-Persian astrolabe makers from Lahore during the Mughal rule.

In 1647, Muḥammad Ṣādiq ibn Muḥammad Ṣāliḥ (better known as Ṣādiq Iṣfahānī) of Jaunpūr (in Uttar Pradesh, India) is known to have completed a Persian atlas, the *Shāhid-i Ṣādiq*, that consists of thirty-three maps depicting the inhabited quarter of the Earth. The prime meridian in these maps is identified as the Blessed or Fortunate Isles (*al-Jazā$^\circ$ir al-Khālidāt*), sometimes associated with the modern-day Canary Islands. These southwards-oriented maps are an important source of Indo-Islamic world maps of the time. See Schwartzberg (2016) for a description of these maps and a review of the practice of cartography in India of the second millennium.[7]

Verses 26–27

The nychthemeron (the twenty-four-hour day) in the regions between the Equator and the Arctic Circle is such that there is a proportion of daylight and darkness all throughout the year. At the Arctic Circle (66° 30′ N), however, there are two solstitial days in a year where the nychthemeron is alternatively and entirely light and dark; in other words, the length of the day (or the night) at the Arctic Circle alternates between zero and twenty-four hours on the two days of the year. From here where this phenomenon of a polar day (or a polar night) is first observed, the day-length becomes an implicit measure of the latitude. Entering into the Arctic, i.e., beyond 66° 30′ N all the way up to the North Pole (90° N), the day-lengths exceed twenty-four hours with several polar days (and polar nights) in year.

According to Nityānanda, the habitable land extends up to the latitude where the arrangement of day or night (*dyuniśa-vyavasthā*) is 60 *ghaṭī*s (or twenty-four hours)—in other words, up to the Arctic Circle (66° 30′ N).[8] Beyond it lie the inhospitable lands with snow-covered mountains that, according to Nityānanda, are both and frigid and frightening.

In verse 27, Nityānanda describes how the regions towards Kāśmīra become inaccessible to men at the beginning of winter (*hima-āgama*, lit. the arrival of snow), and, accordingly, the northern regions further afar become completely impassable on account of constant snowing. The *Nīlamatapurāṇa* (ca. sixth/seventh century, 276–286) elaborates a story of how men come to live in the Kāśmīra valley for the six month (from spring to autumn), and, after gathering their harvest at the end of summer, leave the valley before the advent of the winter. When winter arrives, fifty million flesh-eating ogres (*piśāca*s), along with their king Nikumbha, return to the Kāśmīra valley having fought the demons (*daitya*s) in the southern desert lands (called *vālukārṇava* 'sea of sand') for the past six months. It is during these winter months that the harshness of the cold and the oppression of the ogres make Kāśmīra a dangerous place (*durga-pradeśa*) for men to live. Nityānanda's rhetorical lament on the impassibility (*durgatara*) of these regions in the winter plays on this Purāṇic story of Kāśmīra.

Verse 28

According to Nityānanda, the line (*rekhā*) joining the terrestrial locations of Mt Sumeru, Rohita, Saṃnihitya, Ujjayinī, Śvetaśaila, Kāñcī, and Laṅkā forms the Indian prime meridian (*madhyarekhā*), and accordingly, these cities have no longitudinal difference (*deśāntara*), i.e., they lie along 0° longitude. Ujjayinī (modern-day Ujjain) is most commonly considered as the city through which the Indian prime meridian passes, and even though most Sanskrit *siddhānta*s list the names of various other places on the prime meridian, their modern-day longitudes are often divergent from Ujjain's longitude. See Subbarayappa and Sarma (1985, pp. 24–25) and Chatterjee (1981, foonote 42 on pp. 28–29, Translation and Notes) for the lists of cities on the Indian prime meridian according to various Siddhāntic authors and commentators. Nityānanda's list of cities can be approximately identified as follows:

1 Mt Sumeru (or Mt Meru) is the mythical central mountain (the *axis mundi*) that serves as the fixed north pole of Earth; see Section 1.3.3.5.

2 Rohita (or Rohitaka) is the modern-day city of Rohtak (28° 54′ N, 76° 35′ E) in Haryana.

3 Sāṃnihitya (variously Saṃnihitya or Sāṃnihitī) is mentioned in the *Mahābhārata* (III.83, 190) as the water tank located in Kurukṣetra (also identified as Sthāneśvara). In the modern day, a sacred water reservoir called Sannihit Sarovar (29° 58′ N, 76° 50′ E) is located in Thanesar in the Kurukshetra district of Haryana.

4 Ujjayinī (variously Mālavanagara or Avantī) is believed to be the standard prime meridian in India. In the modern day, it is identified with Ujjain (23° 11′ N, 75° 47′ E), a city is located on the banks of the Kṣiprā river in the Malva region of Madhya Pradesh.

5 Śvetaśaila (variously Ṣaḍāsya, Sitādri, or Sitavara) 'white mountain' is identified in several Siddhāntic sources as lying along the prime meridian; see, e.g., Bhāskara I's *Mahābhāskarīya* (II.1) or Vaṭeśvara's *Vaṭeśvarasiddhānta* (I.8.1). The word *ṣaḍāsya* 'six-faced' refers to the god Kumāra Kārtikeya, and hence, the Ṣaḍāsya mountain can perhaps be identified with the mythical Krauñca mountain. The location of this mountain is often associated with the modern-day settlement of Krauncha Giri (15° 4′ N, 76° 39′ E) in the Bellary district of Karnataka.

6 Kāñcī (or Kāñcīnagarī) is the modern-day city of Kanchipuram (12° 50′ N, 79° 42′ E) in the Tondaimandalam region of Tamil Nadu.

7 Laṅkā (different from Śrī Laṅkā) is the hypothetical null island (or cupola) of the Earth with coordinates (0° N, 0° E). See Section 1.3.3.1.

Verses 29–30

Figure 5.11 depicts the relative positions of the celestial equator and the celestial poles at various latitudes on the Earth-sphere. At the terrestrial equator (*nirakṣadeśa*, lit. a place with no latitude), the north celestial pole is stationed at the northern horizon while the celestial equator remains fixed to the local zenith (see verse 16).

Moving northwards from the terrestrial equator, the gain in northern latitude is equal to the elevation (*unnati*) of the north celestial pole (*dhruva*, the polestar) above the northern horizon, or equivalently

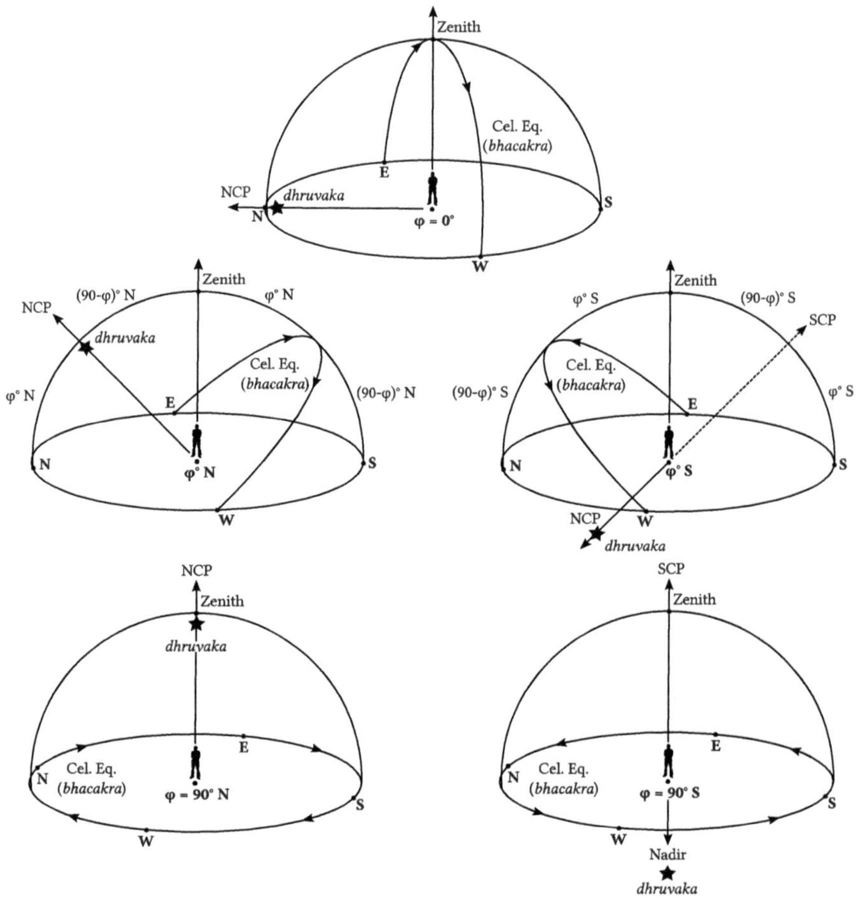

Figure 5.11 The relative positions of the celestial equator and the north celestial pole (directed towards the polestar) at the terrestrial equator ($\varphi = 0°$, top-centre), a northern latitude ($\varphi°$ N, top-left), the north pole ($\varphi = 90°$ N, bottom-left), a southern latitude ($\varphi°$ S, top-right), and the south pole ($\varphi = 90°$ S, bottom-right) on the Earth-sphere.

the depression (*namra*, lit. bowing or inclining) of the celestial equator (*vaiṣavavṛtta*) southwards of the local zenith (*nabhastala*, lit. the sky surface). At the north pole, located at a distance of a quarter of the Earth's circumference (*bhū-caraṇa*) or 90° from the terrestrial equator, the north celestial pole (or the polestar) is directly overhead (*śiras-sthita*) at the zenith while the celestial equator is simply identified with the terrestrial equator.

Reciprocally, moving southwards from the terrestrial equator, the gain in southern latitude is equal to the depression (*nata*) of the north celestial pole (or the polestar) below the northern horizon, or equivalently the depression of the celestial equator northwards of the local zenith. At the south pole, located at a distance of a quarter of the Earth's circumference or 90° from the terrestrial equator, the north celestial pole (or the polestar) is directly underneath (*adhas-sthita*, lit. situated under [the observer's feet]) at the nadir, while the celestial equator is simply identified with the terrestrial equator.

Verses 31–32

According to Nityānanda, an observer on the surface of the Earth (*kṣamā-upari*) remains stationed above it (*upariṣṭha*), i.e., stands on the surface of the Earth-sphere, on account of being pulled towards the centre of the Earth-sphere (*dharaṇī-garbha-karṣita*). In verse 7, the Earth-sphere (*bhūgola*) is described as a self-supporting motionless sphere that supports the firmament (and all things on it). The tendency of the earth element *ku* (one of the five elemental constituents of the Earth) is to gravitate towards the centre of the Earth-sphere, and, in doing so, maintains its pull on all things on the surface of the Earth-sphere. The sphericity of its shape allows terrestrial observers standing anywhere on the surface of the Earth-sphere to consider the distance of a quarter of the Earth (*bhū-caturtha*) or 90° from their location as their terrestrial horizon and the distance of half the Earth or 180° from their location as their antipode (*adhas-sthala*, lit. the place below). See Figure 5.9.

Earlier Siddhāntic scholars have also addressed the question on why men simply do not fall off the surface of the Earth. Āryabhaṭa I, in his *Āryabhaṭiya* (IV.7), compared the Earth-sphere to the 'bulb of a *kadamba* flower' (*kadamba-puṣpa-granthi*) covered on all sides by blossoms (*kusuma*), and hence able to sustain all kinds of terrestrial (*sthalaja*) and aquatic (*jalaja*) creatures. Lalla, in his *Śiṣyadhīvṛddhidatantra* (XVII.7–8), described the movement of people in the Earth's southern hemisphere akin to the forward movement of the house lizard (*graha-godhika*) resting underneath the cross-beams (*upatulā*) of the ceiling of a dwelling (*bhavana-bhāratulā*). Bhāskara II, in his *Siddhāntaśiromaṇi* (II.3.6), provides an explanation closer to what Nityānanda describes in verse 31. According to Bhāskara II, the power of attraction (*ākṛṣṭi-śakti*) is an inherent power (*sva-śakti*) of Earth (*mahī*) with which is attracts any unsupported (*kha-stha*, lit. situated in the sky) heavy object towards it.

Verses 33–34

The division of the surface of the Earth into zones (called *climes*) based on the variation in daylight hours first finds mention in Ptolemy's works from the second century.[9] Following Ptolemy, several medieval European and Islamicate authors also described the geographic regions of the Earth as climes. To the best of my knowledge, among Sanskrit Siddhāntic authors, Nityānanda is the first to describe the parallels of latitudes on the Earth-sphere as the seven climes (*sapta-khaṇḍas*).

According to Nityānanda, going northwards from the equator—the equator being the place where the nychthemeron is equally distributed with twelve hours of day and twelve hours of night—the beginning of the first clime is marked by a place with *vyaṣṭhāṃśadantonmitanāḍikābhis* '[a length in *nāḍikās*] measuring 32 *nāḍikās* minus an eighth part of 1 *nāḍikā*'—in other words, 31;52,30 *nāḍikās* of daylight. These hours of daylight correspond to the longest day of the year—in other words, the summer solstitial day. With a *nāḍikā* (or *ghaṭīkā*) approximately equal to twenty-four minutes (see endnote 8 on page 324), 31;52,30 *nāḍikās* of daylight is approximately $12^{h}45^{m}$.

Beyond this, the beginnings of each of the seven successive climes (going northwards) are marked by places where the daylight hours increase by an additional 1;15 *nāḍikās* (about 30^{m}). For instance, the second clime begins at a place where the summer solstitial day-length is 31;52,30 *nāḍikās* increased by 1;15 *nāḍikās*, i.e., 33;7,20 *nāḍikās* (about, $13^{h}15^{m}$).

The degrees of latitude corresponding to the hours of daylight (on the summer solstitial day) are related to the hour angle of the Sun and its declination. Figure 5.12 shows the position of the Sun S^* in the sky for a northern observer O. The hour-angle ω is the angular separation (measured in units of time) between the observer's local meridian and the hour circle passing through the celestial object. For the Sun, ω measures the deviation of local solar time from solar noon and is considered negative at sunrise and positive at sunset. In Figure 5.12, $\sphericalangle S^*PZ$ is the hour-angle ω of the Sun S^* (with declination δ and solar altitude a) for an observer O at a terrestrial latitude of $\varphi°$ N. The spherical triangle ΔS^*PZ has sides $S^*P = 90 - \delta$, $S^*Z = 90 - a$, $PZ = 90 - \varphi$ with the hour-angle $\sphericalangle S^*PZ = \omega$, and hence, the spherical rule of cosines gives

$$\cos\omega = \frac{\cos(90-a) - \cos(90-\delta)\cos(90-\varphi)}{\sin(90-\delta)\sin(90-\varphi)} = \frac{\sin a - \sin\delta\sin\varphi}{\cos\delta\cos\varphi}$$

At local sunrise (pt. B) on the eastern horizon, the altitude a of the Sun is zero, and hence the sunrise equation can be expressed as $\cos\omega = -\tan\delta\tan\varphi$.[10]

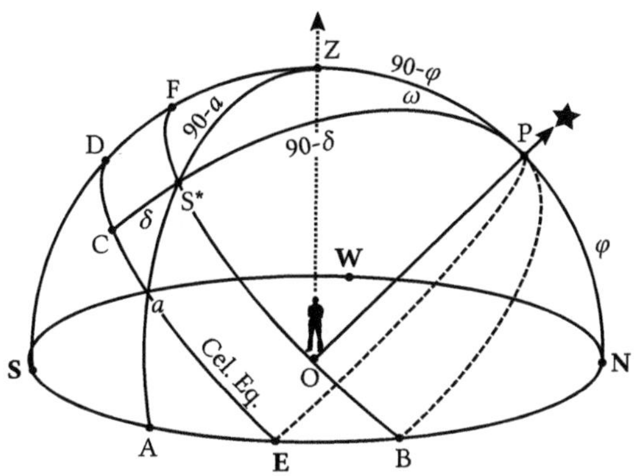

Z	Zenith
P	North Celestial Pole
O	Observer
S*	Sun
arc **NESW**	Local Horizon
arc **NZS**	Local Meridian
arc PS*C	Hour Circle
arc ZS*A	Altitude Circle
arc **PE**	Six o'clock Circle
arc PB	Hour Circle (at sunrise)
arc BS*F	Day/Diurnal Circle
arc **E**CD	Celestial Equator

Figure 5.12 The celestial sphere for a northern observer O (at a latitude φ N) showing the Sun S* in the sky (at a declination of δ) and the hour angle ω of the Sun S* measured as the angle between the local meridian (arc NZS) and hour circle of the Sun (arc PS*C).

At the beginning of the first climes with $12^{h}\,45^{m}$ of daylight, the hour angle ω at sunrise is $6^{h}\,22^{m}\,30^{s}$ (arc BF). Expressed in angles (with $1^{h} = 15°$), ω is $95°37'30''$. On the day of the summer solstice in the northern hemisphere, the declination δ of the Sun is $23°30'$ N (Nityānanda gives this value in verse 76), and hence, the terrestrial latitude φ N corresponding to $12^{h}\,45^{m}$ of daylight is

$$\cos 95°37'30'' = -\tan\varphi \tan 23°30' \Rightarrow \varphi \sim 12°42'12''.$$

Table 5.2 lists the parallels of latitudes corresponding to the increasing hours of daylight (in 30^{m} intervals) in the seven northern climes corresponding to six different values of the summer solstitial declination of Sun (i.e., the ecliptic obliquity). The values of the ecliptic obliquity, taken from Houtsma ([1913–36] 1993, p. 504), are as follows:

1 24° (standard value in most Sanskrit Siddhāntic texts)

2 23° 51′ 20″ (Ptolemy)

3 23° 33′ (Islamicate astronomers from the ninth century, particularly, those around Baghdād during the reign of the Abbasid caliph al-Maʾmūn)

4 23° 35′ (Islamicate astronomers from the ninth century onwards like Bānū Mūsā, al-Battānī, Abu l-Wafāʾal-Būzjānī, al-Bīrūnī, and others)

Table 5.2 The parallels of northern latitude corresponding to successively increasing hours of daylight h (measures on the summer solstitial day) in the seven climes, computed using different values of the ecliptic obliquity ε (i.e., the solar declination on the summer solstitial day)

Climes (h)	Parallels of latitude of the seven climes					
	Obliquity of the ecliptic ε					
	24°	23° 51′ 20″	23° 33′	23° 35′	23° 30′ 17″	23° 30′
I ($12^h 45^m$)	12° 24′ 56″	12° 29′ 51″	12° 40′ 27″	12° 39′ 17″	12° 42′ 2″	12° 42′ 12″
II ($13^h 15^m$)	20° 5′ 45″	20° 13′ 19″	20° 29′ 34″	20° 27′ 46″	20° 32′ 0″	20° 32′ 15″
III ($13^h 45^m$)	27° 1′ 21″	27° 10′ 50″	27° 31′ 9″	27° 28′ 55″	27° 34′ 12″	27° 34′ 31″
IV ($14^h 15^m$)	33° 6′ 14″	33° 16′ 56″	33° 39′ 51″	33° 37′ 20″	33° 43′ 17″	33° 43′ 38″
V ($14^h 45^m$)	38° 20′ 59″	38° 32′ 22″	38° 56′ 41″	38° 54′ 1″	39° 0′ 19″	39° 0′ 41″
VI ($15^h 15^m$)	42° 49′ 44″	43° 1′ 24″	43° 26′ 15″	43° 23′ 32″	43° 29′ 58″	43° 30′ 21″
VII ($15^h 45^m$)	46° 38′ 6″	46° 49′ 49″	47° 14′ 37″	47° 11′ 54″	47° 18′ 19″	47° 18′ 42″

5 23° 30′ 17″(Ulugh Beg, ca. 1425, and following him Mullā Farīd in his *Zīj-i Shāh Jahānī* in ca. 1629/30, and Nityānanda in his *Siddhāntasindhu*, ca. early 1630)

6 23° 30′ (Nityānanda in his *Sarvasiddhāntarāja*, 1639).

The use of 30^{m} increments to define the boundaries of seven successive climes starts with Ptolemy (in his *Almagest*) and then continues in the practices of Islamicate mathematical geographers. Al-Bīrūnī, in his *al-Qānūn al-Mas*ᶜ*ūdī* (*The Mas*ᶜ*ūdic canon*), has tabulated the sizes of the different climes (in Arabic miles and *farsakhs*) and given the boundaries of the successive climes of 30^{m} increments. According to al-Bīrūnī, the first clime starts at a place with 12 ¾ hours of daylight (around 12° 39′ of latitude) and extends all the way to the beginning of the seventh clime with 15 ¾ hours of daylight (around 47° 11′ of latitude) (Mercier 1992, p. 175). Interestingly, the geographic positions of places found in Ulugh Begh's *Sūlṭānī Zīj* (ca. 1440) are not arranged by climes, but rather by regions. Mullā Farīd's *Zīj-i Shāh Jahānī* (ca. 1629/1630), being based on Ulugh Beg's work, also groups various terrestrial places based on regions (and not climes). This arrangement then carries into Nityānanda's Sanskrit translation of the *Zīj-i Shāh Jahānī*, the *Siddhāntasindhu* (ca. early 1630).

Introducing the cosmography of the Purāṇas

In the next thirty-one verses, Nityānanda discusses Purāṇic cosmography with lists of Earth's islands (*dvīpas*), oceans (*samudras*), mountains (*parvatas*), continents (*khaṇḍas*), etc. These lists follow a standard register of Purāṇic geography repeated in almost all Sanskrit *siddhāntas*. Accordingly, Nityānanda declares that his descriptions will follow an established order (*saṃsthā*), alluding to the order of appearance of these topics in previous Siddhāntic texts. Also, in contrast to his own compositions (*racanā*), these verses merely express the opinion of the Purāṇas (*paurāṇika-mata*) and not necessarily his own critical opinions.

Verse 35

According to the Purāṇas, the Earth (*pṛthivī*) is an entirely flat disk-like structure, whereas the Sanskrit *siddhāntas* regard the Earth as a sphere. Nityānanda offers this verse as a way of reconciling Purāṇic and Siddhāntic views; see Plofker (2009, p. 269) and Minkowski (2004, p. 24). According to him, although the Earth is spherical, it appears locally flat (*samavat*) in a vicinity of a hundredth part (*śata-aṃśaka*) of its circumference (*paridhi*). Other Siddhāntic authors have previously made the same observation; e.g., Jñānarāja, in his *Siddhāntasundara* (II.1.28), explicitly states that *mukuratalanibhatvaṃ yat purāṇapradiṣṭaṃ tad avaniśatabhāgasyaiva no bhūmigole* 'the likeness to the surface of a mirror mentioned in the Purāṇas [applies only] to a one-hundredth part of the Earth, not the [entire] sphere of the Earth' (T. L. Knudsen 2014, p. 54). The Earth appears locally flat due to the exceedingly small size of man when compared to the enormous size of the Earth. Bhāskara II, in his *Siddhāntaśiromaṇi* (II.3.13), says *pṛthvī ca pṛthvī nitarāṃ tanīyān naraś ca* 'the Earth is extremely extensive, and a man exceedingly small (in comparison)' (L. Wilkinson and B. D. Śāstrī 1861, p. 114). This disproportion in scale was also noted in the *Sūryasiddhānta* (XII.54) that declares the *alpakāyatayā lokāḥ* 'littleness of the bodies of men' gives them the impression of a *vṛttām apy etām cakrākārāṃ vasundharām* 'circular wheel-like Earth, despite its spherical nature' (Gangooly 1935, p. 289). Like his predecessors, Nityānanda reinterprets the Purāṇic statement on an entirely flat Earth, as pronounced by many scholars of the Purāṇas (*purāṇapaṇḍita*), to refer to its local flatness that can be seen everywhere on the surface on the Earth (within a local field of view).

Verses 36–37

The visible geographic horizon (*dṛṣṭi-bhūja*) is the circular perimeter on the surface of the Earth (*bhū-pṛṣṭha*) where an observer's line of sight (*dṛṣṭi-sūtra*), unimpeded by any obstacles, touches the surface of the Earth tangentially (*tiryak*). This geographic horizon forms the edge of the dome of the sky (*vyoman*) visible to a terrestrial observe, and is smaller than the terrestrial horizon (conceived as the circumference of the Earth); see Figure 5.9.

Purāṇic cosmography identified this circular perimeter of the visible geographic horizon as the Lokāloka mountain. The *Śrīmad Bhāgvatam* (V.20.34, 36) describes the Lokāloka mountain as *lokālokanāmācalo lokālokayor antarāle parita upakṣiptaḥ* 'a mountain named Lokāloka, which divides the countries that are full of sunlight from those not lit by the sun' and *lokāloka iti samākhyā yad anenācalena lokālokasyāntarvartināvasthāpyate* 'between the lands inhabited by living entities and those that are uninhabited stands the great mountain which separates the two and which is therefore celebrated as Lokāloka' (Prabhupāda 1975, pp. 305–306 in Volume II).

Lalla, in his *Śiṣyadhīvṛddhidatantra* (XIX.30), identifies the Lokāloka mountain as the orbit of the sky that is the periphery of the cavity of the egg of Brahmā (*brahmāṇḍakuṇḍa*) (Chatterjee 1981, footnote 1 on p. 263). Bhāskara II, in his *Siddhāntaśiromaṇi* (II.3.67), claims that Purāṇic scholars have declared the periphery of the universe (*brahmāṇḍa*) to be the Lokāloka mountain (L. Wilkinson and B. D. Śāstrī 1861, p. 126). In contrast, in verse 36, Nityānanda identifies the Lokāloka mountain as the visible geographic horizon which appears more reconciliatory as it simply assigns a mythological name to a geographic feature.

In verse 37, Nityānanda objects to another Purāṇic belief that the Lokāloka mountain is the periphery of the sky (*nabha-kakṣa*)—in other words, a perimeter of the visible universe like a celestial horizon. The Purāṇas describe the Lokāloka mountain (also called the Dṛśyādṛśya 'visible and invisible' mountain) as the outermost border of the three worlds (*bhūloka, bhuvarloka,* and *svarloka*) up to which the rays of the Sun extend into the universe; see, e.g., *Śrīmad Bhāgvatam* (V.20.37). This description naturally allows the Lokāloka to be identified as the perimeter between the visible (*dṛśya*) and the invisible (*adṛśya*) parts of the universe.

The main objection to the Lokāloka mountain being a celestial horizon is in its inability to then explain the diurnal motion of planets. The rising or setting of planets requires the presence of a visible geographic horizon against which the rising and setting can be seen. When the entire universe (along with the planets) is contained within the celestial sphere, and the Lokāloka mountain forms its outermost perimeter (or celestial horizon), it cannot be said that the mountain separates the seeing (rising) and not-seeing (setting) of a planet.

The question of a planet's visibility (at rising and setting), and the role of identifying the appropriate horizon, is often discussed in Sanskrit astronomical texts. For example, Parameśvara in his *Goladīpikā* (II.5–6) states that in the 'view of the *śāstras*', planets become visible when they rise above the terrestrial horizon around the Earth's centre and are invisible below it on account of being hidden by the Earth's body (K. V. Sarma 1957, p. 73). However, he adds that only those planets that rise above the surface of the Earth (*kṣitipṛṣṭhāt ūrdhvagā gṛhā*), i.e., above the visible geographic horizon, are seen by an observer, whereas the planets that are below the surface of the Earth (*kupṛṣṭhato'dhogatā gṛhā*), i.e., below the visible geographic horizon, are not visible.

Verse 38

According to Purāṇas, the Sun always rises from behind the eastern Udaya mountain (*udayācala* or *udayagiri*), which, in effect, makes this mountain the fixed horizon to the east. However, by considering the Udaya mountain as the *acala-kṣitija* or fixed terrestrial horizon (as opposed to one's local geographic horizon), it is implied that the Sun rises identically at all places on a planar Earth. In contrast, Siddhāntic astronomy accounts for even the smallest differences (*alpatara-calana*) in the rising times of the planets due to the longitude differences (*deśāntara*) at different places in one's own country. The incongruity (*asaṃgati*) between the Siddhāntic and the Purāṇic narrative is evident in Nityānanda's rhetorical statement in this verse.

Verse 39

The Purāṇas claim that the Sun revolves around the middle (hip) of Mt Meru, the golden mountain (*suvarṇa-śaila*) of enormous height (*vipula-unnata*) that is the *axis mundi* of the world, and the nychthemeron (*dyuniśa*, lit. day and night) for the gods (*suparvans*) living on the eastern face of Mt Meru is twelve months (one year), with their day and their night both being six months long equally. These two views, however, appear contradictory from a Siddhāntic point of view. If the daily revolution of the Sun around Mt Meru brings about the day and night (with the enormous face of the mountain occulting the Sun at night as it travels behind the mountain), then the day-length would be the same for both the gods on Mt Meru and the humans on the surface of the Earth. The half-yearly day-length of the gods at the north pole (Mt Meru) requires the orbit of the Sun to be significantly larger than the girth of any mountain.

In fact, this is one of the false views (*mithyājñāna*) that Lalla addresses in his *Śiṣyadhīvṛddhidatantra* (XX.10–15). According to Lalla, the shadow of the Earth causes nights and not the Sun being occulted by Mt Meru every day. Bhāskara II, in his *Siddhāntaśiromaṇi* (II.3.12), also attacks this Purāṇic view in his own laconic manner.

Verse 40

The *bhūmaṇḍala* of the Purāṇa is a flat disk-like surface of enormous extent. It includes concentric rings of seven annular island-continents and seven oceans of increasingly large dimensions, extending radially outwards from the central island-continent of Jambūdvīpa surrounded by the first ocean of salt water. Nityānanda describes this first saline ocean (*kṣāra-ambudhi*) following a typical Purāṇic narrative; however, he attributes it a size (*vyāsa*) of one hundred *yojana*s, a number significantly smaller than what the Purāṇas claim. For example, the *Śrīmad Bhāgvatam* (V.1.32–33) states, *teṣāṃ parimāṇaṃ pūrvasmāt pūrvasmād uttara uttaro yathāsaṅkhyaṃ dviguṇamānena bahiḥ samantata upakḷptāḥ* 'each island is twice as large as the one preceding it' and *jaladhayaḥ sapta dvīpaparikhā ivābhyantaradvīpasamānā ekaikaśyena yathānupūrvaṁ saptasv api bahir dvīpeṣu pṛthak parita upakalpitās teṣu jambvādiṣu* 'all the islands are completely surrounded by these oceans, and each ocean is equal in breadth to the island it surrounds' (Prabhupāda 1975, pp. 59–61 in Volume I). The central island-continent of Jambūdvīpa was believed to be a hundred thousand *yojana*s in extent (see, e.g., *Viṣṇu Purāṇa*, II.3.28), and hence accordingly, the first saline ocean would also be a hundred thousand *yojanas* in size—a measure thousand times larger than Nityānanda's estimate.

Like Nityānanda assertion in this verse, earlier Siddhāntic authors have also claimed that the entire northern hemisphere of the Earth (*udac-kudala*) makes up the central island-continent of Jambūdvīpa, and the saline ocean surrounding it is the equatorial ocean resembling a girdle around the hips of the Earth. The remaining six island-continents (and their encircling oceans) simply constitute the entire southern hemisphere of the Earth (*apara-kudala*); see, e.g., Bhāskara II's *Siddhāntaśiromaṇi* (II.3.21) or Jñānarāja's *Siddhāntasundara* (II.1.41).

Verses 41–43

Nityānanda's descriptions of the seven oceans (*sapta-abdhi*), each of a different kind, can be found in most Siddhāntic texts; see, e.g., Lalla's *Śiṣyadhīvṛddhidatantra* (XIX.1–3), Śrīpati's *Siddhāntaśekhara* (XV.30, 32), Bhāskara II's *Siddhāntaśiromaṇi* (II.3.17, 22–23), or Jñānarāja's *Siddhāntasundara* (II.1.39). These seven concentric oceans extended from the largest equatorial saline ocean (*kṣāra-sindhu*) that encircled the innermost Jambūdvīpa continent (the entire northern hemisphere) to the smallest outermost sweet-water ocean (*svādu-sindhu*) that encircled the southern polar region of Vaḍavānala (in the southern hemisphere). Vaḍavānala was believed to be the southern polar region, antipodal to Mt Meru in the north pole and forming the base of that mythical mountain (see Section 1.3.3.6).

There are two main differences between Nityānanda's description of these seven oceans and what is conventionally attested in the Purāṇas (e.g., in the *Śrīmad Bhāgvatam*, V.1.33, or in the *Viṣṇu Purāṇa*, II.4). The first difference concerns the ordering of these concentric oceans extending outwards: Nityānanda orders them as the oceans of salt water (*kṣāra*), milk (*payas*), yoghurt (*dadhi*), clarified butter (*ghṛta*), sugarcane juice (*ikṣu*), wine (*madya*), and sweet water (*svādu*), whereas the *Śrīmad Bhāgvatam* and the *Viṣṇu Purāṇa* order them as the oceans of salt water (*kṣāra*), sugarcane juice (*ikṣu*), wine (*surā*), clarified butter (*sarpis*), emulsified yoghurt (*dadhi-maṇḍa*), milk (*kṣīra*), and sweet water (*svādudaka*).

The second difference is a difference in the sizes of the oceans. The Purāṇas declare the size of each successive ocean (going outwards from the innermost saline ocean) to be twice the extent of the preceding one. This makes the outermost sweet-water ocean sixty-four times the size of the inner salt-water ocean. In contrast, Nityānanda calls the outermost ocean of sweet water exceeding small (*atyanta-alpa*) in comparison with the innermost saline ocean (flowing along the Earth's equator). The accommodation of Purāṇic geography in Siddhāntic discourse required

Purāṇic dimensions to be significantly reduced to fit the geometrical constraints of a spherical Earth. Even as Nityānanda's discourse on the Earth-sphere acknowledges Purāṇic cosmography as a valid subject matter, he, like his fellow Siddhāntic authors, modifies Purāṇic sizes and distances to agree with Siddhāntic requirements. At the end of verse 43, Nityānanda briefly describes the Purāṇic netherworld (*pātāla-loka*) located in the concavity of the Earth (*dharaṇī-puṭa*), i.e., deep inside the Earth-sphere in the southern polar region of Vaḍavānala, and brilliantly illuminated by serpentine jewels. The Purāṇas describe the netherworld (Pātāla) as the last in a series of seven subterranean realms; the other six being Atala, Vitala, Nitala, Gabhastimat, Mahātala, and Sutala (*Viṣṇu Purāṇa*, II.5.2). The netherworld is also the abode of the serpentine race of demons (called *nāga*) who are the guardians of the treasures stored there. According to the *Śrīmad Bhāgvatam* (V.24.12), *yatra hi mahāhipravaraśiromaṇayaḥ sarvaṃ tamaḥ prabādhante* 'many great serpents reside there with gems on their hoods, and the effulgence of these gems dissipates the darkness in all directions' (Prabhupāda 1975, p. 383 in Volume I). Other Siddhāntic authors have similarly identified the nether worlds are being inside the Earth-sphere, e.g., Bhāskara II in his *Siddhāntaśiromaṇi* (II.3.23–24) or Jñānarāja in his *Siddhāntasundara* (II.1.61).

Verses 44–45

Like the seven concentric oceans described in verses 41 and 42, the seven annular island-continents (*dvīpa*s) between each pair of oceans are also progressively smaller (*alpa-alpaka*) in size. The innermost island-continent of Jambūdvīpa is the largest (*mahat-para*) encompassing the entire northern hemisphere of the Earth-sphere, while the other six continents, each succeeding smaller than the previous one, extend and occupy the other half of the Earth-sphere (*para-ilā-dala*), i.e., the southern hemisphere.

According to the Purāṇas, the cumulative sum of the extent of the island-continents and their surrounding oceans on the Earth-disk (*bhū-maṇḍala*) is around 500 million *yojana*s (see, e.g., *Viṣṇu Purāṇa*, V.4.96). This incredibly large value has been repeatedly challenged as a false view (*mithyājñāna*) by several Siddhāntic authors using proportionality arguments; see, e.g., Lalla's *Śiṣyadhīvṛddhidatantra* (XX.30), Śrīpati's *Siddhāntaśekhara* (XV.24), and Bhāskara II's *Siddhāntaśiromaṇi* (II.3.14–16).

Interestingly, Jñānarāja, in his *Siddhāntasundara* (II.1.76), offers, for the very first time, an argument to reconcile the Purāṇic and the Siddhāntic points of view on differing dimensions. According to him,

paurāṇikaiḥ samuditāḥ pṛthivīgraharkṣa-
saṃsthānamānagatayaḥ paramārthatastāḥ |
kalpāntare tu kila samprati kālabodha-
śāstroditāḥ samutibhiḥ pariveditavyāḥ ||

> The shapes, measures, and motions of the Earth, planets, and stars given by the followers of the Purāṇas, which are ultimately true, are indeed for another *kalpa*. Now, [in this present *kalpa*,] the [shapes, and so on of the Earth, and so on] given in the treatises that give knowledge of time [i.e., *jyotiḥśāstra*] are to be thoroughly studied by the wise (T. L. Knudsen 2014, p. 69).

This accommodation relies on the idea of *kalpa-bheda* (or *kalpāntara*), i.e., an exegetic difference between phenomena in different cyclical time periods or *kalpas*. In the present instance, the incommensurability of Purāṇic geography and Siddhāntic geometry is due to the fact that the dimensions of the Earth were different in different time cycles. In each aeon (*kalpa*), creation repeats itself anew; however, the details and dimensions of the objects created are adjusted to suit the comprehensibility of the age.

Nityānanda avoids going into detail about the actual sizes of the annular island-continents. Instead, he simply lists them down following similar lists in other Siddhāntic texts, e.g., in Lalla's *Śiṣyadhīvṛddhidatantra* (XIX.1–3), Śrīpati's *Siddhāntaśekhara* (XV.33), Bhāskara II's *Siddhāntaśiromaṇi* (II.3.25), or Jñānarāja's *Siddhāntasundara* (II.1.42). While some Siddhāntic authors order these island-continents (going radially outwards from the innermost Jambūdvīpa) slightly differently, they all derive the names of these island-continents from Purāṇic sources. The Purāṇas have named these island-continents based on the eponymous trees, grass, flowers, mountains, or birds that were believed to proliferate in these place.

1 Jambūdvīpa is named after the black plum tree (Jambū);

2 Śākadvīpa is named after the teak (or pine) tree (Śāka);

3 Śalmaladvīpa is named after the silk cotton tree or Bombax tree (Śalmala);

4 Kauśadvīpa is named after the salt reed-grass (Kuśa);

5 Krauñcadvīpa is named after the bird species curlew (Krauñca), or perhaps, the mountain called Krauñca;

6 Gomedadvīpa, also called Plakṣadvīpa, is named after the Plakṣa tree (*ficus religiosa*); and

7 Puṣkaradvīpa is named after the blue lotus (*puṣkara*), or perhaps, the Maple tree.

Verses 46–51

In these six verses, Nityānanda describes the boundary mountains (*giris*) and the physiographic regions (*varṣas*) enclosed by their intermediate spaces (*vivara-deśas*) on the central island-continent Jambūdvīpa. To situate the Purāṇic flat disk-like Earth (*bhūmaṇḍala*) on a Siddhāntic Earth-sphere (*bhūgola*), the Jambūdvīpa is believed to encompass the entire northern hemisphere and is surrounded on all sides by the equatorial saline ocean (*kṣārāmbudhi*). In essence, the planar Jambūdvīpa is now folded over the convex hemisphere of the Earth-sphere in a manner that Mt Meru remains central and fixed as the north pole (as the *axis mundi*) while the other geographical features are placed in relation to the equatorial cities of Laṅkā, Romaka, Yamakoṭi, and Siddhapurī.

Figure 5.13 depicts a planar view (stereographic projection) of the Jambūdvīpa (northern hemisphere) with the relative arrangement of its different mountain ranges and interior regions around the central Mt Meru (north pole). Most Purāṇic descriptions of Jambūdvīpa adhere to this configuration, with perhaps some minor variations in the names and ordering of the mountains and regions in the different Purāṇas (and also the *Mahābhārata*, VI.7–9). Earlier Siddhāntic authors have also described the physical geography of the Jambūdvīpa near-identically; see, e.g., Lalla's *Śiṣyadhīvṛddhidatantra* (XIX.8–13) or Bhāskara II's *Siddhāntaśiromaṇi* (II.3.26–29).

Verses 52–53

Nityānanda's statements in verses 52 and 53 follow theomorphic cosmography of Vaiṣṇava Purāṇas. Most Purāṇas describe Ilāvṛta, the central *varṣa* of Jambūdvīpa, as the land surrounding Mt Meru, the celestial abode of the gods (*surālaya*). The *Mārkaṇḍeya Purāṇa* (I.4.12 and I.10.10) describes the topography of Jambūdvīpa as elevated in the middle (*madhye tuṅga*) and slopping towards the south and the north (*dakṣinottarato nimna*), with the elevated central Ilāvṛta *varṣa* being the saucer-like extension in the body of Mt Meru (*śarāvākārasaṃstāro merumadhye ilāvṛte*). Other Siddhāntic authors before Nityānanda have also described Ilāvṛta *varṣa* as the beautiful golden earth surrounding a central Mt Meru, the bejewelled golden mountain abode of the immortals gods; see, e.g., Lalla's *Śiṣyadhīvṛddhidatantra* (XIX.14), Śrīpati's *Siddhāntaśekhara* (XV.45), or Bhāskara II's *Siddhāntaśiromaṇi* (II.3.30–31).

In the Gauḍīya Vaiṣṇava tradition, Garbhodakaśāyī Viṣṇu is the plenary expansion of the supreme spirit (*virāj-puṃs*) or Viṣṇu that impregnates and permeates all universe(s) and gives rise to diversity during

Figure 5.13 A schematic representation of the physical geography of the island-continent of Jambūdvīpa with its different boundary mountains and enclosing physiographic regions (*varṣa*s).

the process of material creation. The Vaiṣṇava Purāṇas often describes Brahmā, the creator god, as the first being born on the elevated pericarp (*karṇika*) of the lotus that sprang from the navel of Garbhodaka-śāyī Viṣṇu (because of which Viṣṇu is also called Padmanābha 'the one with the navel-lotus'). This elevated pericarp is Mt Meru and the Earth is the very navel-lotus (*nābhi-saroruha*). In his universal form, Viṣṇu is regarded as the Viśvarūpa who pervades all existence, and the different realms (*loka*s) are identified as the six limbs (*ṣaṣ-aṅga*) of his cosmic body.

Verses 54–57

Figure 5.14 depicts the relative arrangement of the mountains, forests, and lakes in the Ilāvṛta *varṣa* as described by Nityānanda in these verses. According to the Purāṇas, each of the four mountain ranges

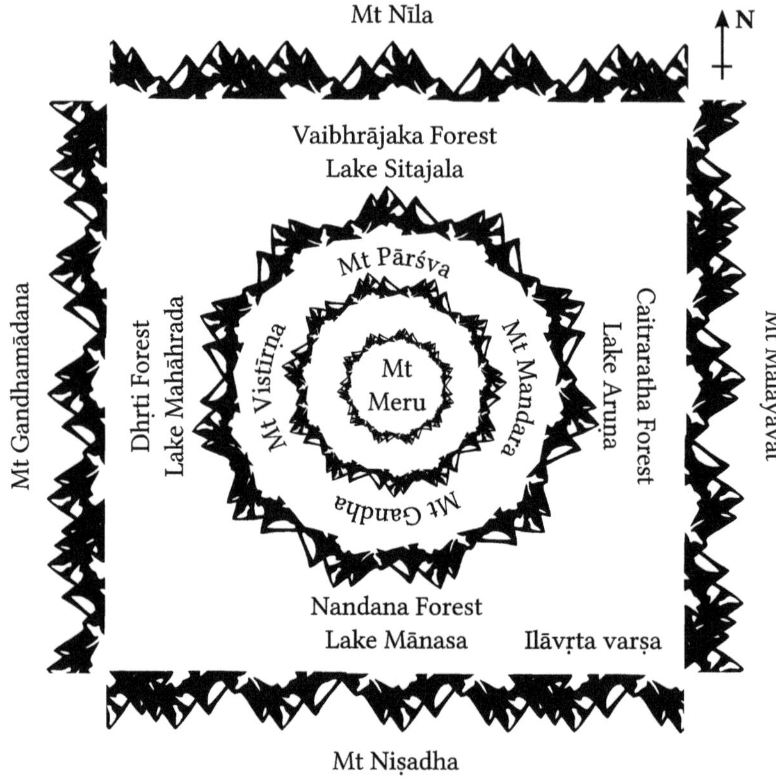

Figure 5.14 A schematic representation of the mountains, lakes, and forests of Ilāvṛta *varṣa*, the central region of the island-continent of Jambūdvīpa.

that act as buttress mountains to Mt Meru have particular types of trees that proliferate on them. The eastern Mandara mountains have burflower (Nīpa or Kadamba) trees (*Neolamarckia cadamba*), the southern Gandha mountains have the black plum (Jambū) trees (*Syzygium cumini*), the western Vistīrṇa mountains have the Indian fig or Banyan (Vaṭa) trees (*Ficus benghalensis*), and northern Pārśva mountains have the sacred fig (Pippala or Aśvatha) trees (*Ficus religiosa*). A similar description of Mt Meru's four buttress mountains can also be found in earlier Siddhāntic texts, e.g., Lalla's *Śiṣyadhīvṛddhidatantra* (XIX.15, 18), Śrīpati's *Siddhāntaśekhara* (XV.40), or Bhāskara II's *Siddhāntaśiromaṇi* (II.3.32).

The forests (or gardens) that are found on these mountains have peculiar attributes, e.g., Bhāskara II, in his *Siddhāntaśiromaṇi* (II.3.34–35), describes Caitraratha to be of varied brilliance, Nandana to be the delight of the celestial nymphs, Dhṛti as the giver of refreshments to the gods, and Vaibhrājaka to be resplendent (L. Wilkinson and B. D. Śāstrī 1861, p. 118). In fact in these same verses, Bhāskara II also adds that the lakes (or reservoirs) found in these forests (gardens) have waters where *saraḥsu rāmāramaṇaśramālasāḥ surā ramante jalakelilālasāḥ* 'the celestial spirits, when fatigued with their dalliance with their fair goddesses, love to disport themselves'. The beauty of these mountain lakes, with the amorous sportings of celestial beings in their waters, has also been described by other Siddhāntic authors like, e.g., Śrīpati in his *Siddhāntaśekhara* (XV.42) or Lalla in his *Śiṣyadhīvṛddhidatantra* (XIX.16, 18).

In verse 55, Nityānanda praises the exquisite *Jambūnada* gold that is found in the sandy soil along the embankment of the Jambū river that flows from the top of the Gandha mountain. The story of *Jambūnada* gold is a story of Purāṇic alchemy. According to the Purāṇas, an enormous Jambū tree (111 *yojanas* in height according to the *Brahma Purāṇa*, XVIII.23–24) stands at the crown of the Gandha mountain. The juice of the fruits of this tree oozes and forms the source of the Jambū river. The *Viṣṇu Purāṇa* (II.2.19–20, 24) describes the fruits of this Jambū tree to be the size of elephants, and hence the nectar that flows from these gigantic fruits ripening and falling onto the ground gushes forth like a river. Wherever this nectarine stream mixes with loam and earth, it transforms the soil into the extraordinary *Jambūnada* gold worthy of being made into the ornaments of divine beings. A similar praise of the Jambū river and its divine *Jambūnada* gold can also be found in Bhāskara II's statements in his *Siddhāntaśiromaṇi* (II.3.33).

Verse 58

Nityānanda follows the Purāṇas in situating the various divinities on Mt Meru (or Mt Sumeru). According to Purāṇas, Mt Meru has three golden summits where the trinity (*trimūrti*) of the Hindu pantheon reside. Brahmā (*virañci*), the creator, lives in Brahmaloka (also called Satyaloka); Viṣṇu (*hari*), the preserver, lives in Viṣṇuloka (also called Vaikuṇṭha); and Śiva (*śambhu*), the destroyer, lives in Śivaloka (also called Kailāsa). Below these three mountain peaks are the cities of the eight regents of the directions:

1 Indra (also called Śakra or Vāsava), the leader of the gods (*devas*) and the guardian of the East, lives in Indraloka or Svarga;

2 Agni, the god of fire and guardian of the south-east, lives in Agniloka;

3 Kāla (also called Yama), the god of time, death, and justice, and the guardian of the south, lives in Yamaloka;

4 Nairṛta (also called Alakṣmi or Nirṛti), the goddess of darkness, poverty, and corruption, and the guardian of the south-west, lives in Nairṛtyaloka;

5 Aṃbupa (also called Varuṇa), the god of water and guardian of the west, lives in Varuṇaloka;

6 Vāyu (also called Pavana or Prāṇa), the god of wind and the guardian of the north-west, lives in Vāyuloka;

7 Soma (also called Kubera or Candra), the lord of wealth and guardian of the north, lives in Somaloka or Alakapura; and

8 Īśāna (also called Śiva Rudra), the lord of prosperity and knowledge, and guardian of the north-east, lives in Īśānaloka.

Verses 59–60

According to Purāṇic lore, the river Gaṅgā (also called Surasarit 'the river of the gods') entered this world through a tear in its covering pierced by the big toenail of the left feet of Viṣṇu. Her celestial waters wash the feet of Viṣṇu upon entering the world and hence she is often called Bhagavatpadī or Viṣṇupadī. The Purāṇas describe her descent as she first arrives on the golden mountain Mt Meru, and, from there, divides herself into four streams and flows in the four cardinal directions: her distributary rivers Sītā to the east, Alakanandikā to the south, Cakṣu to the west, and Subhadrā to the north. In verses 59 and 60, Nityānanda extols the river Gaṅgā and describes these four rivers that distribute from it in the four cardinal directions (*catur-diś*) of Mt Meru. Each of these rivers flows onto the peaks of the buttress mountains (*viṣkambha-acala*) of Mt Meru, and, beyond that, enters into the four lakes (*saras*) in the four forests in these mountains. In his composition, Nityānanda praises the tumultuous and turbulent descent of Gaṅgā through mountains and caves creating waves of rolling water (rapids) with the full finesse of his poetic art.

In fact, Nityānanda's epithets (*upākhyā*) for river Gaṅgā in verse 60 draw on several mythological tropes from Purāṇic stories. For example, Gaṅgā's descent on Earth was broken by Śiva as he held her in his matted locks to save the Earth from the full force of her flow. This gave Śiva the name Gaṅgādhara 'he who holds the river Gaṅgā', while Gaṅgā, the stainless (*vimalā*) jewel in the crown of Śiva's head, became his head ornament (*mastaka-bhūṣaṇa*) like a pearl necklace (*muktāvalī*). Or, how Gaṅgā, having washed the lotus feet (*caraṇa-ambuja*) of Viṣṇu, spread through the glistening three words (*vilasat-lokatraya*) and became known as the destroyer of all evil (*pāpaugha-vidhvaṃsinī*). In fact, Nityānanda's exaltation (*stutivāda*) of Gaṅgā follows in the light of Bhāskara II's statements on her benevolent powers in his *Siddhānta-śiromaṇi* (II.3.39).

Verses 61–63

In these three verses, Nityānanda describes the nine regions and the seven principal mountains of Bhārata *varṣa*. An identical description can also be found in earlier Siddhāntic texts, e.g., Lalla's *Śiṣyadhīvṛddhidatantra* (XIX.24–25), Śrīpati's *Siddhāntaśekhara* (XV.46–47), and Bhāskara II's *Siddhāntaśiromaṇi* (II.3.41–42). Ali (1983, pp. 126–132) provides an excellent overview of the different interpretations of these nine regions in Purāṇic sources. Following Ali's description (p. 130), the nine regions (*nava-khaṇḍas*) of Bhārata *varṣa* can be (tentatively) identified as follows:

1 Aindra *khaṇḍa*, identified with the trans-Brahmaputra region in the north-east of India;

2 Kaśeru *khaṇḍa*, identified with the coastal plain between the deltas of the rivers Godāvarī and Mahānadī on the eastern coast of India;

3 Tāmraparṇa, identified with the Indian peninsular south of the river Kāverī in the south-east of India (associated with the river Tāmraparṇī);

4 Gabhastimādana *khaṇḍa*, identified with the hilly regions between the rivers Narmadā and Godāvarī in the central heartland of India towards the south-east;

5 Kumārī *khaṇḍa*, loosely identified with the central Indo-Gangetic plain in the north-central region of India;

6 Nāga *khaṇḍa*, loosely identified with the kingdoms and colonies between the Vindhya and Satpurā mountain ranges in the south-central region of India towards the west, and extending up to the Chhota Nagpur plateau to the east;

7 Saumya *khaṇḍa*, identified with the coastal regions west of the river Indus;

8 Vāruṇa *khaṇḍa*, identified with the western coast of India; and

9 Gandharvaka *khaṇḍa*, identified with the cis-Indus region to the north west of India.

The Purāṇas describe the Kumārī *khaṇḍa* of Bhārata *varṣa* as the only region where the four-fold class division of society is obeyed. Men are born into one of the four principle classes (called *varṇas*): the Brāhmins, priests, and teachers who receive training in Vedas; the Kṣatriya kings and warriors; the Vaiśya merchants and artisans; and the Śudra labourers and men who serve the other three *varṇas*. Elsewhere in the other eight regions live the foreign tribes (*mlecchas*, *yavanas*, *kirātas*, etc.) that are not bound to this social order and are considered to be outcasts of non-Aryan (*anārya*) descent.

The Purāṇas also describe the seven principal mountain ranges (*kula-giris*) that lie in Bhārata *varṣa*. These mountains were identified by the clan names (*kula-nāma*) of the tribes that were thought to live there. A (tentative) identification of these mountain ranges, based on Ali (1983, pp. 112–113), is as follows:

1 Mahendra *giri*, identified with the Eastern Ghats of India, extending from Ganjam in Odhisa to Tirunelveli in Tamil Nadu;

2 Malaya *giri*, identified with the Western Ghats (Sahyādri) along the Malabar coast, extending from the Nīlagiri hills to around Kanyakumari (south of the river Kāverī);

3 Sahya *giri*, identified with the Western Ghats (Sahyādri) from the river Tāptī to the Nīlagiri hills along the west coast of India;

4 Śukti *giri*, identified with the ring of mountains that encircle the Mahānadī river basin in the eastern part of central India;

5 Ṛkṣaka *giri*, identified with the modern-day Vindhya mountains (in central India) from the source of the river Sunar to the eastern ranges that form the catchment area of the river Son;

6 Vindhya *giri*, identified with the Satpurā mountains (south of river Narmadā), the Mahādeo hills, the Kaimur range (east of the modern-day Vindhya mountains), the hills of the Chhota Nagpur plateau, and the Rajmahal hills (and their spurs towards the river Gaṅgā) that spread across central India; and

7 Pāriyātra *giri*, identified with the Aravalī and western Vindhya mountains in the north-western part of India.

Verses 64–65

Purāṇic cosmography describes seven worlds (*sapta-lokas*) that constitute the terrestrial and celestial realms, with each realm inhabited by a particular class of divine beings. According to the *Viṣṇu Purāṇa* (II.7.3–20), these realms (sometimes understood as concentric orbs) are as follows:

1 Bhūrloka, the terrestrial realm or Earth;

2 Bhuvarloka, the atmospheric realm extending between the Earth and the Sun, and inhabited by *munis*, *siddhas*, etc.;

3 Svarloka, the celestial realm above the Sun extending up to the Polestar (*dhruva*);

4 Maharloka, the celestial realm above the Polestar inhabited by Bhṛgu and the other saints;

5 Janaloka, the celestial realm inhabited by Sanatkumāra, Sanaka, Sanātana, and Sanandana (the four adopted sons (*manas-putra*s) of Brahmā);

6 Taparloka, the celestial realm of penance inhabited by deified mendicants (*vairāgin*s); and

7 Brahmaloka (or Satyaloka), the celestial realm of truth and the abode of the creator Brahmā.

In his efforts to adapt Purāṇic cosmography to a Siddhāntic context, Nityānanda limits the first three worlds to the Earth-sphere. According to him, the Bhūrloka is the southern hemisphere of the Earth (*ilā-dala*) below the terrestrial equator (*vyakṣadeśa*, lit. a place with no latitude) and includes the six island-continents (*dvīpa-ṣaṭka*) previously described in verse 44. The Bhuvarloka is situated to the north (*saumya*) of the Bhūrloka—in other words, the entire northern hemisphere that forms the island-continent of Jambūdvīpa. The Svarloka is then identified with Mt Meru, perhaps, just the base of the mountain on the surface of the Earth (as the Brahmaloka, abode of Brahmā, is already mentioned to be on the peak of Mt Meru; see verse 58). The four remaining celestial realms of Mahar, Jana, Tapas, and Satya are simply located in the heaven (*vyoman*), perhaps, also in conjunction with the enormous Mt Meru that extends into the sky.

Towards the end of verse 65, Nityānanda includes the seven subterranean regions (see verse 43) with the aforementioned seven superterranean worlds to make a total of fourteen worlds (*caturdaśa-loka*) described in Purāṇic sources. The *Viṣṇu Purāṇa* (II.5.2–3) classifies the seven nether worlds according to the colour of its soil: Atala with its white soil, Vitala with its black soil, Nitala with its purple soil, Gabhastimat with its yellow soil, Mahātala with its sandy soil, Sutala with its stony soil, and Pātāla with its golden soil. Various different kinds of demons (*dānava*s, *daitya*s, *yakṣa*s, and *nāga*s) inhabit these subterranean regions. An identical description of Purāṇic cosmography can be found in most Siddhāntic texts, e.g., Lalla's *Śiṣyadhīvṛddhidatantra* (XIX.28–29), Śrīpati's *Siddhāntaśekhara* (XV.48), and Bhāskara II's *Siddhāntaśiromaṇi* (II.3.43).

Verse 66

Having completed his disquisition on Purāṇic cosmography (in verses 35–65), Nityānanda now resumes to his discussions on the Earth-sphere. In verse 66, he describes the relative time of the day at the equatorial cities of Laṅkā, Yamakoṭi, Siddhapurī, and Romaka. The arrangement of these cities on the equator (at a mutual separation of 90° see verses 13–17) implies that sunrise (*udaya*) at Laṅkā corresponds to midday (*madhyāhna*) at Yamakoṭi (to the east of Laṅkā), sunset (*asta*) at Siddhapurī (antipodal to Laṅkā), and midnight (*rajanidala*) at Romaka (to the west of Laṅkā). Bhāskara II, in his *Siddhāntaśiromaṇi* (II.3.44), has also described the time of the day in these equatorial cities (corresponding to sunrise at Laṅkā) in an almost identical verse.

Verse 67

The method to determine the true north-south direction (meridian) at a place relies on simple geometry. Drawing two arcs centred on an eastern point (*pūrva-diś*) corresponding to the rising (*udaya*) of the Sun (on the eastern horizon) and on a western point (*pratyak-diś*) corresponding to the setting (*asta*) of the Sun (on the western horizon), the perpendicular bisector of the east-west line determines the north-south direction as shown in Figure 5.15.

Geometrically, this perpendicular bisector passes through the two points of contact where the two circular arcs meet, enclosing a figure within that resemble the shape of a fish. With this geometric construction, it is easy to see how the local north-south direction (meridian) at any place on the Earth-sphere would point northwards towards Mt Meru. Most Siddhāntic texts prescribe this method

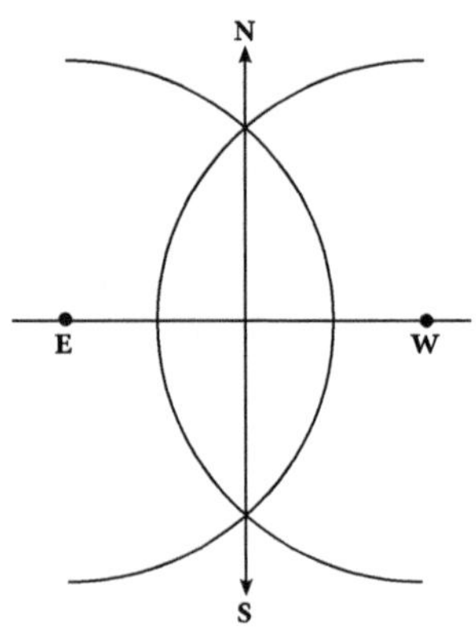

Figure 5.15 The fish-like intersecting circular arcs to determine the perpendicular bisector of the east-west line, i.e., the local north-south meridional direction.

to determine the local meridian line (pointing northwards towards Mt Meru); e.g., Bhāskara II description in his *Siddhāntaśiromaṇi* (II.3.45) is exactly identical to Nityānanda's statement in this verse.

This method of determining the perpendicular bisector using intersecting circular arcs that resemble the outline of a fish (*matsya* or *timi*) can be found in the mathematical Vedāṅga texts from as early as the first millennium BCE. The *Śulbasūtra*s 'Rules of the cord', a part of the Śrautasūtra or texts dealing with the performance of various religious ceremonies and sacrifices, discusses geometrical methods for the construction of fire altars (*agnicayana*). The *Kātyāyana Śulbasūtra* (ca. mid-fourth century BCE) describes the method of determining the east-west line using the forenoon and afternoon shadows of a gnomon along a locally flat surface, while the older *Baudhāyana Śulbasūtra* (ca. 800–500 BCE) describes the method of intersecting circular arcs creating the illusion of a fish-shaped figure (*matsya-ākāra*) to find the perpendicular bisector along the north-south direction (Sarasvati Amma [1979] 2007, pp. 22–24).

Verses 68–69

In verse 28, Nityānanda listed the cities of Laṅkā, Kāñcī, Ujjayinī, Saṃnihitya, Rohita, and Mt Meru as those lying on the Indian prime meridian (0° longitude). In verse 68, he adds the following two other regions to the list:

1 Kurujāṅgala, an area covering the modern-day cities of Rohtak (28° 54′ N, 76° 35′ E), Hansi (29° 6′ N, 75° 59′ E), and Hisar (29° 9′ N, 75° 43′ E) in Haryana.

2 Śrī Gargarāṭ, a prime meridional city mentioned by several Siddhāntic authors, e.g., Śrīpati in his *Siddhāntaśekhara* (II.96) or Vaṭeśvara in his *Vaṭeśvarasiddhānta* (I.8.1). Its modern-day location, however, is uncertain. It is perhaps related to the Śrī Karkarājeśvara temple (23° 10′ N, 75° 45′ E) on the banks of the Kṣipra river close to Ujjain.[11] The Tropic of Cancer is believed to pass through this temple, with its deity being the regent lord (*rājeśvara*) of the zodiacal sign Cancer (*karka*).

Moreover, in describing the relative positions of the equatorial cities in verses 13 and 17, Nityānanda placed the cities of Yamakoṭi and Romaka to the east and the west of the central city of Laṅkā respectively, both being at a distance of a quarter of the Earth (*ku-caturtha-bhāga*), i.e., 90° from Laṅkā in their respective directions. The spherical shape (*gola-ākṛti*) of the Earth-sphere would imply that all cities on the Indian prime meridian would have Yamakoṭi and Romaka as the pole of their great

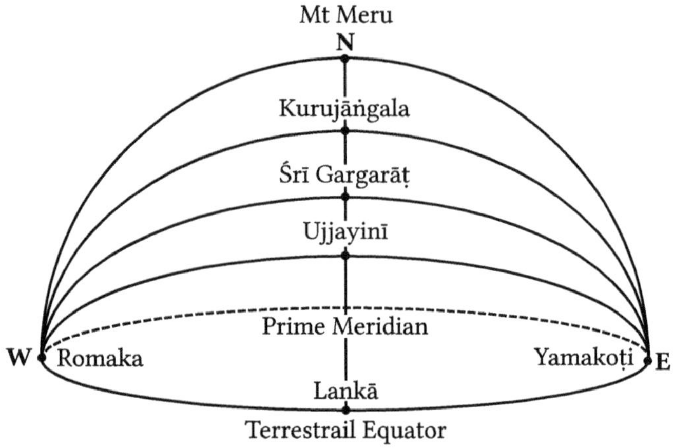

Figure 5.16 The sphericity of the Earth demonstrated by the cardinal orientations of the equatorial and the non-equatorial cities along the Indian prime meridian.

circle—in other words, always separated from these two cities by 90°—see Figure 5.16.

However, the only city eastwards and westwards of the cities of Romaka and Yamakoṭi respectively is the city of Laṅkā along the equator. The other non-equatorial cities lie in the intercardinal directions with respect to Romaka and Yamakoṭi. Bhāskara II, in his *Siddhāntaśiromaṇi* (II.3.46–47), states

> *yathojjayinyāḥ kucaturthabhāge prācyāṃ diśi syadyamakoṭireva* |
> *tataśca paścānna bhaveda laṅkaiva tasyāḥ kakubhi pratīcyām* ||
>
> *yat syāt prācyāṃ tat astan na bhavet pratīcyām* |
> *nirakṣadeśāditaratra tasmāt prācīpratīcyau ca vicitrasaṃsthe* ||
>
> Only Yamakoṭi lies dues east from Ujjayinī, at a distance of 90° from it; but Laṅkā and not Ujjayinī lies due west from Yamakoṭi.
>
> The same is the case everywhere; no place can lie west of that which is to its east except on the equator, so that east and west are strangely related.
>
> (L. Wilkinson and B. D. Śāstrī 1861, p. 120–121)

Essentially, Indian astronomers (*jyotirvids*) regarded due east (*prācī*) and due west (*pratīcī*) of any terrestrial place to lie along the prime vertical passing through the place (see, e.g., the arcs of the great circle passing through the different cities and converging at Yamakoṭi in the east and Romaka in the west in Figure 5.16). The cities that were located at an angular separation of 90° from a particular place along its parallel of latitude were not considered eastwards (or westwards) of that location. This strange arrangement (*vicitra*) of the directions (*kakubha*) is what prompted them—and in the present case, prompted Nityānanda—to consider the Earth (*dhātrī*) as a sphere (*gola*) as opposed to being mirror-like (*darpaṇa*), i.e., the flat disk-like Earth described in the Purāṇas. Incidentally, a comparison of flatness of an object to a mirror (*darpaṇa*) is an old Siddhāntic analogy; e.g., Lalla, in his *Śiṣyadhīvṛddhidatantra* (XX.34), refutes the flatness of the Earth by first saying *yadi darpaṇavat samā mahī*...'if the earth is plane like a mirror...' (Chatterjee 1981, p. 275).

Verse 70

According to Nityānanda, the surface area of a clime on the surface of the Earth (*bhūmi-khaṇḍa*) is the product of the circumference (*pariṇāha*) of the Earth and the height (*śara*) of the clime, both measured in *yojanas*.[12] Nityānanda states this fact without providing any explanation on why this is so. (See notes to verses 33 and 34 for Nityānanda's description of climes.)

The method of calculating the spherical surface area of a segment (or zone) on a sphere can be traced back to Archimedes' Hat-Box theorem, described in his work *On the Sphere and Cylinder*, II.5 (*Περὶ σφαίρας καὶ κυλίνδρου*, ca. 225 BCE). As Figure 5.17 shows, when a sphere is enclosed by a cylinder and sliced by two parallel planes perpendicular to the axis of the exterior cylinder, the Hat-Box theorem equates the surface area

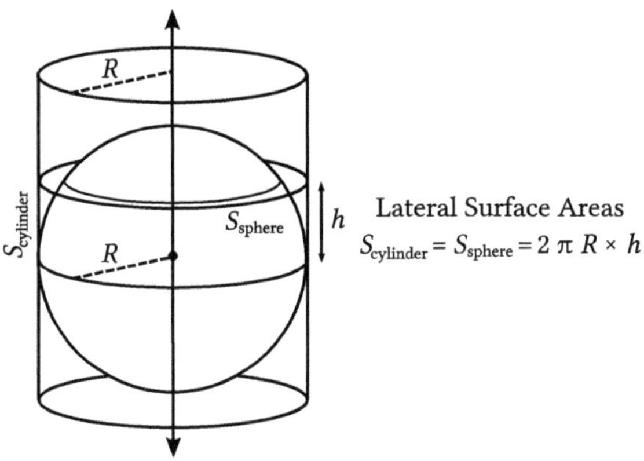

Figure 5.17 Archimedes' Hat-Box theorem describing the equivalence of the lateral surface areas of a spherical segment and cylindrical strip of equal height.

S_{sphere} of the spherical segment on a sphere (of radius R) and the surface area $S_{cylinder}$ of the spherical strip of the enclosing cylinder (also of radius R) of equivalent height h. In other words, $S_{sphere} = S_{cylinder} = 2\pi Rh$.

To my knowledge, this particular formula to compute the surface area of a clime is not found in any Siddhāntic texts prior to Nityānanda's *Sarvasiddhāntarāja*; however, earlier mathematical texts in Sanskrit have described the method to calculate the surface area of a sphere by integrating the area of trapezoidal strips folded out of the parallel circular segments on a sphere's surface; see, e.g., Jyeṣṭhadeva's method in his *Gaṇita Yuktibhāṣā* (ca. 1530) (Sarasvati Amma [1979] 2007, pp. 213–215).

Several Islamicate mathematicians like Thābit bin Qurra (826–901), Isḥāq ibn Ḥunayn (ca. 830–910), Ibn al-Haytham (965–ca. 1041), Abū Sahl l-Qūhī (fl. 970), al-Bīrūnī (972–1048), and even Naṣīr al-Dīn Tūṣī (1201–1274) have discussed the completeness of Archimedes' propositions extensively in their works. For instance, al-Qūhī's treatise *On Filling the Lacuna in the Second Book of Archimedes* (*fī suddi l-khalal fī l-maqālati l-thāniya min kitāb Arshimīdis*) discusses Archimedes' Hat-Box theorem as it tries to construct spherical segments of a sphere equal in surface area to one another (Berggren 1996). Nityānanda's familiarity with Islamicate astronomy (perhaps, mediated by his interaction with Mullā Farīd) could explain his knowledge of this formula; however, at the time of writing, I have not been able to identify any textual sources that may have informed him.

Verse 71

For a non-equatorial terrestrial location, the corrected circumference of the Earth (*sphuṭa-ku-paridhi*) is essentially the parallel of latitude of that place. As Figure 5.18 shows, the location D with northern latitude $\varphi°$ N ($\angle$COD or smaller $\widehat{\mathrm{CD}}$), or co-latitude $\overline{\varphi}°$ ($\angle$**N**OD or smaller $\widehat{\mathbf{N}\mathrm{D}}$), has a circle of latitude of radius $R_e \cos\varphi$ where R_e is the Earth's radius. The measure of the corrected circumference of the Earth at D, i.e., the circumference of the parallel of latitude ⊙ADQB, is $2\pi R_e \cos\varphi$. The parallel of latitude is marked (*aṅkita*) with 360°, or, alternatively, 60 *nāḍikā*s (24 hours) in units of time. The time units of a parallel of latitude indicate the relative positions of celestial objects when seen from different terrestrial places along that circle of latitude, and at different times of the day. (Compare the positions of the Sun at different equatorial cities at different times of the day in verse 66.)

The latitude-corrected circumference of the Earth or the parallel of latitude, measured in *yojana*s, also helps in calculating the longitudinal correction (*deśāntara*) to the mean daily longitude of planets (*kheṭa*s) for a terrestrial observer (at that latitude) at a particular time of the day. To understand this, consider as follows (with Figure 5.18):

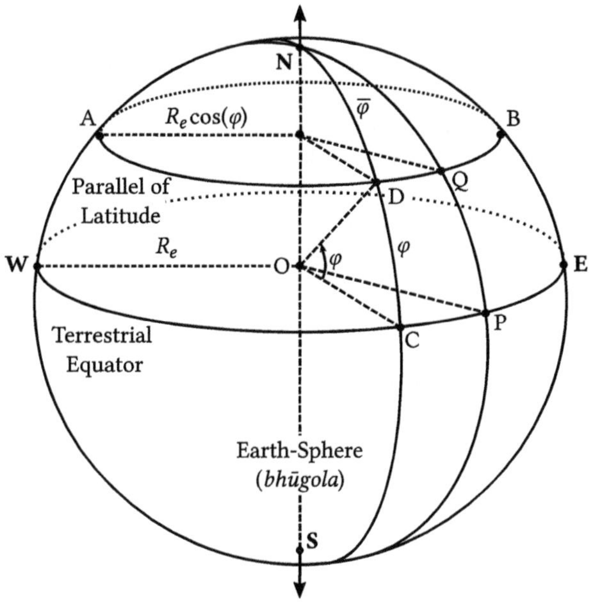

Figure 5.18 The corrected circumference of the Earth (parallel of latitude) for a non-equatorial terrestrial location of latitude φ° N.

1 For two terrestrial positions D and Q along the same parallel of latitude ⊙ADQB, the local meridians are represented as $\widehat{\mathbf{NDCS}}$ and $\widehat{\mathbf{NQPS}}$ respectively. The circumference of the parallel of latitude (passing through D and Q) is $2\pi R_e \cos\varphi$ *yojanas* where $\cos\varphi \in [0,1]$. For φ being 0°, the parallel of latitude is coincident with the terrestrial equator ⊙**WCPE** of circumference $2\pi R_e$ *yojanas*, and for φ being 90°, the parallel of latitude is simply the north pole **N**.

2 Thus, the measure of the circumference of the parallel of latitude is smaller than the measure of the equatorial circumference of the Earth, i.e., $2\pi R_e \cos\varphi < 2\pi R_e \; \forall \; \varphi > 0^\circ$. Accordingly, the distance of arc $d(\widehat{\mathrm{DQ}})$ (in *yojanas*) between two observers at D and Q along the same parallel of latitude is smaller than the corresponding distance $d(\widehat{\mathrm{CP}})$ (in *yojanas*) for two observers at C and P at the Earth's equator.

3 Now, a 360° rotation of the celestial sphere in 60 *nāḍikās* (24 hours) corresponds to one full rotation of the parallel of latitude. Hence, the measure of arc (in *yojanas*) along the parallel of latitude ⊙ADQB corresponding to 1 *nāḍikā* (about 24 minutes)

is $\frac{2\pi R_e \cos\varphi}{60}$. Whereas, at the terrestrial equator ⊙WCPE, the measure of arc (in *yojana*s) corresponding to 1 *nāḍikā* is $\frac{2\pi R_e}{60}$.

4 The difference in times δt between an event, e.g., sunrise at the locations D and Q, can be calculated as

$$\frac{\delta t}{60 \textit{ nāḍikās}} = \frac{d(\widehat{DQ})}{2\varpi R_e \cos\varphi} \Rightarrow \delta t = \frac{d(\widehat{DQ})}{2\pi R_e \cos\varphi} \times 60 \textit{ nāḍikās}.$$

This time difference is called *deśāntarakalā* and is applied positively or negatively to account for the eastward or westward position of a terrestrial location (along the same parallel of latitude) respectively. Typically, the measure is added for a location east of the prime meridian and subtracted for a place west of it. With this adjustment, the proper time of a celestial event at a particular location can be determined (knowing its time of occurrence at a place along the prime meridian and on the same latitude).

5 For a planet with a mean daily longitude $\Delta\theta$ (in 60 *nāḍikā*s), the amount of deviation in $\Delta\theta$ when seen at two places, e.g., D and Q, with a time difference of δt is

$$\frac{\delta\theta}{\delta t} = \frac{\Delta\theta}{60} \Rightarrow \delta\theta = \frac{\delta t\,\Delta\theta}{60} = \frac{d(\widehat{DQ})\,\Delta\theta}{2\pi R_e \cos\varphi}.$$

The correction $\delta\theta$ is then applied to determine the mean daily longitude $\Delta\theta$ of the planet. For observers west of the prime meridian, this correction is additive, whereas for observers east of it, the correction is subtractive. This method of correcting the mean daily longitude of a planet due to the difference in an observer's location is called *deśāntara-saṃskāra* in Indian astronomy (see Ramasubramanian and Sriram 2011, pp. 41–42).

Bhāskara II, in his *Siddhāntaśiromaṇi* (II.4.25), also describes a small circle (the parallel of latitude) on the Earth-sphere that represents the rectified circumference of the Earth (*sphuṭa-bhūparidhi*) for one's own latitude. This circle (on the sphere) is also centred on Mt Meru and passes through one's own location (at a known distance in *yojana*s from Mt Meru).

End of the section on the Earth-sphere

Nityānanda ends the discussions on the Earth-sphere (*bhūgola*) with a prosaic colophon. Here, the word *racanā* 'composition' alludes to both the content of the Earth-sphere and Nityānanda's own arrangement of the topics, in metrical verses, discussed here in this section on the Earth-sphere.

Beginning the section on the sphere of asterisms

The second section in Nityānanda's *golādhyāya* discusses the sphere of asterisms (*bhagola*). The *bhagola* is the right celestial sphere that is an extension of the Earth-sphere to include the heavens. From the perspective of a geocentric observer, the north and south poles of the Earth-sphere extended to become the north and south celestial poles (*dhruva-dvaya*) respectively, and the terrestrial equator extends to become the celestial equator (*viṣuvatvṛtta*) (as the celestial horizon). See Section 1.3.4.

Once again, Nityānanda's use of the word *racanā* (in *bhagola-racanā* 'composition of the sphere of asterism') alludes to the natural configuration of the right celestial sphere, as well as his own act in arranging the topics that discuss aspects of the *bhagola* in metrical verses. The geometrical configuration of the heavens is often a topic discussed under the subject of *ʿilm al-hayʾa* 'science of the figure (of the heavens)' or *ʿilm al-falak* 'science of the sphere' in Islamicate astronomy—it is differentiated from *ʿilm al-nujūm* 'science of the stars' that discusses topics on astrology. Hence, comparatively, the study of the configuration (*hayʾa* in Arabic, *racanā* in Sanskrit) of the sphere (*falak* in Arabic, *gola* in Sanskrit) of the heavens is a separate topic in both Sanskrit and Islamicate astronomy.

Verse 72

The right celestial sphere (*bhagola*) is the celestial sphere that extends in all directions, from above the Earth's atmosphere, i.e., above the outermost winds (*vāyu*) of the Earth (*bhūmi*), up to the periphery (*avadhi*) of the heavens (*vyoman-kakṣa*). Nityānanda describes this space shaped like a sphere (*gola-ākāra-mūrti*) and permeated by the Pravaha wind.

According to Purāṇas, seven celestial winds formed the wheels of the chariot of Agni, the Vedic deity of fire. These seven winds, namely, Āvaha, Pravaha, Udvaha, Samvaha, Suvaha, Parivaha, and Parāvaha were responsible for terrestrial (meteorological) and celestial motions. The flow of Pravaha wind impelled the sphere of the heavens to rotate westwards perpetually. Śrīpati, in his *Siddhāntaśekhara* (XV.51–52), and Bhāskara II, in his *Siddhāntaśiromaṇi* (II.4.1–3), both list the names of these mythical winds, and then go onto declare the Pravaha wind as the cause of the eternal westward movement of the celestial sphere.

In addition to its shape and movement, Nityānanda describes the physical form of the *bhagola*. According to him, the sphere of asterisms is a globular object (*piṇḍībhūta*), transparent (*svaccha*) like a crystalline pitcher (*sphaṭika-ghaṭa*), clear like the colour of deep water (*nīra*), and enclosing an expanse of space (*sa-avakāśa*). Other Siddhāntic authors have described a wind-driven westward rotating celestial sphere; however, the physical description of this sphere as a crystalline translucent object containing an expansive space up to the outermost orbit of the stars is novel to Nityānanda's text.[13] In fact, in the chapter on computations (*gaṇitādhyāya*) of his *Sarvasiddhāntarāja* (I.3.180–200), Nityānanda dedicates twenty-one verses to the formation of the universe (*brahmāṇḍa-nirmāṇa-darśana*). There, he describes the universe as concentric spherical shells, beginning with the spheres of the four primal elements and extending outwards up to the ninth sphere or the *Primum Mobile*. In talking about the ninth sphere, Nityānanda says (in verses 189 and 190),

> *atrāṣṭāv eva golāḥ syur navamaḥ svayam eva hi |*
> *etān aṣṭau dyusad golāṃścālayen nijavegataḥ ||*
>
> *ratyakpravahavāyusthaḥ sadā gatyaikayaiva hi |*
> *svodarasthāṃs tathā saptagolāṃś candrādi khe sadā ||*
>
> Here, there are just eight spheres. The ninth, just on its own, based in the heavens, causes the eight spheres to move, by its own motion.
>
> The ninth sphere, positioned in the westwards blowing Pravaha wind, always [moving] with uniform motion (causes the) seven spheres contained within it to move through space.

The underlying Aristotelian cosmology in this description points to the Islamicate influence in Nityānanda's *Sarvasiddhāntarāja*.[14]

Verse 73.1

Reading 1 in MSS Bn.I, Np, Pb, and Sc

The sphere of asterisms (*bhagola*) contains three important reference circles:

1 the celestial equator (*viṣuvavalaya*) in the middle;

2 the diurnal circle (*dyuvṛtta*) produced on the either side (*pārśvajāta*) of the celestial equator that marks the diurnal (westwards) movement of a celestial object (an asterisms or a star-planet); and

3 the ecliptic (*bhacakra*) that forms the background zodiac (constellations) against which the slow daily (eastwards) movement of a celestial object can be mapped.

Nityānanda describes the diurnal circles in the sphere of asterisms to be parallel (*sama*) to the celestial equator but slightly unfastened (*ślatha*) to it. The superposition of the slow eastward movement of a celestial object (along its own orbit) coupled with its daily westward motion (impelled by the westwards-rotating celestial sphere) traces spirals along its path of motion; see Figure 5.19. These spiral-like diurnal circles are thought to be loosely bound to the plane of the celestial equator with an elevation (*tuṅga*) or depression (*pāta*) towards the northern or southern celestial poles.

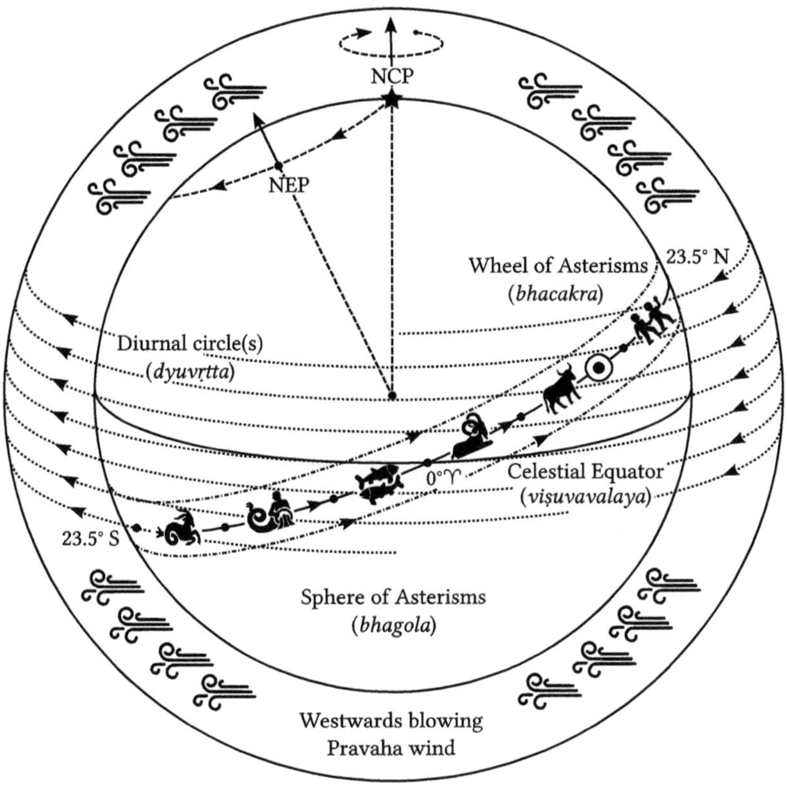

Figure 5.19 The sphere of asterisms with the celestial equator, the wheel of asterisms, and the diurnal circles. The westwards blowing Pravaha wind (in the outermost sphere) impels the diurnal westward motion of the Sun even as the Sun advances eastwards along the ecliptic (in the inner sphere of asterisms) every day. Here, the vernal equinox is coincident with the beginning of the zodiac (*Meṣādi* or 0° ♈); the westward regression of the vernal equinox (with respect to *Meṣādi*) corresponds to the precession of the North Ecliptic Pole (NEP) in the same direction.

The annual eastward migration of the Sun set against the sidereal background of constellations traces a path in the celestial sphere called the ecliptic, or, more simply, the zodiacal circle. In the present reading of the verse, Nityānanda identifies the *bhacakra*, lit. the wheel of asterisms, as this zodiacal girdle-like ring of the celestial sphere. According to him, the *bhacakra* bears all the star-planets (*sakala-khacara*) within it, with each star-planet moving eastwards with its own daily motion (*sva-bhukti*) along its own orbit (inclined to the ecliptic). At the same time, the *bhacakra* itself moves eastwards at an extremely slow pace (*mṛdu-gati*). This eastward motion of the zodiac (beginning at *Meṣādi* or 0° ♈), or equivalently the relative westward regress of the vernal equinox, is what gives rise to the precession of the equinoxes. In Indian astronomy, the motion of the equinox (*ayana-calana*) is described by a related phenomenon called trepidation where the equinox is thought to alternatively oscillate eastwards and westwards of *Meṣādi* by up to 24° (obliquity of the ecliptic) (see Pingree 1972).

Verse 73.2

Reading 2 in MSS Bn.II, Br, Pm, and Rr

In an alternative reading of verse 73, the sphere of asterisms (*bhagola*) is again described as containing the celestial equator (*viṣuvavalaya*) in the middle with the diurnal circles (*dyuvṛtta*) on its sides (*pārśva-jāta*). In this alternative reading of the verse, the movements of celestial objects along their *dyuvṛtta*s are considered analogous to fish (*jhaṣa*) swimming in water (*antarjala*), suggesting that the interior of the sphere of asterisms is like water. This interpretation is in keeping with the description of the sphere of asterisms in verse 72. The daily motions of the asterisms (*uḍu*s) and the stars-planets (*khecara*s) in the sphere of asterisms are simultaneous, each moving in their own orbit and with their own speed. These motions are set against the backdrop of the sidereal constellations that form the zodiacal girdle, more commonly called the zodiacal circle or the ring/wheel of asterisms (*bhavalaya*). Sometimes, this zodiacal circle is identically and nominally called the circle of asterisms (*bhagola*).

Verses 74–76

In verse 74, Nityānanda defines the ecliptic as the path (*mārga*) travelled by the Sun in its eastward journey in the celestial sphere. The path of the Sun, seen against the backdrop of the sidereal constellations, forms the zodiacal circle (*rāśicakra*) that intersects the celestial equator (*viṣuvavalaya*) at the two equinoctial points. As Figure 5.20 shows, the northward movement (*uttarāyana*) of the Sun from 270° (*Mṛgādi* or beginning of Capricorn) to 90° (*Karkādi* or beginning of Cancer) passes through 0° Aries (*Meṣādi* or beginning of Aries) marking the point of vernal equinox in the tropical (*sāyana*) zodiac. Similarly, the southward movement (*dakṣiṇāyana*) of the Sun from 90° (*Karkādi*) to 270° (*Mṛgādi*) passes through 180° (*Jūkādi* or beginning of Libra) marking the autumnal equinox of the tropical zodiac (in the northern hemisphere). The northern and southern parts of the ecliptic (with respect of the celestial equator) are the six zodiacal signs beginning with Aries (*meṣa*) and the six zodiacal signs beginning with Libra (*jūka*) respectively.

The ecliptic forms an important reference circle of the celestial sphere and according to what is measured, it is synonymously called (i) the circle of declination (*krāntivṛtta*) that measures the elevation of the Sun with respect to the celestial equator, (ii) the circle of lunar mansions (*bhavanavalaya*) that traces the path of the Moon across the constellations (fixed houses or stations), or simply (iii) the wheel of asterisms (*bhacakra, bhavalaya*, or *bhavṛtta*).

Also, as Figure 5.20 shows, the ecliptic poles are on the two sides of the ecliptic to the north and the south. The ecliptic pole (*kadamba*) rotates

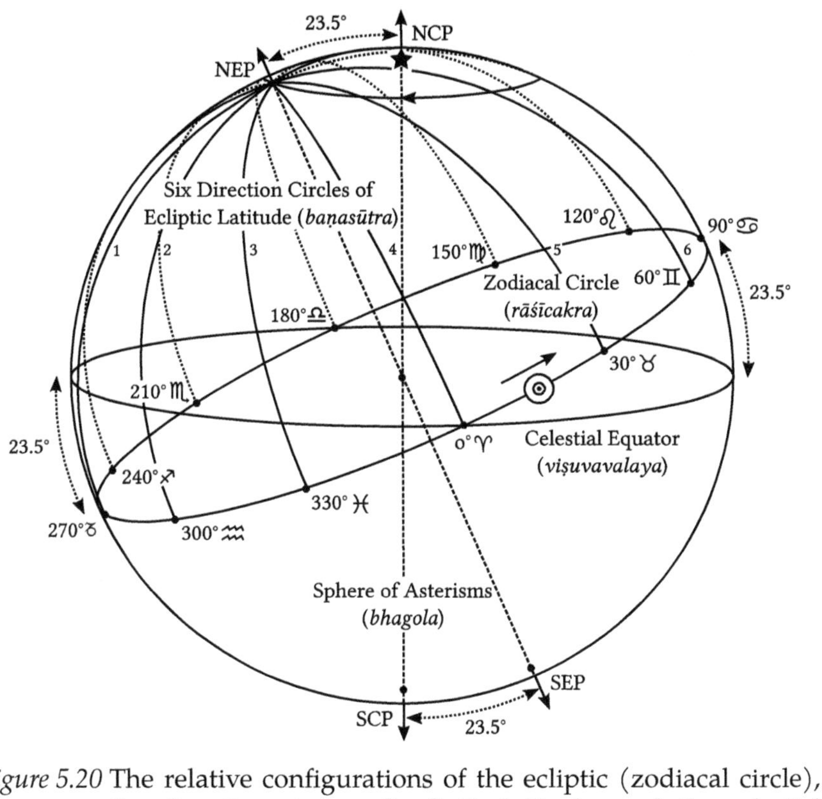

Figure 5.20 The relative configurations of the ecliptic (zodiacal circle), the direction circles of ecliptic latitude, and the celestial equator in the sphere of asterisms.

around the polestar (*dhruva*), i.e., the north celestial pole, everyday. The ecliptic itself is divided into twelve equal parts (*tulya-aṃśa*) by the six direction circles of ecliptic latitude (*bāṇasūtras*, lit. bow-strings). These reference circles form the secondary circles to the ecliptic that converge at the ecliptic poles and help determine the ecliptic latitude (*bāṇa*, lit. arrow) of celestial objects. The twelve divisions of the ecliptic mark out the twelve zodiacal constellations (*rāśi*s) or lunar mansions (*bhavana*s) beginning with *Meṣādi* (0° ♈), the first point of Aries (*aja*).

The solstitial colure, i.e., the circle passing through the pairs of celestial and ecliptic poles, intersects the ecliptic at two points. These are the two solstitial (*āyana*) points, indicated in Figure 5.20 as 90° ♋ (the summer solstitial position of the Sun in the tropical zodiac for the northern hemisphere) and 270° ♑ (the winter solstitial position of the Sun in the tropical zodiac for the northern hemisphere).

And lastly, the ecliptic is inclined to the celestial equator to the north and the south by 23° 30′, an amount that equals the obliquity of the ecliptic. This is equivalent to the tilt of the pair of ecliptic poles (*kadamba-yugma*) to the pair of celestial poles (*dhruva-dvaya*). An ecliptic obliquity of 23° 30′ is of Islamicate origin, whereas most Sanskrit Siddhāntic texts take this value to be 24°; see notes to verses 33 and 34.

Verse 77

The planets, beginning with the Moon, experience two different kinds of motion, namely, a diurnal westward motion impelled by the westwards-blowing Pravaha wind and a slow, self-initiated eastward motion (see verse 73.1). The diurnal circles (*dyuvṛtta*s) appear as spirals of motion (*śaṅkhāvarta*, lit. convolution of a shell) due to the cumulative effect of both these types of movement. Accordingly, the orbit (*adhvan*) of a planet is the apparent path created by successive conjunction (*apara-yoga*) of these diurnal spirals, slowing graduating eastwards after every revolution.

For each planet, its orbit (projected onto the sphere of asterisms) is inclined to the ecliptic/zodiacal circle (*rāśimaṇḍala*), intersecting it twice at its ascending and descending nodes going northwards and southwards of the ecliptic/zodiacal plane respectively; see Figure 5.21. Nityānanda refers to these projected orbits as latitudinal circles (*bāṇavṛtta*s). The orbit of a planet, being inclined to the ecliptic, is indicative of the planet's ecliptic latitude (*bāṇa*) at any given instance. Hence, referring to these projected orbits as latitudinal circles is akin to calling the ecliptic the declination circle (*krāntivṛtta*) as it indicates the declination (*krānti*) or inclination of a planet with respect to the celestial equator. The latitude of a planet is measured on a secondary circle to the elliptic (passing through the planet)—the direction circles of ecliptic latitude (*bāṇasūtra*) described in verse 75.

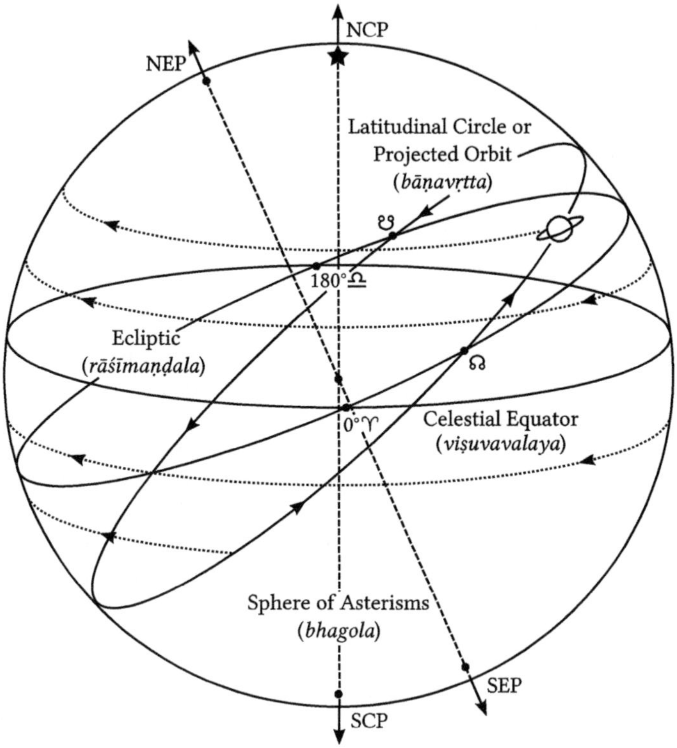

Figure 5.21 The latitudinal circle or projected orbital path of a planet in relation to the ecliptic (zodiacal circle) in the sphere of asterisms. The projected orbit is the apparent path of a planet's motion traced by the cumulative conjunctions of its diurnal westward spirals and slow eastward migration.

From as early as the fifth century, Siddhāntic texts have described the same order of the planets starting from the closest planet, the Moon, and extending outwards towards Saturn. For example, Varāhamihira, in *Pañcasiddhāntikā* (V.13.39), states the order of the planets (going progressively higher from a central Earth towards the farthest regions of the stars) as the Moon, Mercury, Venus, the Sun, Mars, Jupiter, and Saturn. Nityānanda, in the chapter on computations (*gaṇitādhyāya*) of his *Sarvasiddhāntarāja* (I.3.185–186), describes the progressively higher orbits of the planets, beginning from the Moon, as spheres (or orbs) by saying,

atha tasyopariṣṭāc ca budhakakṣākhyagolakaḥ |
candravat parato'nyeṣāṃ śukrārkāṅgārapūrvataḥ ||

grahāṇāṃ kakṣikā golā vijñeyāḥ śilpavittamaiḥ |
rāśīnām aṣṭamo golo bhānām api nibadhyate ||

> And then, situated above it [i.e., above the sphere of the orbit of the Moon] is the sphere known as the orbit of Mercury similar to [the sphere of the orbit of the] Moon. Above it are the other [orbital spheres], beginning with Venus, the Sun, and Mars.
>
> The orbits of the planets are to be known by scholars of the arts as their spheres. The eighth sphere of zodiacal signs is composed of the very stars [in the heavens].

Verse 78

The eastward motion of a planet along its own orbit is in opposition to its diurnal westward motion (impelled by the Pravaha wind). Nityānanda describes the planet's orbital circle (*gola*) as being upwards (*ūrdhva*) and downwards (*adhas*) with respect to the ecliptic plane in equal measure (*samatva*). Like the Sun, the planets themselves journey eastwards on account of their own motion, and along this path, they reach their apogee (*ucca* or *uccatā*) and perigee (*nīca*). Figure 5.22 depicts the actual orbit of a planet around the Earth with the planet being situated at its apsidal

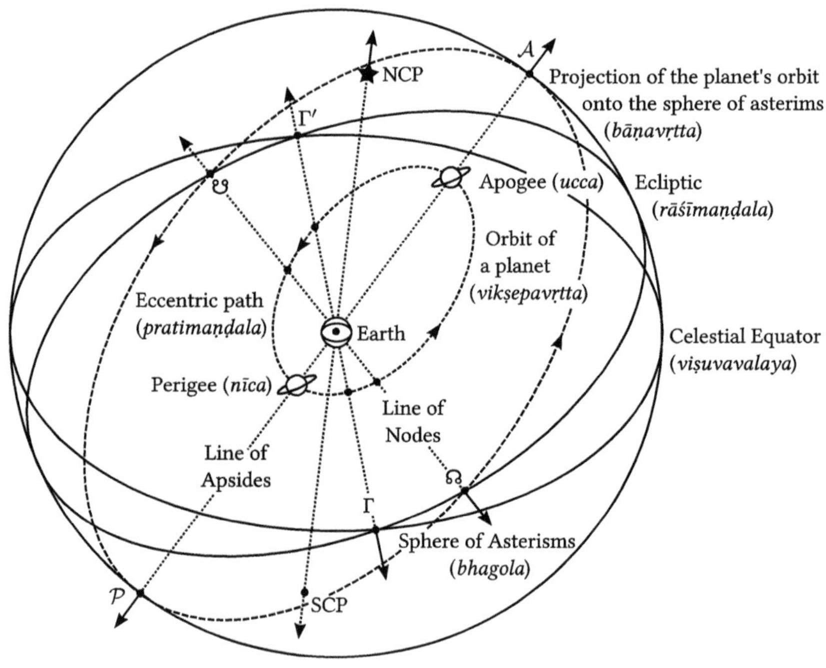

Figure 5.22 The actual orbit of a planet around the Earth with its apsidal and nodal lines extended to meet its projected orbit on the sphere of asterisms.

extremities. The orbit of the planet, when projected onto the sphere of asterisms, traces the planet's course as a latitudinal circle (*bāṇavṛtta*). The apsidal line of its orbit intersects the *bāṇavṛtta* at points $\mathcal{A}$ (apogeal point) and $\mathcal{P}$ (perigeal point), while the line of nodes (marking the intersection of the planet's orbital plane and the ecliptic plane) is indicated by the ☊ (ascending node) and ☋ (descending node). The direction of the vernal equinox is indicated by the equinoctial line $\Gamma\Gamma'$; in the tropical zodiac (*sāyana*), the vernal equinox (Γ) is considered coincident with *Meṣādi* (0° ♈).

Although the words *ucca* and *nīca* refer to the apogee and the perigee of a planet's orbit respectively, in a technical sense, these words are also used to refer to the higher and lower apsides in the epicyclic (*ucca-nīca-vṛtta*) model of planetary motion. In the equivalent eccentric model as depicted in Figure 5.22, the *ucca* and *nīca* simply refer to the farthest and nearest points along the eccentric path (*pratimaṇḍala*), also known as the planet's orbit (*vikṣepavṛtta*), along which the planet moves with respect to a central Earth-sphere.

Verse 79

When a planet's position on its actual orbit is projected onto the outermost sphere of asterisms, its projected position on the latitudinal circle (*bāṇavṛtta*) helps ascertain its elevation. This elevation can be measured with respect to two different planes of references. As Figure 5.23 shows, the elevation of a planet—in other words, the elevation of its projected position on its *bāṇavṛtta*—has two great circles passing through it that act as the secondary circles to the celestial equator (*viṣuvavalaya*) and the ecliptic (*rāśimaṇḍala*) accordingly.

1 The circle passing through the pair of ecliptic poles (*kadamba-yuga*) and the planet is called the direction [circle] of ecliptic latitude (*śilīmukhasūtraka*). This is identical to the *bāṇasūtra*s described in verse 75. The ecliptic latitude (*śilīmukha* or *bāṇa*) is then the measure of the planet's elevation with respect to the ecliptic plane, measured on its direction circle perpendicular to the ecliptic.

2 The circle passing through the pair of celestial poles (*dhruva-yugma*) and the planet is called the direction circle of declination (*apakramasūtraka*). The declination (*apakrama* or *krānti*) is the measure of the planet's elevation with respect to the equatorial plane, measured on its direction circle perpendicular to the celestial equator.

Nityānanda uses *sūtraka*, lit. string or thread, to refer to a circular arc on the sphere of asterisms that is oriented in a particular direction. The *bāṇasūtra* or *śilīmukhasūtraka* is the direction along which the ecliptic

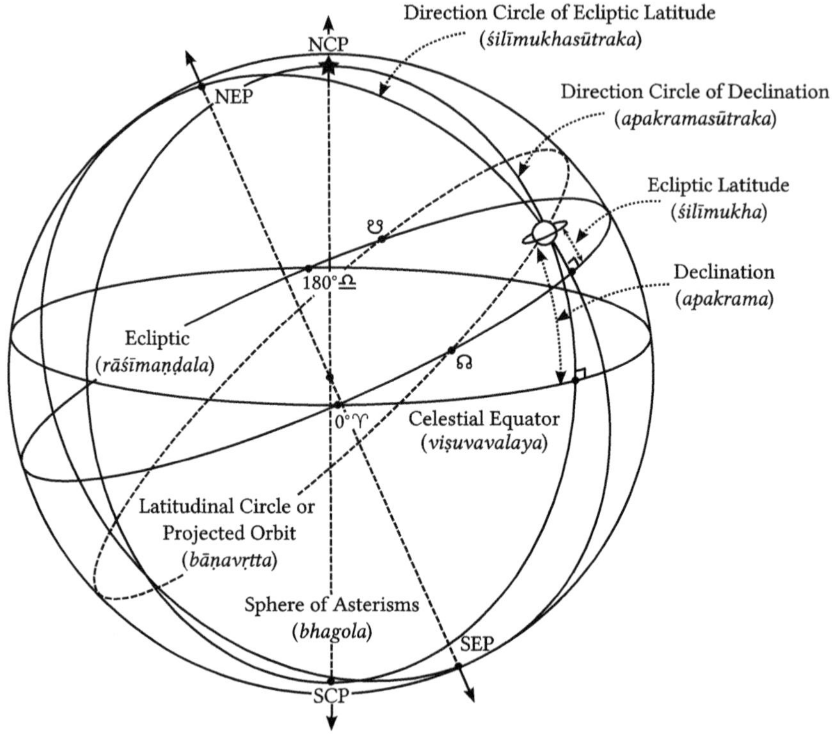

Figure 5.23 The direction circles of ecliptic latitude and declination corresponding to a planet's position on its latitudinal circle in the sphere of asterisms.

latitude is measured, just as *apakramasūtraka* is the direction for measuring the proper declination. The directional secondary circles are not commonly named in Siddhāntic literature; however, some secondary studies on Sanskrit astronomy do mention the direction circle of declination as *dhruvasūtra*, lit. fixed string, or *dhruvaprotavṛtta*, lit. fixed string circle (see, e.g., Somayaji 1971, Appendix B on pp. v, vii). The practice of naming secondary circles is commonplace in most Islamicate texts; for example, Maḥmūd al-Jaghmīnī describes the declination circle (*dā*ᵓ*irāt al-mayl*) and the latitude circle (*dā*ᵓ*irāt al-*ᶜ*arḍ*) in his *al-Mulakhkhaṣ fī al-hay*ᵓ*a al-basīṭa* (ca. thirteenth century, I.3.10–11) (Ragep 2016, pp. 110–113).

Verses 80–81

Nityānanda states that according to the tradition of the foreigners (*yavana-āgama*), there are 48 constellations with 1022 stars in them. (The numbers 48 and 1022 are expressed as the *bhūtasaṁkhyā* numerals *gaja-veda* and *sahasraṃ nayana-akṣi-yukta*.) Of these 48 constellations, 12 constellations constitute the ecliptic (*bhacakra*), i.e., the 12 signs (*rāśis*) of the zodiac, 21 constellations are north of the ecliptic, while the remaining 15 constellations are to the south of the ecliptic.

This particular distribution of the 48 constellations relative to the zodiac is identical to what is found in Ptolemy's catalogue in his *Almagest* (Peters and Knobel 1915, Table 1 on p. 16). Ptolemy's catalogue lists 1028 objects that constitute these 48 constellations (see, e.g., Peters and Knobel 1915, pp. 27–50); however, among them, three stars are shared between constellations and hence repeated twice and three other stars are in fact double clusters. This reduces the total number to 1022 stars. Over the course of the history of the transmission of Ptolemy's star catalogue, Islamicate scholars extensively discussed these numbers.[15] As Ptolemy's inventory of stars passed through the revisions of Islamicate astronomers, various emended version were successively produced (see, e.g., Kunitzsch 2008). Of particular note is the version by ᶜAbd al-Raḥmān al-Ṣūfī (903–986) in his *Ṣuwar al-Kawākib al-Thamāniyah wa al-Ārbaᶜīn* 'The 48 constellations', also known as *Kitāb al-Kawākib al-Thābitah* 'The Book of the Fixed Stars' (ca. 964). al-Ṣūfī's work included 48 constellations and 1016 stars. The 21 northern constellations included a total of 359 stars, the 12 zodiacal constellations included a total of 349 stars, and 15 southern constellations included a total of 308 stars. These totals included the stars that constituted the figures of the constellations as well as those that were outside them. The difference of six stars between Ptolemy's list of 1022 stars and al-Ṣūfī's list of 1016 stars is due to al-Ṣūfī excluding the six additional stars Ptolemy included in his count for the last southern constellation Piscis Austrinus (Hafez 2010, pp. 249–253).

al-Ṣūfī's work formed the basis of the star catalogue in Ulugh Beg's *Sūlṭānī Zīj* (epoch 1437) that used a Persian translation of al-Ṣūfī's text prepared in Marāgha by Naṣīr al-Dīn al-Ṭūsī around 1250. Ulugh Begh's *Sūlṭānī Zīj* describes 1018 stars (Knobel 1917), with some sources counting 1022 stars. Nityānanda's *Siddhāntasindhu* (ca. early 1630), a translation of Mullā Farīd's *Zīj-i Shāh Jahānī* (itself a rendition of Ulugh Beg's *Sūlṭānī Zīj*), specifies a list of 1018 fixed stars along with their polar coordinates (*dhruvaka* and *śara*), their northern or southern orientation (*diś*) with respect to ecliptic plane, and their measures of brightness (called *kadara* in Sanskrit, from the Arabic *qadr*). These descriptions follow the opinions of Ptolemy (*bhatallamūjūsa-mata*) and ᶜAbd al-Raḥmān al-Ṣūfī (*iva-suphī-mata*) who modified Ptolemy's list of stars to make them more relevant to Samarqand's time and place.[16] It is perhaps likely that Nityānanda's statement of 1022 stars derives from one of the copies of Ulugh Beg's *zīj* in circulation during his time.

Notably though, most Siddhāntic texts only list twenty-nine junction stars (*yogatāras*) that divide the ecliptic based on the daily transit of the Moon in a synodic month. These stars, sometimes referred to as constellations (*nakṣatras*) themselves, are mentioned in the *Paitāmaha-siddhānta* (III.30) summarised by Varāhamihira in his *Pañcasiddhāntikā*.[17] Nityānanda describes eighty-nine stars (along with their associated parameters) in his section on the conjunction of the planets with the stars (*saṃkrantyādi* or *bhagrahayuti*) in the chapter on computations (*gaṇitādhyāya*) of his *Sarvasiddhāntarāja* (I.8.5–56). This compilation of eighty-nine stars is not found in any other Siddhāntic text currently known to us. However, the exact relationship between the descriptions of the eighty-nine stars from Nityānanda's *Sarvasiddhāntarāja* and the 1018 (or 1022?) stars in Ulugh Beg's *Sūlṭānī Zīj*, Mullā Farīd's *Zīj-i Shāh Jahānī*, or even Nityānanda's *Siddhāntasindhu* remains to be investigated.

Verse 82

The circles of day-radius of a planet (*dyujyāmaṇḍala*) are parallel small circles to the plane of the celestial equator that trace the daily westward movement of the planet. In verse 73, Nityānanda described the diurnal circles (*dyuvṛttas*) as parallel circles that remain loosely attached to the plane of the celestial equator and extend northwards and southwards with their appropriate elevation and depression. Moreover, as verse 77 indicated, these diurnal circles are, in fact, spirals of motion (*śaṅkhāvarta*) that slowly regress eastwards tracing what is known as the latitudinal circle (*bāṇavṛtta*)—the eastward orbit of the planet project onto the outermost sphere of asterisms. The circles of day-radius resemble the diurnal circles; however, they are not spirals; they are fixed circles like the parallels of latitude on Earth. On any particular day, the circle of day-radius of a planet is a measure of the elevation of the planet above the celestial equator—in other words, a circle of constant declination (*apakrama*) of the planet for that day—see Figure 5.24. Every day, the circle of day-radius of a planet intersects its latitudinal circle (*bāṇavṛtta*) at a different point, with that point of intersection slowly regressing eastwards.

An alternative way to slice up the latitudinal circle (*bāṇavṛtta*) of a planet is to construct parallel small circles to the plane of the ecliptic (*rāśimaṇḍala*) going all around the ecliptic pole (*kadamba*). These parallels of ecliptic latitude (*śarāgramaṇḍala*, lit. circle [traced by] the tip of the ecliptic latitude), like the circles of day-radius, are also parallel small circles that intersect the latitudinal circle of the planet at a different point daily. Each parallel of latitude indicates a small circle of fixed ecliptic latitude (*śilīmukha* or *śara*) in much the same way as each circle of day-radius is a function of the fixed declination. Figure 5.24 depicts the parallels of ecliptic latitude and the circles of day-radius of a planet intersecting its latitudinal circle.

Verse 83

Nityānanda declares all the aforementioned circles (*vṛttas* or *maṇḍalas*) on the sphere (*gola*) as imaginary (*kalpita*). These circles represent the

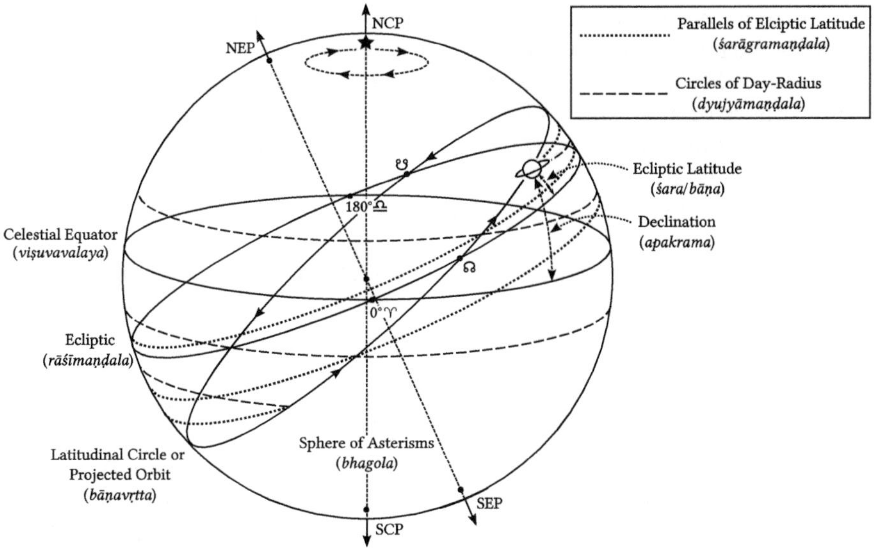

Figure 5.24 The circles of day-radius and parallels of ecliptic latitude (corresponding to constant declinations and the ecliptic latitudes respectively) of a planet intersecting its latitudinal circle on the sphere of asterisms.

movement of celestial objects in the sphere of asterisms, and, as such, are mere cognitive tools to gain a geometrical perspective. He compares these circles to the whirling circles of firebrand (*ālāta-cakra-bhrami*) made by children in the darkness (*andhakāra*) as play. The purpose of his analogy is to remind the reader that these circles are mere illusions created by prolonged perception. Unlike the celestial objects (stars, planets, constellations, etc.) which are real, these imaginary circles simply illuminate the science behind the computations (*gaṇita*) by offering the light of geometrical reasoning.

The analogy of firebrand circles (*ālāta-cakras*) as fictitious impressions of persistent objects created solely by the fallacy of perception is a figurative device employed commonly in Buddhist and Advaita Vedāntic texts (see, e.g., R. King 1995, pp. 117–179). Being a Gauḍa Brahmin himself, it is not surprising to find Nityānanda use this Vedāntic analogy in explaining illusory representations.

Verse 84

The orbit of a planet (*dyusad-kakṣa*), represented by its orbital circle (*vikṣepavṛtta*) as seen in Figure 5.22, is considered fastened in the sphere of asterisms. The geocentric orbits of the different star-planets are progressively larger, starting from the innermost orbit of the Moon and extending outwards to the orbit of Saturn. Beyond that lies the outermost boundary, namely, the circle of zodiacal signs (*rāśis*), that forms the backdrop against which all celestial motions are measured. In verse 77, Nityānanda has described these planetary orbits projected onto the sphere of asterisms as latitudinal circles (*bāṇavṛttas*). These projected orbits intersect the ecliptic at their respective nodes and at different ecliptic inclinations. Both here and in verse 77, the order of the planets is suggested as beginning from the Moon (*vidhu*) (without explicitly stating the order).

Varāhamihira, in *Pañcasiddhāntikā* (V.13.40–41), describes the orbits of the star-planet as having increasing larger linear extension of the *rāśis*; in other words, the complete revolutions (measure with respect to the background zodiac) of the star-planets that are closer to the central Earth-sphere are faster than the revolutions of those star-planets that are further afar. This description is akin to viewing the spokes on an oil-press wheel that are closer towards the central navel and more widely spread towards the rim of the wheel. As the wheel spins, the revolution period of the part of the spokes closer to the navel appears faster than the parts closer to the rim. In other words, the faster the revolution period, the shorter the orbit of the celestial object.

MS Rr (f. 35v) of Nityānanda's *Sarvasiddhāntarāja* (*gaṇitādhyāya*) includes a drawing depicting the geocentric orbits of the star-planets (up until the outermost zodiac) stacked one above the other with their circumferences in *yojanas* written within. Figure 5.25 is a digital rendering of the original drawing from the manuscript. In the text surrounding the image in MS Rr, there are five verses (vv. 12–16) from the section on lunar eclipse (*candragrahaṇa*) that lists the circumferences (in *yojanas*) of the planetary orbits seen in the image, along with the diameters (*vistṛtis*) of the planet's spheres (*bimbas*) in *yojanas* and minutes of arc (*kalā*). The measures of the orbital circumferences, as *yavanair niruktā* 'declared by the *yavanas*', are enumerated in *bhūtasaṁkhyā* numerals as follows: the Earth (*bhūkakṣa*) 6000 *yojanas*; the Moon (*candrakakṣa*) 352,500 *yojanas*; Mercury (*budhakakṣa*) 666,109 *yojanas*; Venus (*śukrakakṣa*) 3,567,000 *yojanas*; the Sun (*sūryakakṣa*) 8,284,252 *yojanas*; Mars (*bhaumakakṣa*) 36,570,000 *yojanas*; Jupiter (*gurukakṣa*) 85,280,700 *yojanas*; Saturn (*śani-kakṣa*) 128,035,500 *yojanas*; and the farthest orbit of the constellations (*nakṣatrakakṣa*) 150,086,000 *yojanas*.

Beginning the section on the oblique celestial sphere

The third section in Nityānanda's *golādhyāya* is on the oblique celestial sphere (*khagola*). The *khagola* is the oblique celestial sphere that forms an extension of the Earth-sphere similar to the *bhagola*; however, the north and south celestial poles (*dhruva-dvaya*) are now inclined to the local zenith by an amount equal to the co-latitude (*lambaka*). Accordingly,

Figure 5.25 A digital reproduction of the original drawing of concentric orbital spheres of the star-planets, beginning from the central Earth-sphere and extending outwards towards the outermost orbit of the constellations, seen in MS Rr RORI, Alwar 2619, f. 35v.

the celestial equator is also inclined to the celestial horizon by an equal measure. When the co-latitude is zero, i.e., at the north or south pole of the Earth, the *bhagola* (right celestial sphere) and *khagola* (oblique celestial sphere) are identical. See Section 1.3.4.

Verse 85

Nityānanda describes the orientation of a geocentric observer (*kugarbhasthita nṛ*) as being radial—in other words, extending from Earth's centre (*bhūgarbha*) to the Earth's surface (*kṣitipṛṣṭha*). It is in this

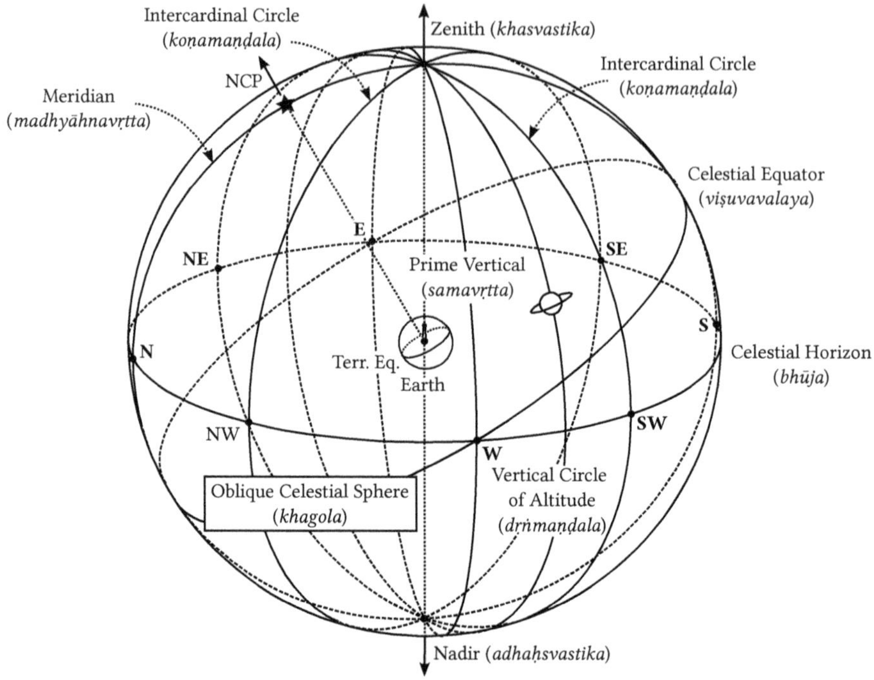

Figure 5.26 The oblique celestial sphere with its zenith, nadir, and celestial horizon, as well as the prime vertical, the meridian, the vertical circle of altitude, and the intercardinal circles.

radial direction, isotropically (*sarvatas*), the geocentric oblique celestial sphere (*khagola*) extends towards the edge of the visible heavens. As Figure 5.26 shows, the oblique celestial sphere is marked by its equatorial circle (*vṛtta*) called the celestial horizon (*bhūja*), and its two fixed poles (*dhruva-dvaya*) called the zenith (*khasvastika*) and the nadir (*adhaḥsvastika*). Nityānanda describes this sphere as eternal (*ananta*) in that it never ceases to exist.[18]

Verses 86–87

Nityānanda describes the following the great circles (*vṛtta*s or *maṇḍala*s) on the oblique celestial sphere, all of which intersect at the local zenith (*khasvastika*); see Figure 5.26.

1 The prime vertical (*samavṛtta*) passing though the due east and due west points on the celestial horizon, and traversing the local zenith.

2 The meridian (*madhyāhnavṛtta*) passing through the due north and due south points on the celestial horizon, and traversing the local zenith.

3 The vertical circle of altitude (*dṛṅmaṇḍala*) passing through the direction of a celestial object (along the celestial horizon), its reversed or antipodal direction, and traversing the local zenith.[19]

4 The intercardinal circle (*koṇamaṇḍala*), a special case of the vertical circle of altitude, passing through antipodal intercardinal directions on the celestial horizon, and traversing the local zenith.

All these great circles are perpendicular to the celestial horizon, with the prime vertical and the meridian being mutually orthogonal to one another. The vertical circle of altitude passes through the body of a celestial object, and, as its name implies, it measures the altitude or elevation of the object from the plane of celestial horizon (in other words, the co-altitude or depression of the object from the local zenith). If the celestial object is positioned along an intercardinal directions, i.e., if its vertical circle of altitude intersects the celestial horizon at a due intercardinal point, the circle is called an intercardinal circle.

Nityānanda's construction of the compound words to define these great circles is noteworthy. In every instance, the antipodal locations of a circle are separated by their central position. In other words, for two antipodal points A and B of a circle, with point C being the central point along its circumference at a distance of 90° from A to B, Nityānanda's compound words arrange these points in the sequence A–C–B or A–B–C with an appropriate terminal endings. For example, the prime vertical (*samavṛtta*) is described by the compound *prāk-khasvastika-paścima-upari* 'east-zenith-west-over'. This is equivalent to the modern-day (geometrical) identification of circles with three points along its circumference.

Verse 88

From the perspective of the oblique celestial sphere, the celestial equator (*viṣuvavalaya*) is inclined to the celestial horizon (*bhujā*) by an amount equal to the co-latitude (*lamba*) of a geocentric observer. Moreover, the ecliptic (*rāśimaṇḍala*) is itself inclined to the celestial equator by an amount that equals the obliquity of the ecliptic, and hence, it is also inclined to the celestial horizon as Figure 5.27 shows. In verse 88, Nityānanda defines two great circles in the oblique celestial sphere in relation to the ecliptic and celestial poles:

1 the secondary circle to the ecliptic (*dṛkkṣepakamaṇḍala*) passing through the zenith (*khasvastika*) and the ecliptic pole (*kadamba*); and

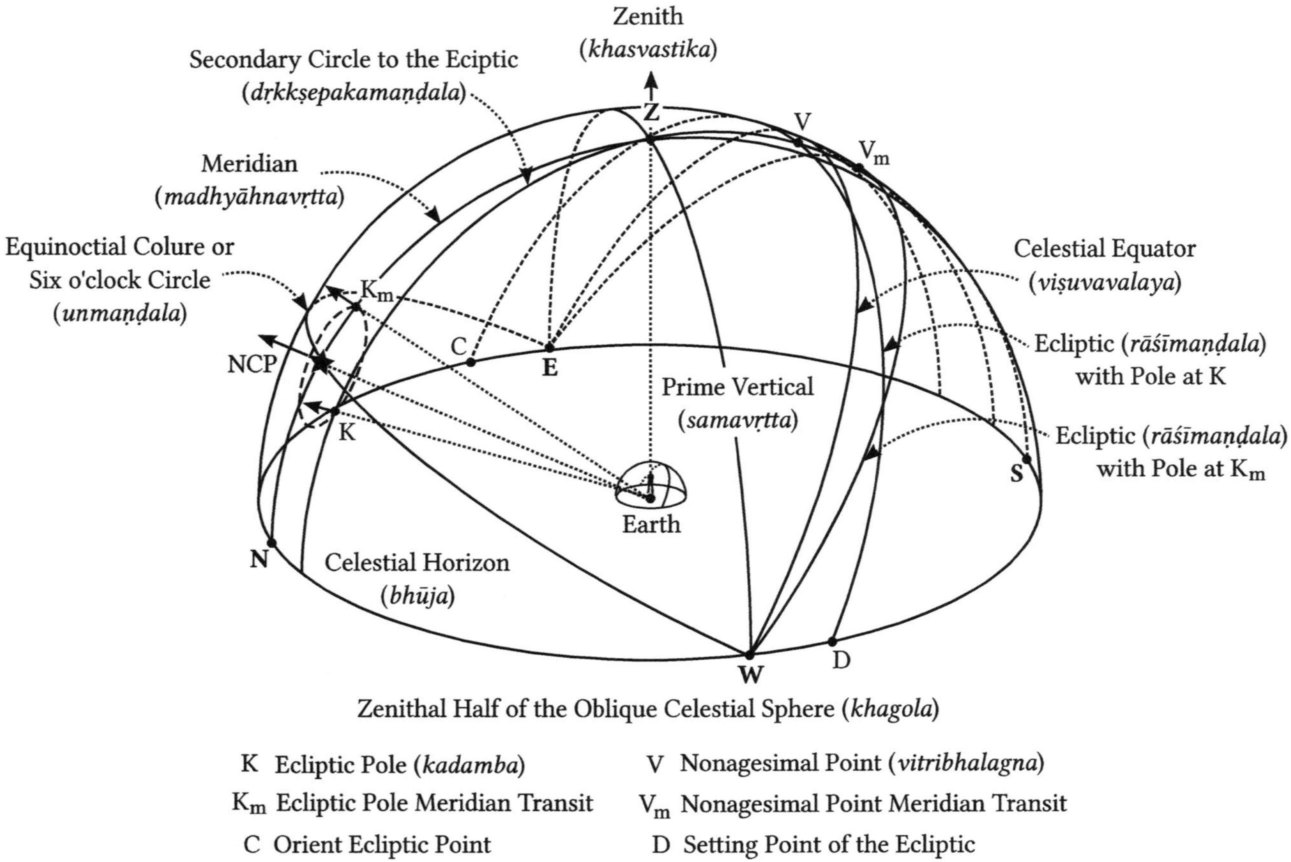

Figure 5.27 The zenithal half of the oblique celestial sphere with the secondary circle to the ecliptic and the equinoctial colure, also called the six o'clock hour circle.

2 the equinoctial colure or six o'clock hour circle (*unmaṇḍala*) passing through the due east and due west point on the celestial horizon, and traversing the celestial pole.

The secondary circle to the ecliptic intersects the ecliptic at a point called the nonagesimal point (*vitribhalagna*), sometimes called the central ecliptic point. As Figure 5.27 shows, the nonagesimal point is a point on the ecliptic exactly 90° from the ecliptic point ascending on the eastern horizon (i.e., 90° from the orient ecliptic point). The daily rotation of the ecliptic pole (*kadamba*) around the celestial pole (*dhruvaka*) implies that the secondary circle to the ecliptic (along with the ecliptic itself) is not fixed but changes orientation throughout the day. When the ecliptic pole is on the meridian, at that instant, the nonagesimal point lies on the meridian (i.e., the nonagesimal point is coincident with the meridian ecliptic point).

The secondary circle to the ecliptic and the nonagesimal point help determine the parallaxes in longitude (*lambana*) and latitude (*nati*) of a planet in relation to the ecliptic. The equinoctial colure or six o'clock hour circle is important to determine the ascensional difference (*cara*) in the rising of celestial objects at different terrestrial latitudes. The pole of the equinoctial colure is the meridian ecliptic point. The intersection of the diurnal circle of a celestial body with this six o'clock hour circle helps determine its visibility in the sky (or day-length); the eastern and western points of intercept of its diurnal path with the six o'clock hour circle correspond to the object's six o'clock rising (in the morning) and six o'clock setting (in the evening) respectively. The meridian transit of the object marks its midday position.

At the Earth's equator, the polestar lies along the horizon (to the north), and hence, the six o'clock hour circle is coincident with the horizon; in other words, the six o'clock hour circle is simply the equatorial horizon at the terrestrial equator. At the Earth's poles, the polestar is directly overhead, and hence, the six o'clock hour circle passes through the due east and due west points on the horizon and the zenith; in other words, the six o'clock hour circle is the equinoctial colure or prime vertical at the terrestrial poles.

Verses 89–91

The circles of day-radius (*dyujyāvṛttas*) in the oblique celestial sphere (*khagola*) resemble the circles of day-radius (*dyujyāmaṇḍalas*) in the right celestial sphere (*bhagola*). Nityānanda described the circles of day-radius in the *bhagola* in verse 82 as parallel small circles above and below the plane of the celestial equator. These parallel circles of day-radius of a planet intersect its latitudinal circle (*bāṇavṛtta*) at a different point everyday, and that point of intersection slowly regresses eastwards every day;

see notes to verse 82. In the *khagola*, the circles of day-radius of a planet are simply inclined to the plane of the celestial horizon by an amount equal to the constant declination (*apakrama*) of the planet (on the day) plus the local co-latitude (*lambaka*) of the place; see Figure 5.28 (right). By treating the declination of the planet on a particular day as constant, the circle of day-radius (in the *bhagola* or the *khagola* perspectives) can be regarded as stationary (*sthira*). This contrasts the diurnal circles (*dyu-vṛttas*) described in verse 73 as loosely attached (*ślatha*) spirals to the plane of the celestial equator.

In addition to this circle of day-radius, Nityānanda defines two great circles and two small circles in the *khagola* in verses 89–91 that relate to the (instantaneous) position of a planet. Figure 5.28 (left) depicts the two great circles, the secondary circles to the meridian and the prime vertical, that pass through the body of a planet.

1 The *samasūtra* (also called *samaprota*, lit. fixed in parallel or equally) is a great circle that passes through the body of the planet and meets the celestial horizon at its due north and due south points. In other words, the *samasūtra* is the secondary circle to the prime vertical (*samavṛtta*). Just as the direction circles of ecliptic latitude (*bāṇasūtra*) described in verse 75 divide the ecliptic into zodiacal signs, the *samasūtras* also divide the prime vertical into different parts. The division of the ecliptic or the prime vertical (both into twelve parts) helps determine the twelve astrological houses according to different systems of domification (more on this in verses 95–97).

2 The *madhyasūtra* is a great circle that passes through the body of the planet and meets the celestial horizon at its due east and due west points. This makes the *madhyasūtra* the secondary circle to the meridian (*madhyāhnavṛtta*), also known as the central meridian of a planet.

Figure 5.28 (right) also depicts two small circles, the parallels of altitude and meridian, passing through the body of the celestial object (like a star-planet).

1 The *madhyavṛtta*, lit. middle circle, is a small circle that passes through the body of the celestial object, parallel to the plane of the meridian, and is centred on the due east or due west points on the celestial horizon. It is called the parallel of meridian and it resembles the circle of day-radius in the *khagola* (which is parallel to the plane of the celestial equator and centred on north celestial pole). The circle of day-radius is a small circle of constant declination of the celestial object on a particular day, i.e., a circle of radius Cos δ where δ is the declination of the celestial object on that day

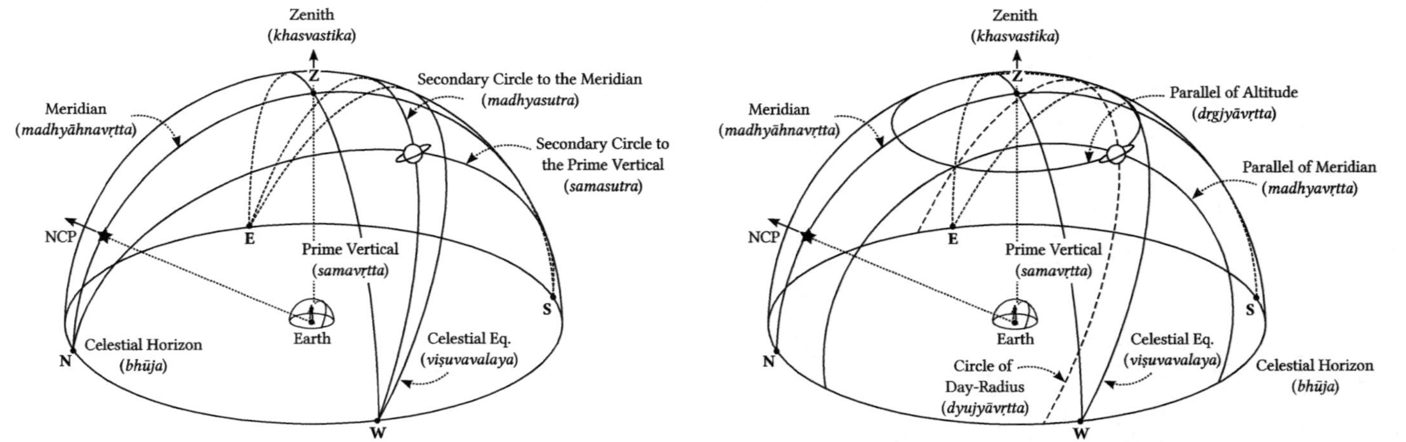

Great Circles in the Zenithal Half of the Oblique Celestial Sphere (*khagola*)

Small Circles in the Zenithal Half of the Oblique Celestial Sphere (*khagola*)

Figure 5.28 The zenithal half of the oblique celestial sphere with the secondary circles to the meridian and the prime vertical (great circles, left), and the parallels of altitude and meridian (small circles, right), passing through a planet.

(taken as a constant). Similarly, the parallel of meridian is a small circle corresponding to the zenith distance of the celestial object at a given instant of the day, i.e., a circle of radius Cos z where z is the zenith distance of the celestial object at a particular time of the day.

The parallel of meridian is a non-standard circle in Sanskrit Siddhāntic astronomy. Nityānanda's description of this object (in verse 90) appears to be the first instant where such an object is defined in a Siddhāntic text. Moreover, he does not directly refer to this circle as the *madhyavṛtta*; instead, he identifies it by the name (*abhidhā*) that was previously (*pūrva*) used to describe the secondary circle to the meridian (*madhyasūtra*) in the first two *pāda*s of verse 90 immediately preceding the description of the parallel of meridian.[20]

2 The *dṛgjyāvṛtta*, lit. circle of the Sine of the zenith distance, is a small circle that passes through the body of the celestial object, parallel to the plane of the celestial horizon, and is centred on the zenith. In essence, these parallels of altitude indicate the altitude (or complement of zenith distance) of the celestial object, and hence they are also called the circles of zenith distance.[21] In other words, for a celestial object at a distance of z from the local zenith, the parallel of altitude is a circle of radius Sin z. The position of the parallels of altitude circle above (or below) the celestial horizon indicates the vertical elevation (or depression) of the celestial body.

Verse 92

According to Nityānanda, the ecliptic circle is divided into twelve arcs (*bhūja*s, understood as *bāhu*s 'sides') by the twelve semicircles of the six secondary circles to the prime vertical (*samasūtra*); six arcs above and six arcs below the plane of the celestial horizon (*kṣitija*). These twelve arcs of the ecliptic form the twelve astrological houses (*bhāva*s) whose positions remain fixed with respect to the celestial horizon.[22] The first house, identified as *dhanādya*, lit. immediately preceding the second house of wealth, should be understood as the one in which the Ascendant (*lagna*) is located. Strictly speaking, the division of the ecliptic by the secondary circles to the prime vertical creates unequal houses, i.e., houses of unequal arc-measures, as long as poles of the ecliptic and prime vertical are not coincident. Figure 5.29 depicts this division of the ecliptic into unequal houses by the secondary circles to the prime vertical.

In the practice of genethlialogy or natal astrology, the twelve houses signify various aspects of human life and its concerns. Based on one's

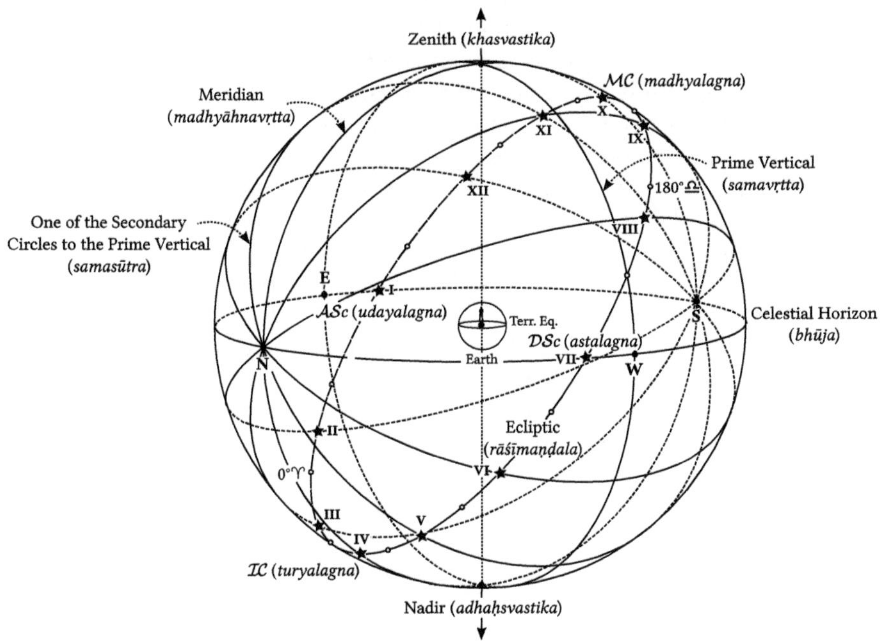

Figure 5.29 The oblique celestial sphere depicting the twelve (unequal) astrological houses (indicated with roman numerals along the ecliptic) created by twelve semicircular arcs of the secondary circles to the prime vertical. The ecliptic is also marked at its Ascendant ($\mathcal{A}Sc$), the Medium Coeli ($\mathcal{M}C$), the Descendant ($\mathcal{D}Sc$), and the Imum Coeli ($\mathcal{I}C$) positions. In the equal house division of the zodiac, the asterisms (part of individual zodiacal signs) mark the beginnings (cusps) of the houses. These are indicated in the figure above as solid stars, whereas the beginnings (zero degrees) of the zodiacal signs themselves are indicated with circular dots.

natal chart, the identification of a particular zodiacal sign and a particular house affects these aspects in several ways. The *praśnādhyāya* in the *Yavanajātaka* of Sphujidhvaja (ca. 269/270) categorises the astrological houses (*bhāva*s) as

1 cardinal or angular houses (*kendra*s, from the Greek *κέντρα*) numbered first, fourth, seventh, and tenth houses[23];

2 succedent houses (*pāṇaphara*s, from the Greek *ἐπαναφοραι*) numbered second, fifth, eight, and eleventh; and

3 cadent houses (*āpoklima*s, from the Greek *ἀποκλίματα*) numbered third, sixth, ninth, and twelfth (Pingree 1981b, p. 82).

Varāhamihira, in his *Bṛhajjātaka* (I.15), lists the significance of the twelve astrological houses in the circle of houses (*bhāvacakra*). According to him, these houses, beginning from the house of Ascendant, represent body (*tanu*), family (*kutuṃba*), brothers (*sahottha*), relations (*bandhu*), sons (*putra*), enemies (*ari*), wife (*patni*), death (*maraṇi*), deeds of virtue (*śubha*), avocation (*āspada*), gain (*aya*), and loss (*riḥpha*, from the Greek ῥιφή) (Aiyar 1905, p. 12). However, according to later texts, e.g., the popular *Bṛhatpārāśarahorā*, ascribed to Parāśara, the ordering of the twelve house is slightly different. The *Bṛhatpārāśarahorā* (VII.37) lists them as body (*tanu*), wealth (*dhana*), brothers (*sahaja*), relatives (*bandhu*), sons (*putra*), enemy (*ari*), young maiden (*yuvatī*), imperfection (*randhra*), virtue (*dharma*), livelihood (*karma*), gain (*lābha*), and loss (*vyaya*) (Santhanam 1984–88, p. 102).

Nityānanda does not list the names of the houses in this verse; he simply refers to the first house as the one preceding the (second) house of wealth (*dhana*). The names and the order of astrological houses would have been common knowledge among natal astrologers (especially among court astrologers) and therefore it is reasonable to assume Nityānanda left them out in favour of brevity.

Verses 93–94

In the Indian *bhāva*-system of natal astrology, the ordering of the houses (*bhāva*s) is determined by the position of the orient ecliptic point or Ascendant (*udayalagna*). The Ascendant is the degrees of a zodiacal sign (*bha-aṃśa*, lit. degree/part of an asterism), beginning with 0° Aries (*Meṣādi* or 0° ♈), rising on the eastern horizon at a particular time and place. Concurrently, the degrees of the zodiacal sign setting on the western horizon marks the occident (setting) ecliptic point or Descendant (*astalagna* or *astatanu*); the degree of the zodiacal sign transiting the local meridian (upper culmination towards the zenith) marks the meridian ecliptic point or Medium Coeli (*madhyalagna*); and the degrees of the zodiacal sign transiting the local meridian (lower culmination towards the nadir) marks the fourth ecliptic point or Imum Coeli (*turyalagna*). The houses are numbered anticlockwise; however, the beginning (cusp) of a house, and also its length, is dependant on the system of domification chosen (see notes to verses 95–97). Figure 5.29 depicts the arrangement of (unequal) houses where the cusp of the first house begins with the Ascendant, the fourth house cusp being the Imum Coeli, the seventh house cusp being the Descendant, and the tenth house cusp being the Medium Coeli. References to these ecliptic positions can also be found in other Siddhāntic texts, e.g., Lalla's *Śiṣyadhīvṛddhidatantra* (XV.16, 32), Vaṭeśvara's *Vaṭeśvarasiddhānta* (II.3.16), or Bhāskara II's *Siddhāntaśiromaṇi* (II.7.26).

The remaining eight intermediate houses are arranged in four pairs between these four angular/cardinal houses (*kendras*). As Figure 5.29 shows, each pair is placed between successive pairs of cardinal houses in the following order:

1 The house cusps of the second and third houses are between the Ascendant (first house cusp) and the Imum Coeli (fourth house cusp);

2 the house cusps of the fifth and sixth houses are between the Imum Coeli (fourth house cusp) and the Descendant (seventh house cusp);

3 the house cusps of the eighth and ninth houses are between the Descendant (seventh house cusp) and the Medium Coeli (tenth house cusp); and

4 the house cusps of the eleventh and twelfth houses are between the Medium Coeli (tenth house cusp) and the Ascendant (first house cusp).[24]

Typically, the lengths of the houses in the Indian system of house divisions of the zodiac (*rāśimaṇḍala*) are all equal (measuring 30° each) with a whole sign constituting a house. The first house is simply the entire zodiacal sign of the Ascendant, beginning at zero degrees of the rising sign. Each subsequent house is then simply the next zodiacal sign in its entirety. Alternatively, in the equal house system, as depicted in Figure 5.29, the houses are also 30° in longitudinal length; however, they are measured beginning with the exact degrees of the rising sign (i.e., the longitude of the Ascendant) acting as the cusp of the first house. The subsequent houses then advance by 30° into succeeding signs accordingly. See endnote 25.

Nityānanda's statement in verse 93 on the first house being the Ascendant sign, beginning with Aries (*kriya*) (i.e., the *Meṣādi* or 0° ♈), alludes to the house division of the zodiac following the system of whole signs. Indian astrology employs the sidereal (*nirāyana*) zodiac where the vernal equinoctial point (*mahāviṣuva*) is identified with 0° Aries (*Meṣādi*), i.e., a zodiac without any precession (*niś-ayana*). This implies that on the day of the vernal equinox, at sunrise, the degree of the Ascendant is exactly 0° Aries marking the beginning of the first house in both the whole sign and equal house systems of diving the zodiac.

Verses 95–97

The division of the sphere of the heavens into twelve sectors—in other words, the demarcation of the zodiacal circle into twelve astrological houses—is called domification. Historically, the domification of the oblique celestial sphere was done with respect to one of three great circles (and its associated poles). Al-Bīrūnī, in his *al-Qānūn al-Masᶜūdī* (II.1), discusses, at length, three methods of domification. (See North 1986, pp. 32–33 and pp. 46–47.)

1 **The well-known (*mashhūr* method, based on the division of the celestial equator**

First, the celestial equator is divided into twelve equal arcs (of 30° each) with respect to the celestial poles: six arcs above the equinoctial colure (or six o'clock hour circle) and six below. Then, six projection (great) circles are drawn passing through the due north and due south points of the celestial horizon, and six antipodal pairs of points of the celestial equator marking by its twelve equal arcs. The intersections of these projection circles with the ecliptic demarcate the cusps of the houses of unequal lengths. This is depicted in Figure 5.30 to the left.

In this method of domification, the angular or cardinal house cusps correspond to the longitudes of the Ascendant (first house cusp), the Imum Coeli (fourth house cusp), the Descendant (seventh house cusp), and the Medium Coeli (tenth house cusp). The inclination of the celestial equator to the celestial horizon results in opposing cardinal sectors being equal even as all four quadrants are not equal. The cusps of the intercardinal houses correspond to trisected arcs in each of the four cardinal sectors of the ecliptic, with opposing arcs having equal (sidereal) rising times.

2 **The preferred method of al-Bīrūnī, based on the division of the prime vertical**

Here, the prime vertical is divided into twelve equal arcs (of 30° each) by the six secondary circles to the prime vertical passing through the due north and due south points of the celestial horizon (i.e., the poles of the prime vertical). Each quadrant of the prime vertical between the celestial horizon and the local meridian is trisected to form three equal arcs. The parts of the ecliptic corresponding to each of these twelve arcs of the prime vertical then form the twelve astrological houses of unequal lengths. This is depicted in Figure 5.30 in the middle.

Like the previous method, the cardinal house cusps are defined by the celestial horizon and the meridian, while the intercardinal houses are parts of the ecliptic between equally trisected arcs of the four cardinal sectors of the prime vertical. In both these methods of domification, the cusps of the houses are fixed for an observer at a particular latitude, and, hence, easily inscribed onto the plate of an astrolabe.

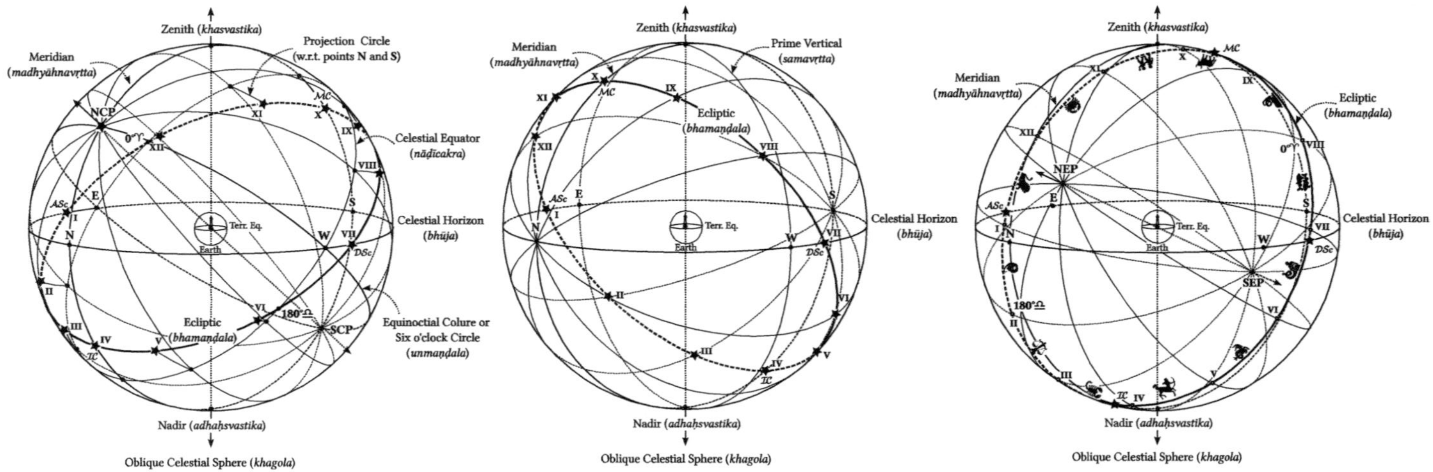

Figure 5.30 The domification of the oblique celestial sphere with respect to the celestial equator (left), prime vertical (figure in the middle), and the ecliptic (right). The house cusps in the methods of domification based on dividing the celestial equator and the prime vertical are the exact asterisms (indicated by solid star) of zodiacal signs, whereas the house cusps in the method based on the division of the ecliptic begin with the zero degrees of the individual zodiacal signs (indicated by circular dots).

3 **The method of the ancestors (*awāʾil*), based on the divisions of the ecliptic**

In this case, the ecliptic is simply divided into twelve equal arcs by twelve secondary circles to the ecliptic, each arc corresponding to the length of a complete zodiacal sign (i.e., 30 degrees). The ordering of the houses begins with the zero degree of the ascending zodiacal sign, regardless of the actual degrees of the Ascendant in that sign. And hence, all houses are of equal length. This is depicted in Figure 5.30 to the right.

In verses 95–97, Nityānanda describes the three-fold (*trividha*) nature of the circle of astrological houses (*bhāvacakras*)—in other words, the three systems of domification. The cardinal sectors—the arcs of great circles between the celestial horizon (*bhūja*) and the local meridian (*madhyāhna*)—corresponding to the celestial equator (*nāḍīcakra*), the prime vertical (*samavṛtta*), and the ecliptic (*bhamaṇḍala*) form four semihemispherical parts of the oblique celestial sphere. In each of these parts, the individual cardinal sectors (*aṅghrigata*, 'quadrants') are trisected (*trisamāna-lava*, lit. three equal parts), and these divisions, when projected onto the zodiacal circle, form the twelve fixed zones (*susthira-valaya*) called astrological houses (*bhāvas*).

The division of the oblique celestial sphere into spherical lunes by the secondary circles (to the celestial equator, the prime vertical, or the ecliptic) is different in each instance. In modern geometry, this division is a twelve-gonal hosohedron with two common antipodal vertices, and hence, the twelve spherical lunes (in any particular type of domification) are oriented symmetrically about the pole of the generating great circle. However, the projection of these symmetrical spherical lunes onto the ecliptic creates unequal (and asymmetrical) houses when the oblique celestial sphere is divided with respect to the celestial equator or the prime vertical. When the ecliptic is divided into twelve parts (complete zodiacal signs) with respect to its own pole, the house lengths remain equal; see Figure 5.30.

Typically, the degrees of a zodiacal sign (*rāśi-lava*) ascending on the eastern horizon—i.e., the longitude of the Ascendant; see verse 92—determines the cusp of the first house (*dravyādi*, lit. immediately preceding the second house of wealth) fixed below the horizon. Subsequently, all twelve houses occupy the four semihemispherical parts in sets of three (*kramataya*) going anticlockwise (*asavya*) from below the horizon till the twelfth house called *riṣpha* (identified with *riḥpha*, from the Greek *ῥιφή*). The twelve astrological houses are individually distinct (*vyasta*) and have fixed position with respect to the celestial horizon: six houses lie above the horizon (towards the zenith) and six of them lie beneath (towards the nadir). In the first two methods of domification, i.e., when the celestial equator or prime vertical is divided into twelve arcs and projected onto the ecliptic, the house cusps are longitude of the Ascendant. However, in the case of the division of the ecliptic into zodiacal signs, the house cusps are typically the zero degrees of the ascending sign, regardless of the actual degrees of the Ascendant in that sign.[25]

Verses 98–102

Nityānanda describes the oblique ascensions of the zodiacal signs and the sidereal day-lengths at the Arctic Circle, i.e., at a terrestrial latitude of 66° 30′ N in these five verses. The Arctic Circle is a place where the terrestrial co-latitude (*lamba*) equals the northernmost maximum declination (*udac-para-krānti*) of the Sun. The ecliptic obliquity (or maximum declination of the Sun) is stated to be 23° 30′ (see verse 76), and accordingly, the terrestrial location with an equivalent co-latitude corresponds to the place with a latitude of (90° − 23°30′) N or 66° 30′ N.

According to the sidereal (*nirāyana*) zodiac, the Sun is situated at the beginning of Cancer (*karka*) on the day of the summer solstice in the northern hemisphere. At the Artic Circle, at midnight of this solstitial day, the Sun rises due north on the celestial horizon just as the beginning of Capricorn (*nakra*) begins to sets due south. This is depicted in Figure 5.31 to the top-left. At that instant, the ecliptic (*bhacakra*), understood as the sidereal zodiac with no *ayanacalana* 'equinoctial displacement', is momentarily coincident with the celestial horizon (*bhūja*) with its vernal equinoctial point (*Meṣādi* or 0° ♈), summer solstitial point (*Karkādi* or 90° ♋), autumnal equinoctial point (*Tulādi* or 180° ♎), and winter solstitial point (*Mṛgādi* or 270° ♑) being coincident with the due east, due north, due west, and due south points of the celestial horizon respectively.

After being horizontal for an instant (during sunrise at midnight), the ecliptic rises in a manner that the six zodiacal signs beginning with Capricorn (*Mṛgādi* or 270° ♑) ascend almost parallel to the eastern celestial horizon, while the six zodiacal signs beginning with Cancer (*Karkādi* or 90° ♋) simultaneously descend along the western side.[26] Thereafter, each of the six signs beginning with Cancer rises along the eastern horizon just as the six signs beginning with Capricorn set to the west. In Figure 5.31,

1 the top-right drawing shows the rising of the sign Leo along the eastern horizon (to the north) just as Aquarius is setting along the western horizon (to the south) in the early hours of the morning (before six o'clock);

2 the bottom-right drawing shows the midday position of the ecliptic where the vernal equinoctial point (*Meṣādi* or 0° ♈), summer solstitial point (*Karkādi* or 90° ♋), autumnal equinoctial point (*Tulādi* or 180° ♎), and winter solstitial point (*Mṛgādi* or 270° ♑) are

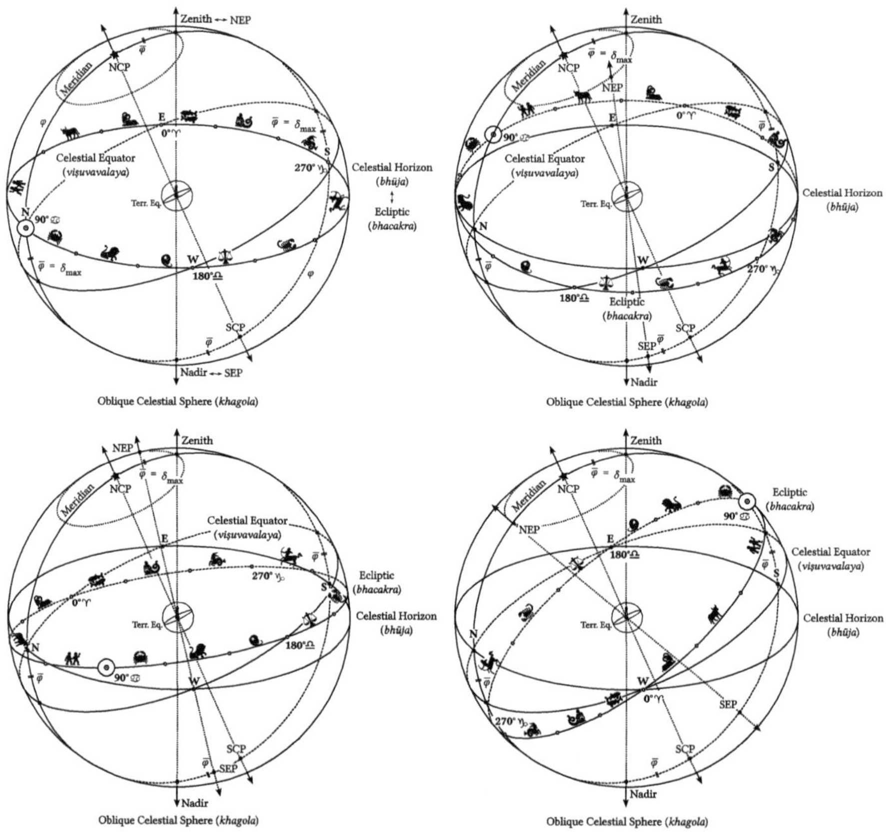

Figure 5.31 The positions of the Sun along the ecliptic (sidereal zodiac with no equinoctial displacement) in the oblique celestial sphere at the Arctic Circle (66° 30′ N) on the day of the summer solstice in the northern hemisphere. The top-left figure depicts the position of the ecliptic at sunrise at midnight; the top-right figure shows the simultaneous rising of six zodiacal signs in the early hours of the morning (before six o'clock); the bottom-right figure depicts the position of the ecliptic at the midday transit of the Sun; and the bottom-left figure shows the simultaneous rising of six zodiacal signs in the late hours of the evening (after six o'clock).

coincident with the due west, southern meridian transit (towards the zenith), due east, and the northern meridian transit (towards the nadir) respectively; and

3 the bottom-left drawing shows the rising of the sign Scorpio along the eastern horizon (to the south) just as Taurus is setting along

the western horizon (to the north) in the late hours of the evening (after six o'clock).

According to Nityānanda, the oblique ascensions of the zodiacal signs beginning with Cancer (*Karkādi* or 90° ♋) at the Arctic Circle are twice their right ascensions (*laṅkodaya*, lit. rising times at Laṅkā) at the terrestrial equator. Various Siddhāntic authors have noted the relation between the ascension times of the zodiacal signs at a terrestrial location and the angle the ecliptic makes to the horizon at that place; see, e.g., Lalla's *Śiṣyadhīvṛddhidatantra* (XVI.13–16), Śrīpati's *Siddhāntaśekhara* (XVI.50–53, 58), and Bhāskara II's *Siddhāntaśiromaṇi* (II.7.16–23). Nityānanda also makes a similar observation (in his discussions on the visible celestial sphere); see notes to verses 127 and 128.

Bhāskara II, in his *Siddhāntaśiromaṇi* (II.7.19), states that each quarter of the ecliptic rises in exactly 15 *ghaṭīkā*s or 6^h. Table 5.3 lists the right ascensions of the zodiacal signs (equalling six hours for each quarter of the ecliptic) following Bhāskara II's statements (from his *Siddhāntaśiromaṇi*, II.7.16–23).

At the Arctic Circle, the ascending signs (Capricorn to Gemini) are more oblique to the horizon than at the equator and hence they rise in shorter periods of time (compared to their equatorial ascension times). Reciprocally, the descending signs (Cancer to Sagittarius) are more upright to the horizon there than at the equator and hence these signs take longer to rise at the Arctic Circle than at the equator. Looking at Table 5.3, we can observe that the total right ascensions of the descending signs are 10,800 *asu*s or 12^h at the equator. On the day of the summer solstice in the northern hemisphere, at the Arctic Circle, the six descending signs beginning with Cancer rise on the eastern horizon in 21,600 *asu* or $2{\times}12^h$. In other words, the six ascending signs beginning with Capricorn set on the western horizon in the same amount of time. Thus, the length of day (*dinamāna*) at the Arctic Circle on the day of the summer solstice (in the northern hemisphere) is 24^h or 60 *ghaṭīkā*s (with 1 *ghaṭīkā* equals 24^m). This implies that is no period of night (*rātripramāṇa*) on this day at the Arctic Circle.

As the Sun advances along the descending half of the ecliptic (*dakṣiṇāyana*), i.e., from the beginning of Cancer to the end of Scorpio, the length of day at the Arctic Circle diminishes with a commensurate increase in the hours of night. Halfway through this journey, the Sun transits through autumnal equinoctial point (*Tulādi* or 0° ♎) where the hours of daylight and darkness are again equal, similar to the Sun's position at the vernal equinoctial point (*Meṣādi* or 0° ♈) in its journey along the ascending half of the ecliptic (*uttarāyana*), i.e., from the beginning of Capricorn to the end of Gemini. Going southwards from the autumnal equinox, i.e., entering the third quarter of the ecliptic from the beginning of Libra to the end of Scorpio, the length of the day at the Arctic Circle further

Table 5.3 The ascension times of the zodiacal signs at the equator, following Bhāskara II's *Siddhāntaśiromaṇi* (II.7.16–23). In Sanskrit astronomy, the ascension times of celestial objects are indicated in units of *asus*, where one *asu* equals four sidereal seconds or a one minute of arc

Zodiacal Sign	*Right Ascension (in units of time)*
Aries (0° to 30° ♈)	1670 *asus* or $1^h51^m20^s$
Taurus (30° to 60° ♉)	1793 *asus* or $1^h59^m32^s$
Gemini (60° to 90° ♊)	1937 *asus* or $2^h9^m8^s$
Cancer (90° to 120° ♋)	1937 *asus* or $2^h9^m8^s$
Leo (120° to 150° ♌)	1793 *asus* or $1^h59^m32^s$
Virgo (150° to 180° ♍)	1670 *asus* or $1^h51^m20^s$
Libra (180° to 210° ♎)	1670 *asus* or $1^h51^m20^s$
Scorpio (210° to 240° ♏)	1793 *asus* or $1^h59^m32^s$
Sagittarius (240° to 270° ♐)	1937 *asus* or $2^h9^m8^s$
Capricorn (270° to 300° ♑)	1937 *asus* or $2^h9^m8^s$
Aquarius (300° to 330° ♒)	1793 *asus* or $1^h59^m32^s$
Pisces (330° to 360° ♓)	1670 *asus* or $1^h51^m20^s$

diminishes as the hours of night proportionally increase. When the Sun reaches its winter solstitial point (*Mṛgādi* or 270° ♑)—in other words, at the end of Sagittarius—the length of the night is 24^h or 60 *ghaṭīkās*, and accordingly, the day-length is zero. In this manner, the distribution of the hours of day and night at the Arctic Circle is completely reversed when the Sun is at its two opposite solstital positions.

Verse 103

The visibility of a planet at a terrestrial location—in other words, the planet being seen above the local horizon (*bhūja*)—is dependent on the true declination (*sphuṭakrānti*) of the planet and the co-latitude (*lambana*) of the place. The true declination of a planet is the arc of the secondary circle to the celestial equator between the visible orb (*dṛśya-tanu*, lit. visible body/form) of the planet and the celestial equator (*viṣuvavalaya*). The co-latitude of a place is the inclination of the celestial equator to the celestial horizon, or, equivalently, the inclination of the celestial poles to the line of joining the local zenith and nadir.

According to Nityānanda, as long as the northern true declination of a planet exceeds the local co-latitude, the planet is always visible in the

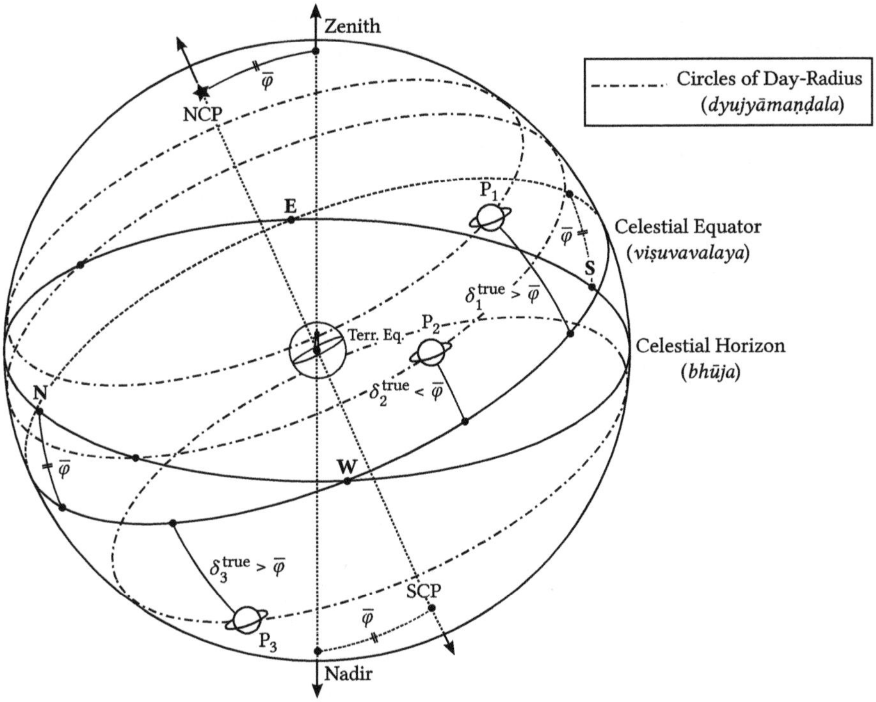

Oblique Celestial Sphere (*khagola*)

Figure 5.32 The oblique celestial sphere showing the visibility of a planet in three positions in relation the celestial horizon, namely, P_1 with the planet being circumpolar in the northern hemisphere, i.e., with its true declination $\delta_1^{\text{true}} > \overline{\varphi}$ towards the NCP where $\overline{\varphi}$ is the co-latitude; P_2 where the planet rises (and sets) above (and below) the celestial horizon, i.e., $\delta_2^{\text{true}} < \overline{\varphi}$; and P_3 where the planet remains below the celestial horizon (seen from the northern hemisphere), i.e., $\delta_3^{\text{true}} > \overline{\varphi}$ towards the SCP.

sky. In Figure 5.32, this is depicted by the planet at position P_1. The motion of the planet is circumpolar with respect to the north celestial pole, remaining above the celestial horizon at all times. Conversely, when the southern true declination of the planet exceeds the local colatitude, the planet remains below the horizon at all times. Position P_3 in Figure 5.32 depicts the planet perpetually set when seen from the northern hemisphere. In this position, the motion of the planet is again circumpolar with respect to the south celestial pole. In an intermediate position, e.g., at position P_2 in Figure 5.32, the true declination of the planet is less than the local co-latitude and hence the planet is visible in the sky for a certain period of time (moving along its day-circle in the zenithal half of the oblique celestial sphere) and remains invisible for the rest of a nychthemeron (when it moves along its day-circle in the nadiral half of the oblique celestial sphere).

Verses 104–106

Entering into the Arctic, i.e., at places where the degrees of terrestrial latitude (*akṣa-bhāga* or *pala-bhāgaka*) is greater than 66° 30′ N, the number of days in a sidereal year with 24^h or 60 *ghaṭīkās* of daylight increases.

1 At places with a latitude of 69° 48′ N, there are already days with 24^h of daylight when the Sun is in Gemini as opposed to just one summer solstitial day of 24^h at the Arctic Circle (when the Sun is situated at the beginning of Cancer, i.e., *Karkādi* or 90° ♋; see verse 100).

In verse 104, Nityānanda claims that at a latitude of 69° 48′ N days with 24^h of daylight begin when the Sun enters the lunar mansion *indubha* (also known as *mṛgaśirṣa*) of Gemini (*yugma*). *Mṛgaśirṣa* is the fifth lunar mansion (*nakṣatra*) that extends across the zodiacal signs Taurus and Gemini from 53° 20′ ♉ to 66° 40′ ♊ (see Pingree 1978a, p. 537).[27] Thus, at a place with a latitude of 69° 48′ N, there is 24^h of daylight on all those days when the Sun lies between the beginning of Gemini (approximately, within the first 6° of Gemini) and the end of Cancer (approximately, within the last 6° of Cancer) of the sidereal zodiac (with no equinoctial displacement); see Figure 5.33.

Correspondingly, when the Sun is between Sagittarius and Capricorn—in other words, situated between the first 6° of Sagittarius and the last 6° of Capricorn of the sidereal zodiac (with no equinoctial displacement)—there is no daylight in places at a latitude of 69° 48′ N; see Figure 5.33. In these locations, the polar night (24^h of darkness) also extends over several days in contrast to the one winter solstitial night of 24^h at the Arctic Circle (when the Sun is situated at the beginning of Capricorn, i.e., *Mṛgādi* or 270° ♑; see verse 102). On the other days of the year, there is a proportional increase (or decrease) in the lengths of the day and night depending on the movement of the Sun northwards (or southwards) along the ecliptic. In this way, the lengths of day and night in a nychthemeron (*dyuniśa*) are symmetrically distributed for different times of the year at these high latitudinal places inside the Arctic Circle.

2 Going further north into the Arctic at a terrestrial latitude of 78°30′ N, the number of days with 24^h of daylight, i.e., a polar day,

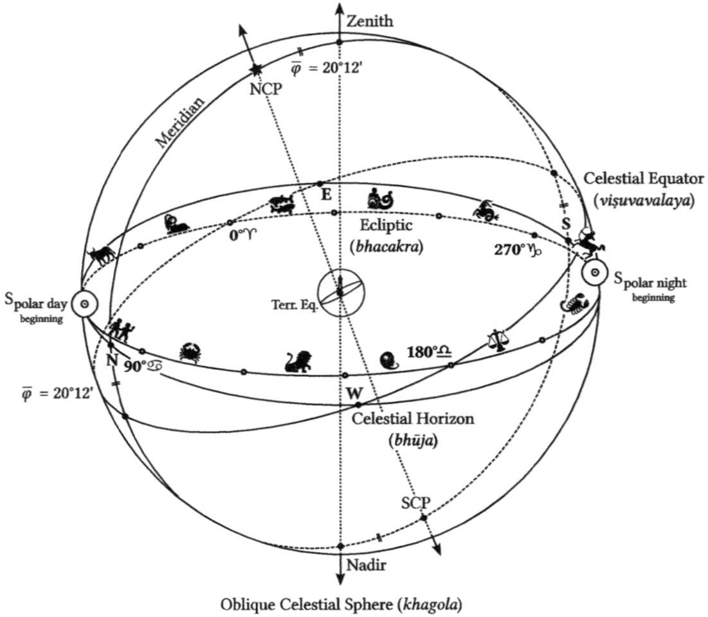

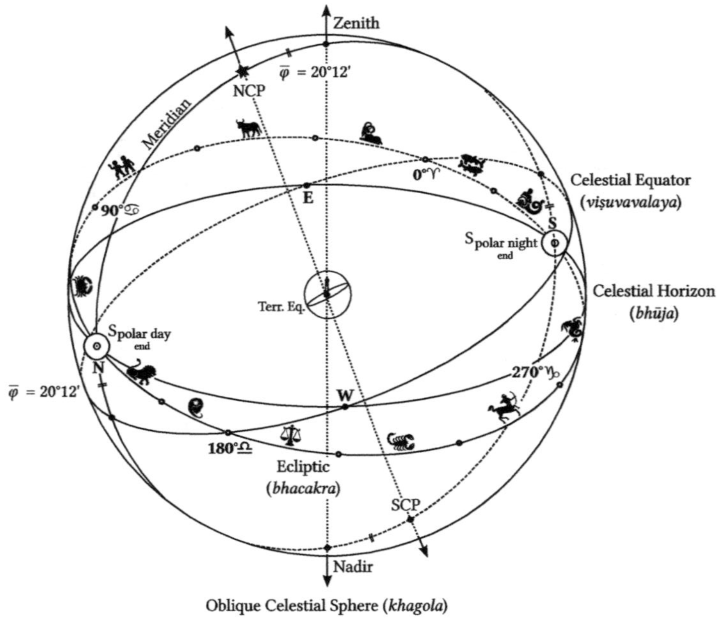

Figure 5.33 The oblique celestial sphere at a terrestrial latitude of 66° 48′ N. The figure to the left depicts the ecliptic positions of the Sun (at midnight) at the beginning of Gemini (start of the polar day) and at the beginning of Sagittarius (start of the polar night), while the figure to the right depicts the ecliptic positions of the Sun (at midnight) at the end of Cancer (end of the polar day) and the end of Capricorn (end of the polar night).

increases and lasts for the entire time it takes the Sun to transit the four zodiacal signs of the sidereal zodiac (with no equinoctial displacement) beginning with Taurus (30° ♉) and ending with Leo (150° ♌). Similarly, when the Sun transits the four signs beginning with Scorpio (210° ♏) and ending with Aquarius (330° ♒), there are an equal number of days with 24^h of perpetual darkness, i.e., a polar night. The positions of the Sun (at midnight) at the beginning and end of a polar day (and polar night) at an Arctic latitude of 78°30′ N are depicted in Figure 5.34.

3 At the north pole (90° N), the hours of daylight and darkness in a year are equal, both lasting a period of six months. The northward journey of the Sun along the sidereal zodiac (with no equinoctial displacement) from the beginning of Aries (*Meṣādi* or 0° ♈), i.e., the vernal equinox, to the beginning of Cancer (*Karkādi* or 90° ♋), i.e., the summer solstice, and then back southwards towards the beginning of Libra (*Tulādi* or 180° ♎), i.e., the autumnal equinox, constitutes a polar day (24^h of daylight) for an observer at the north pole. Reciprocally, the southward journey of the Sun from the beginning of Libra (*Tulādi* or 180° ♎) to the beginning of Capricorn (*Mṛgādi* or 270° ♑), i.e., the winter solstice, and back northwards to the beginning of Aries (*Meṣādi* or 0° ♈) constitutes a polar night (24^h of darkness) for an observer at the north pole. The positions of the Sun (at midnight) at the beginning and end of a polar day (and polar night) at the north pole (90° N) are depicted in Figure 5.35.

In verse 106, Nityānanda identifies observers at the north pole as the gods who reside on Mt Meru; this is in line with his statement in verse 20. According to the Purāṇic view, the length of a day for the gods is understood as twelve months long, with six months of daylight (polar day) and six months of darkness (polar night). In verse 39, Nityānanda rhetorically dismisses the Purāṇic statement that the Sun revolves around Mt Meru once every nychthemeron (24^h) on account of its contradiction with the six-monthly polar days and nights experienced by the gods living at Mt Meru; see notes on verse 39. Here, in verse 106, he offers an astronomical explanation for this polar phenomenon experienced by the gods who live at the north pole. From their perspective, the six-month-long polar day starts (i.e., the polar sunrise begins) with the Sun's entry into Aries (*Meṣādi* or 0° ♈); it is the polar midday when the Sun is at the beginning of Cancer (*Karkādi* or 90° ♋); the polar day ends and the polar night begins (i.e., the polar sunset is marked) with the Sun's entry into Libra (*Meṣādi* or 0° ♎); and it is polar midnight when the Sun is at the beginning of Capricorn (*Mṛgādi* or 270° ♑); see Figure 5.37 to the left.

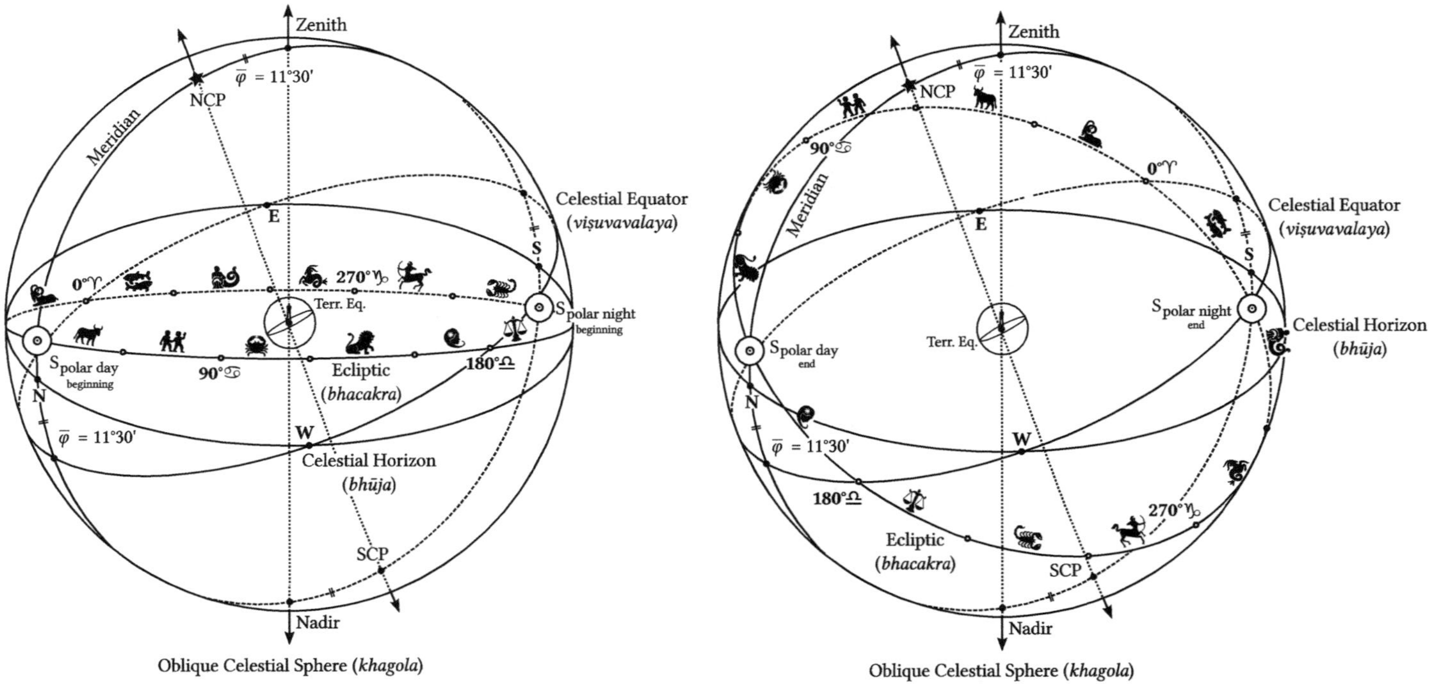

Figure 5.34 The oblique celestial sphere at a terrestrial latitude of 78° 30′ N. The figure to the left depicts the ecliptic positions of the Sun (at midnight) at the beginning of Taurus (start of the polar day) and at the beginning of Scorpio (start of the polar night), while the figure to the right depicts the ecliptic positions of the Sun (at midnight) at the end of Leo (end of the polar day) and the end of Aquarius (end of the polar night).

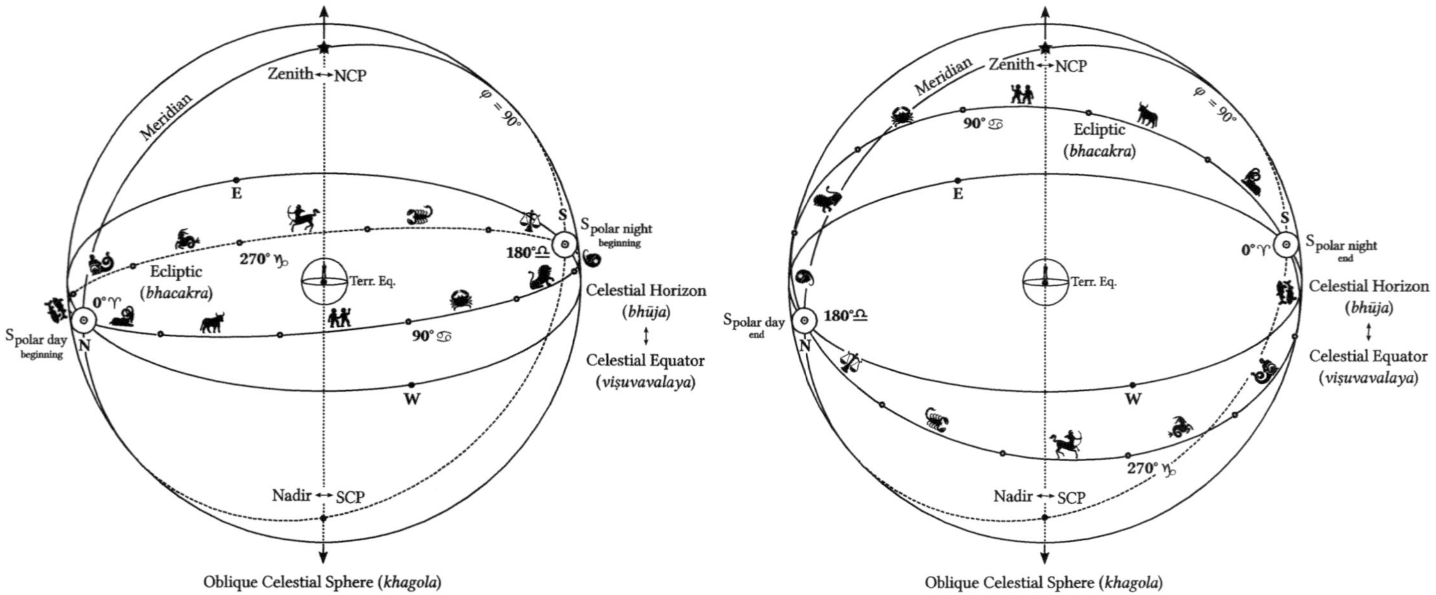

Figure 5.35 The oblique celestial sphere at a terrestrial latitude of 90° N. The figure to the left depicts the ecliptic positions of the Sun (at midnight) at the beginning of Aries (start of the polar day) and at the beginning of Libra (start of the polar night), while the figure to the right depicts the ecliptic positions of the Sun (at midnight) at the end of Virgo (end of the polar day) and the end of Pisces (end of the polar night).

On Nityānanda's choice of the Arctic latitudes 69° 48′ N, 79° 30′ N, and 90° N

1 In verse 103, Nityānanda explains the condition for the visibility of a celestial object above the horizon by comparing its true declination with the co-latitude of the observer. At a latitude of 69° 48′ N, i.e., at a co-latitude $\overline{\varphi}$ of 20° 12′ N, the Sun would always remain visible above the horizon, i.e., a polar day, so long as its declination is greater than 20° 12′ N. The declination δ of a celestial object is related to its ecliptic longitude λ by the expression $\sin\delta = \sin\varepsilon \ \sin\lambda$ where ε is the obliquity of the ecliptic (taken as 23° 30′). For $\delta = +20°12'$, the corresponding ecliptic longitude is approximately 59° 59′ 30″. As the Sun enters Gemini (60° ♊) its declination (approximately, +20° 12′ 6″) is already greater than the Arctic co-latitude of 20° 12′ N implying the start of the polar day. By symmetry, the Sun's declination continues to be greater than the co-latitude till it arrives at the end of Cancer (120° ♋) marking the end of the polar day.

In the antipodal position, just as the Sun enters the first 6° of Sagittarius (240°–246° ♐) or reaches the last 6° of Capricorn (294°–300° ♑), its declination (between −20° 12′ 6″ and −21° 21′ 47″) is greater than a southern co-latitude of 20° 12′ S and accordingly, it always appears to remain below the horizon when viewed from an equivalent northern co-latitude. The transit of the Sun between these positions of the sidereal zodiac (with no equinoctial displacement) marks a polar night at an Arctic latitude of 69° 48′ N.

In other ecliptic positions where the declination of the Sun is less than 20° 12′ N, in other words, when the Sun is between the four signs Aquarius to Taurus (in its northwards ecliptic movement) or between the four signs Leo to Cancer (in its southwards ecliptic movement), the lengths of day and night at an Arctic latitude of 69° 48′ N are proportionally distributed—the day-length increases with the northward journey of the Sun along the ecliptic and correspondingly decreases with its southward return.

2 At a latitude of 79° 30′ N, i.e., at a co-latitude of 11° 30 ′ N, a solar declination greater than 11° 30 ′ N would mean that the Sun will always remain visible above the horizon, i.e., a polar day. A solar declination of 11° 30 ′ N corresponds to an ecliptic longitude of 29° 59′ 56 ″ N, and hence, just as the Sun enters Taurus (30° ♉), its declination (approximately, +11° 30′ 1″) becomes greater than the co-latitude of 11° 30′ N implying the start of the polar day. By symmetry, the Sun's declination continues to be greater than the co-latitude till it arrives at the end of Leo (150° ♌) marking the end of the polar day.

In the antipodal position, when the Sun lies between the six zodiacal signs beginning with Scorpio (210° ♏) and ending with Aquarius (330° ♒), its declination −11° 30′ 1″ is just greater than the southern co-latitude of 11° 30′ S and accordingly, it always appears to remain below the horizon when viewed from an equivalent northern co-latitude. The transit of the Sun between these positions of the sidereal zodiac (with no equinoctial displacement) marks a polar night at an Arctic latitude of 11° 30 ′ N.

3 At the north pole (90° N), there is no co-latitude and hence, so long as the declination of the Sun is positive, i.e., during the transit of the Sun from 0° ♈ to 180° ♎ through 90° ♋, the Sun remains above the horizon, i.e., a six-month-long polar day. Symmetrically, when seen from the north pole, as long as the declination of the Sun remains negative, i.e., during the transit of the Sun from 180° ♎ to 360° ♓ (identified with 0° ♈) through 270° ♑, the Sun remains below the horizon, i.e, a six-month-long polar night.

Earlier Siddhāntic authors have also discussed the effect of increasing terrestrial latitudes on ecliptic visibility. Lalla, in his *Śiṣyadhīvṛddhidatantra* (XVI.20), speaks of the invisibility of Scorpio, Sagittarius, Capricorn, and Aquarius at a terrestrial latitude of 75° N. Śrīpati, in his *Siddhāntaśekhara* (XVI.57), states that Sagittarius and Capricorn are invisible at latitudes of 66° 30′ N while the four signs beginning with Scorpio become invisible at latitudes of 75° N. Bhāskara II, in his *Siddhāntaśiromaṇi* (II.7.28–30), describes the different lengths of day and night in relation to the position of the Sun (on the sidereal zodiac with no equinoctial displacement) at the terrestrial latitudes 69° 20′ N, 78° 15′ N, and 90° N. Interestingly, Bhāskara II's northern terrestrial latitudes are identical to those found in Varāhamihira's *Pañcasiddhāntikā* (XIII.25). Parameśvara, in his *Goladīpikā* (III.50–52), makes a similar statement to Nityānanda's remark in verse 104. However, Parameśvara describes the visibility of zodiacal signs in relation to terrestrial location, i.e., comparing the declination of the signs to terrestrial latitudes, whereas Nityānanda describes the visibility of the signs at Arctic latitudes as day-lengths corresponding to the positions of the Sun on the sidereal zodiac (with no equinoctial displacement).

Verse 107

The elevation of the Sun (with respect to the celestial horizon) changes with a change in the declination of the Sun. At any given time of the day, the degrees of elevation (*unnata-aṃśa*) of the Sun are measured on the vertical circle of altitude (*dṛṅmaṇḍala*) passing through the Sun (see verse 87), while the degrees of declination (*krānti-aṃśa*) of the Sun are measured on the secondary circle to the celestial equator (*viṣuvavalaya*).

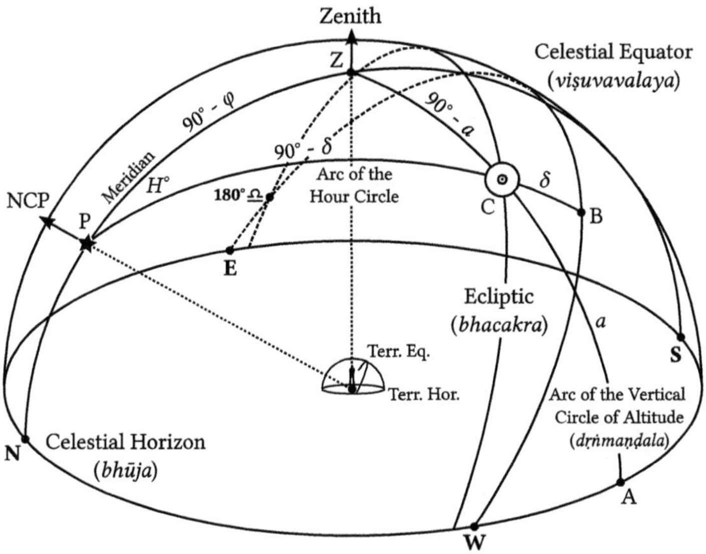

Figure 5.36 The zenithal half of the oblique celestial sphere showing the spherical triangle traced by the arcs of the local meridian, the hour circle, and the vertical circle of altitude corresponding to an arbitrary position of the Sun in the sky.

Figure 5.36 depicts the position of the Sun in the sky at a place with latitude φ (co-latitude $\overline{\varphi}$). In the figure, $\widehat{AC}$ is the altitude or elevation a of the Sun (complement of the zenith distance), $\widehat{CB}$ is the declination δ of the Sun, and $\sphericalangle ZPC$ represents the hour angle H, i.e., the deviation of the secondary circle to the celestial equator passing through the Sun (also known as the hour circle) and the local meridian. Applying the spherical rule of cosines to $\triangle ZPC$, we get

$$\cos\widehat{ZC} = \cos\widehat{PZ}\cos\widehat{PC} + \sin\widehat{PZ}\sin\widehat{PC}\cos\sphericalangle ZPC$$
$$\Rightarrow \cos(90 - a) = \cos(90 - \varphi)\cos(90 - \delta) + \sin(90 - \varphi)\sin(90 - \delta)\cos H$$
$$\Rightarrow \sin a = \sin\varphi\sin\delta + \cos\varphi\cos\delta\cos H$$

The degrees of maximum altitude or elevation (*para-unnata-aṃśa*) of the Sun corresponds to its upper culmination at local noon; in other words, the altitude of the Sun is maximum when it crosses the local meridian at midday (*ahardala*). At this moment, the hour angle $\sphericalangle ZPC$ or H is $0°$, and hence,

$$\sin a = \sin\varphi \cdot \sin\delta + \cos\varphi \cdot \cos\delta = \cos(\varphi - \delta) = \sin\left[90° - (\varphi - \delta)\right]$$
$$\Rightarrow a = 90° - (\varphi - \delta) \text{ with } a \in [0°, 90°]$$

Thus, at a given location (i.e., φ being fixed), the maximum solar altitude a (at midday) depends on the declination δ of the Sun and the local latitude φ.

1 For equatorial observers ($\varphi = 0°$N), at midday on the day of the vernal and autumnal equinoxes (0° ♈ and 180° ♎ respectively), $\delta = 0°$, and hence, $a = 90°$, i.e., the Sun is directly overhead. At the equator, the maximum elevation of the Sun changes through the year from $a = 66°30'$ towards the north on the day of the summer solstice (90° ♋) with $\delta = +23°30'$ to the same amount towards the south on the day of the winter solstice, (270° ♑) with $\delta = -23°30'$.

2 For observers at the Tropic of Cancer ($\varphi = 23°30'$N), at midday on the day of the summer solstice (90° ♋), $\delta = +23°30'$, and hence, $a = 90°$, i.e., the Sun is directly overhead. On equinoctial days (0° ♈ and 180° ♎, with $\delta = 0°$), the Sun culminates towards the south with $a = 66°30'$, whereas at midday on the day of the winter solstice (270° ♑, with $\delta = -23°30'$), the Sun has a maximum elevation, $a = 43°$.

3 For higher northern observers ($\varphi > 23°30'$N), an increase in the northern declination of the Sun, i.e., the northward journey (*uttarāyana*) of the Sun from the winter solstital point (270° ♑, with $\delta = -23°30'$) to the summer solstitial point (90° ♋, with $\delta = +23°30'$) on the sidereal zodiac (with no equinoctial displacement), through the vernal equinox (0° ♈, with $\delta = 0°$), corresponds to an increase in the maximum solar altitude of the Sun at its local meridional transit southwards. The maximum solar elevation at midday on the day of the summer solstice (90° ♋, with $\delta = +23°30'$) corresponds to $a = 113°30' - \varphi°$, where $a < 90°$ for all $\varphi > 23°30'$N. Correspondingly, the maximum solar elevation at midday on the day of the winter solstice (270° ♑, with $\delta = -23°30'$) corresponds to $a = 66°30' - \varphi°$; and the maximum solar elevation at midday on the day of the vernal and autumnal equinoxes (0° ♈ and 180° ♎, with $\delta = 0°$) corresponds to $a = 90° - \varphi°$.

4 At the north pole ($\varphi = 90°$N), at midday of any given day of the year, the maximum solar elevation is simply equal to the declination of the Sun. Accordingly, this value is largest on the day of the summer solstice (90° ♋, with $\delta = +23°30'$) where $a = 23°30'$ for an entire twenty-four hours. On the days of the vernal and autumnal equinoxes (0° ♈ and 180° ♎, with $\delta = 0°$), the Sun moves on the horizon at the north pole without ever rising, whereas on the day of the winter solstice (270° ♑, with $\delta = -23°30'$) the Sun remains set $23°30'$ (maximum depression) below the horizon for twenty-four hours.

Verse 108

For the demons (*suradvis*, lit. enemies of the gods) who live on the south pole (*vāḍavavahni-vāsin*, lit. those who dwell in the regions of the submarine fires; see verse 20)—in other words, the observers at the south pole (90° S)—the nychthemeron is twelve months long with six months of daylight and six months of darkness. However, the arrangement of the polar day and polar night at the south pole is reversed in comparison to what is observed at the north pole (see verses 105 and 106). For observers at the south pole, the polar day begins (i.e., the polar sunrise occurs) when the Sun enters Libra (*Meṣādi* or 0° ♎). It is polar midday when the Sun is at the beginning of Capricorn (*Mṛgādi* or 270° ♑). The polar day ends and the polar night begins (i.e., the polar sunset is marked) with the entry of the Sun into Aries (*Meṣādi* or 0° ♈) and the polar midnight is when the Sun is at the beginning of Cancer (*Karkādi* or 90° ♋). Thus, for both the gods and the demons, the length of a nychthemeron (*divāniśa*, lit. day and night) corresponds to the length of a solar year (*bhāsvat-varṣa*) with a mutual reversal (*viparyaya*) in the proportion (*vyavasthā*) of night and day (*rātri-dina*). Figure 5.37 depicts the positions of the Sun (on the sidereal zodiac with no equinoctial displacement) marking the beginning, the middle, and the end of a polar day and polar night for the northern gods (left) and southern demons (right).

Most Siddhāntic authors have commented on the reversal in the lengths of the polar day and polar night for antipodal polar observers; see, e.g., Āryabhaṭa I's *Āryabhaṭiya* (IV.17), Varāhamihira's *Pañcasiddhāntikā* (XIII.27), Brahmagupta's *Brāhmasphuṭasiddhānta* (XXI.8), Lalla's *Śiṣyadhīvṛddhidatantra* (XVIII.14), Śrīpati's *Siddhāntaśekhara* (XV.58–59), and Bhāskara II's *Siddhāntaśiromaṇi* (II.7.7, 8, 10). Among the Kerala (Niḷā) school of astronomers, Parameśvara (in his *Goladīpikā*, III.43–45) expressly declares that the six-month-long day for the gods commences with the Sun entering Aries (0° ♈, *Meṣādi*) and the six-month-long day for the demons commences with the Sun entering Libra (0° ♎, *Tulādi*), and in this way, the nychthemeron for the gods and demons is essentially an entire solar year. In a similar vein, Nīlakaṇṭha Somayājī, in his *Tantrasaṅgraha* (I.14), contrasts human and divine experience of time by calling a whole year for humans as merely a day for the gods (and the demons).

Verse 109

According to the *Ārdharātrikāpakṣa* (i.e., the system of midnight day-reckoning) of Āryabhaṭa I, the day begins at midnight instead of sunrise (as is conventional in most other systems of Indian astronomy).

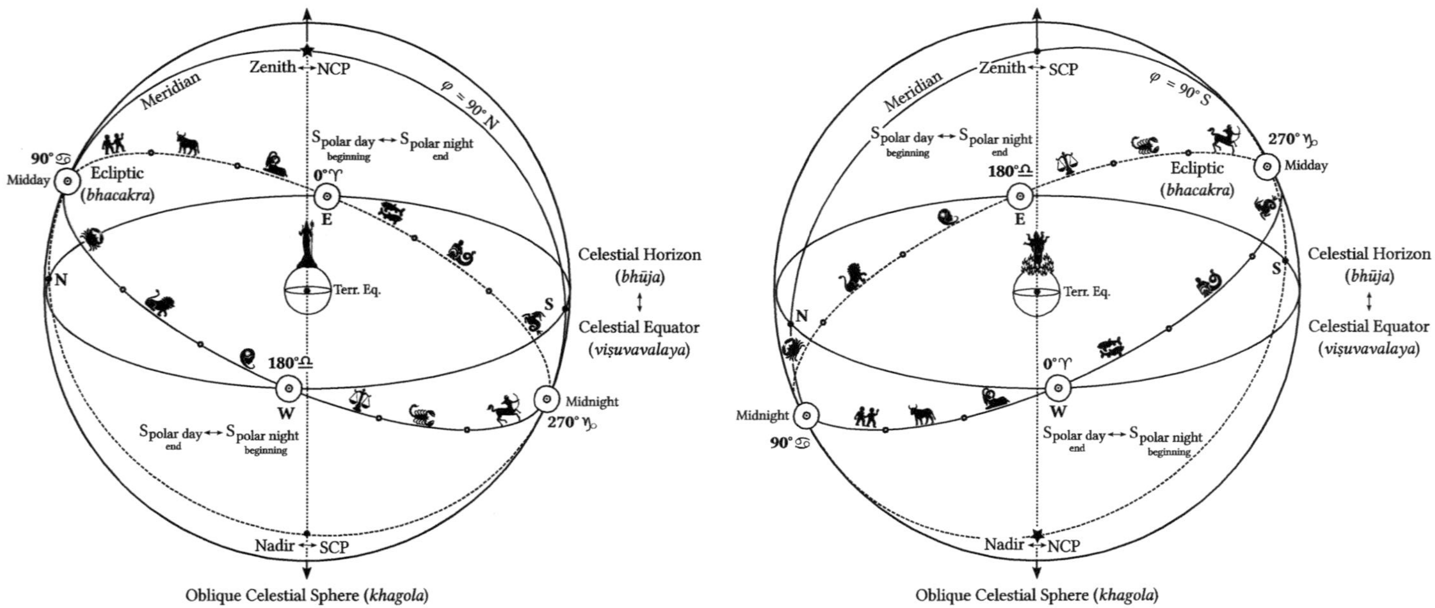

Figure 5.37 The annual ecliptic positions of the Sun in the oblique celestial sphere marking the beginning, the middle, and the end of the six-month-long polar day (and correspondingly, the six-month-long polar night) for the gods who live in the north pole (left) and for the demons who live in the south pole (right).

Āryabhaṭa I is believed to have described this system in his *Āryabhaṭa-siddhānta* that is no longer extant; however, several Siddhāntic authors (and commentators) since then have credited this system to Āryabhaṭa I. For instance, in talking about *ahargaṇa*s (number of mean civil days since the beginning of an epoch) in his *Pañcasiddhāntikā* (XV.18–20), Varāhamihira summarises various historical opinions on when a day begins. According to him,

1 Lāṭācārya, the composer of the old *Sūryasiddhānta* and a pupil of Āryabhaṭa I, considered the day to begin from sunset at Yavanapura (Alexandria in Egypt);

2 Siṃhācārya considered the day to begin with sunrise at Laṅkā;

3 Yavanācārya, the author of a *Yavanajātaka*, thought that the day to commence at the end of the tenth *muhūrta* (a period of two *ghaṭīkā*s or 48^{m}) of night; while

4 Āryabhaṭa I considered the day to begin at different hours according to two systems: the day began at midnight at Laṅkā following the *ārdharātrika* system, whereas the day began with sunrise at Laṅkā in the *sūryodaya* system.

Later Siddhāntic authors like Bhāskara I (in his *Mahābhāskarīya*) and Brahmagupta (in his Khaṇḍakhādyaka) have also attributed the system of midnight (*ārdharātrikā*) day-reckoning to Āryabhaṭa I (Pingree 1981b, pp. 20, 33). See Shukla (1967, 1977) for a fuller discussion on Āryabhaṭa I's *Ārdharātrikāpakṣa* and its position in Siddhāntic astronomy.

In the first part of verse 109, Nityānanda attributes the midnight day-reckoning system to the opinions of learned men of astronomy who follow the *saṃhitā*s (doctrinal texts). In this system, the day begins with midnight (*niśārdha*) and the night begin with midday (*dinārdha*) at any terrestrial latitude. In the latter part of the verse, Nityānanda describes the implication of this midnight day-reckoning system for the gods and demons (i.e., observers at the north pole and south pole respectively): the length of the polar day (or night) at the terrestrial poles is now marked by the movement of the Sun between opposite solstitial positions (instead of between opposing equinoctial positions as described in verses 106 and 108). In other words, the northward (*uttarāyana*) journey of the Sun along the ecliptic from the beginning of Capricorn (*Mṛgādi* or 270° ♑, i.e., the winter solstice) to the beginning of Cancer (*Karkādi* or 90° ♋, i.e., the summer solstice) via the vernal equinox (*Meṣādi* or 0° ♈) corresponds to the polar day of the northern gods, and the southward (*dakṣiṇāyana*) journey of the Sun along the ecliptic from the beginning of Cancer (*Karkādi* or 90° ♋, i.e., the summer solstice) to the beginning of Capricorn (*Mṛgādi* or 270° ♑, i.e., the winter solstice), via the autumnal equinox (*Tulādi* or 180° ☊), corresponds to the polar night of the northern gods. The arrangement of the polar day and polar night is simply reversed from the vantage of the demons who live in the south pole.

The astronomical explanation of day-length at the poles (in verses 106 and 108) corresponding to the Sun's movement between antipodal equinoctial points (i.e., moving north and south of the celestial equator) is different to the astrological (and Purāṇic) explanation of the polar day-length measured between antipodal solstitial points (i.e., the northward and southward journey of the Sun along the ecliptic). With this verse, Nityānanda attempts to reconcile the two different systems of sunrise and midnight day-reckoning, and, in doing so, makes the astronomical (theoretical) and astrological (practical) interpretation of a polar day (or polar night) equivalent.

In fact, he calls the midnight day-reckoning system as efficacious (*phalaprada*)—in other words, having a computational advantage in calculating certain astrological results (*phala*s). Other Siddhāntic authors have also commented on this computational advantage of the midnight day-reckoning system. For example, Bhāskara II, in his *Siddhānta-śiromaṇi* (II.7.11), says

> *dinaṃ surāṇāmayanaṃ yaduttaraṃ niśetarat saṃhitikaiḥ prakīrtitam |*
> *dinonmukhe'rke dinam eva tanmatam niśā tathā tatphalakīrttanāya tat ||*

> As for the doctrine of astrologers, that it was day with the gods at *Meru* whilst the Sun was in the *uttarāyana* (or moving from winter to the summer solstice) and night whilst the Sun was in the *dakṣiṇāyana* (or moving from the summer to the winter solstice), it can only be said in defence of such an assertion, that it is day when the Sun is turned towards the day, and it is night when turned towards the night. Their doctrine has reference merely to judicial astrology and the fruits it foretells.
>
> (L. Wilkinson and B. D. Śāstrī 1861, p. 163)

Verse 110

According to Purāṇic lore, one's deceased forefather or manes (*pitṛgaṇa*) live on the surface of the sphere of the Moon (*indugola*), and from their perspective, the nychthemeron (*ghasra-rātra*, lit. day and night) is believed to last one synodic month with equal lengths of day (*dina*) and night (*rātri*). Figure 5.38 depicts the position of the Moon along its orbit around the Earth in its four cardinal positions, namely, the full moon (F○), the first quarter or middle of the waxing phase (I◐), the new moon (N●), and the third/last quarter or middle of the waning phase (III◑).

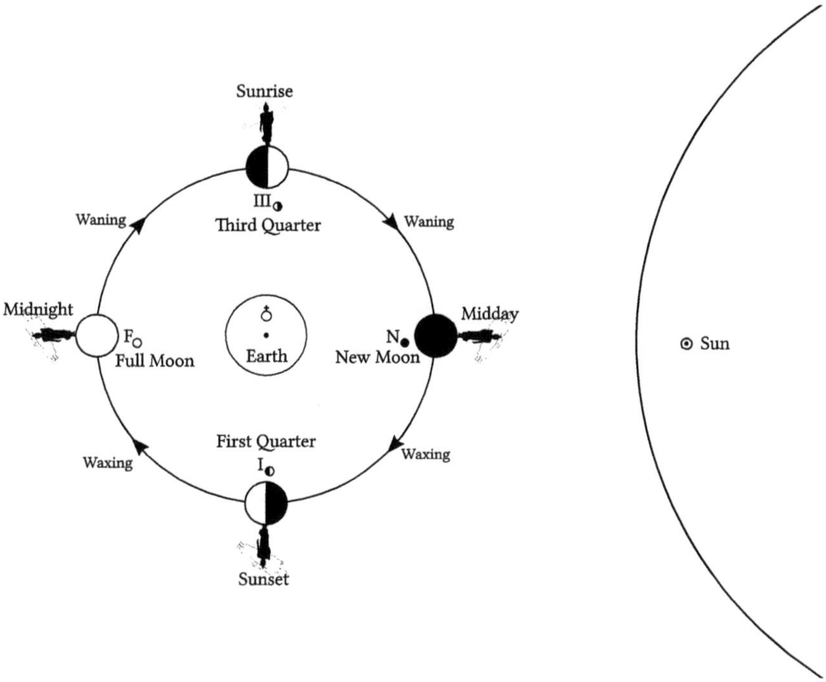

Figure 5.38 The lunar phases in a synodic month constituting the nychthemeron (the twenty-four-hour day) for one's deceased forefathers (manes) living on the surface of the Moon.

On the day of the new moon (*amāntakāla*), i.e., in position N●, the forefathers see the Sun directly overhead (at the zenith) and hence this is their midday (*dinārdha*). In an antipodal position on the day of the full moon (*pūrṇamāsī*), i.e., in position F○, the Sun appears below their feet (at the nadir) and hence it is their midnight (*rajanidala*). Between these two positions, sunrise (*udaya*) for the lunar manes occurs on the eighth day (*aṣṭami*) of the waning fortnight (*kṛṣṇapakṣa*), i.e., at position III◑, and sunset (*asta*) is on the eighth day (*aṣṭami*) of the waxing fortnight (*śuklapakṣa*), i.e., at position I◐. Thus, the period of day and night for the lunar manes lasts an entire synodic lunar month (*candramāsa*).

When seen from the Earth, the day-length for the lunar manes (with the day beginning at sunrise) extends from the third quarter III◑ to the first quarter I◐ through the new moon N●; and correspondingly, the period of night extends from the first quarter I◐ to the third quarter III◑ via the full moon F○. In this way, the nychthemeron of the lunar manes constitutes two fortnightly motions of the Moon between the middle of its waxing (*sita* 'light') and waning (*asita* 'dark') phases.

Alternatively, when the beginning of the day is taken as midnight (following the *ārdharātrikā* system described in verse 109), the waxing phase of the Moon between the new moon $\mathrm{N}_{\bullet}$ and the full moon $\mathrm{F}_{\circ}$ via the first quarter $\mathrm{I}_{\text{◐}}$ forms the day, while the waning phase of the Moon between the full moon $\mathrm{F}_{\circ}$ and the new moon $\mathrm{N}_{\bullet}$ through the third quarter $\mathrm{III}_{\text{◑}}$ constitutes the night. Here, the two waxing (*sita* 'light') and waning (*asita* 'dark') phases of the synodic lunar month together make the nychthemeron of the lunar manes.

Most Siddhāntic authors have also described the nychthemeron of the lunar manes lasting a synodic lunar month; see, e.g., Āryabhaṭa I's *Āryabhaṭiya*, (IV.17), Varāhamihira's *Pañcasiddhāntikā* (XIII.27 and XV.4, 14), Brahmagupta's *Brāhmasphuṭasiddhānta* (XXI.8), Lalla's *Śiṣyadhīvṛddhidatantra* (XVIII.14, 18), Śrīpati's *Siddhāntaśekhara* (XV.61), and Bhāskara II's *Siddhāntaśiromaṇi* (II.7.10, 13, 14). In all of these works, one's deceased ancestors on the Moon are said to enjoy two alternating fortnights of daylight and darkness every synodic lunar month. Nityānanda simply follows his predecessors in restating this Siddhāntic explanation of a Purāṇic lore.

Verses 111–112

With the first two *pāda*s of verse 111, Nityānanda finishes his discussion on the oblique celestial sphere (*khagola*). Like in the previous sections, he uses the word *racanā* 'composition' (in the compound *samasta-racanā-yukta* 'endowed with its entire composition') to allude to the natural configuration of the oblique celestial sphere as well as his own arrangement of the metrical verses discussing the various topics in this section.

Following this, the fourth and final section in Nityānanda's *golādhyāya* is on the visible celestial sphere (*dṛggola*). The *dṛggola* is like the oblique celestial sphere (*khagola*) in having the north and south celestial poles (*dhruva-dvaya*) inclined to the local zenith by an amount equal to an observer's co-latitude (*lambaka*). However, unlike the oblique celestial sphere that is geocentric with a celestial horizon, the visible celestial sphere is centred on the surface of the Earth (where the observer stands) and has a relatively smaller geographic horizon (the visible perimeter where the locally flat Earth appears to meet the sky, ignoring any obstructions along an observer's line of sight). See Section 1.3.4. The distinction between the *khagola* and the *dṛggola*, as acknowledged by men learned in the science of the spheres (*golavijña*), is due to the difference in their centres (*kendra-bhinnatva*). In the remaining verses of this chapter, Nityānanda describes the characteristic names (*lakṣaṇa-nāmaka*) of the individual arcs (*cāpa*s) of the celestial circles (*maṇḍala*s) in this observer-centric local perspective of the celestial sphere. These circles, being centred on the physical observer, are considered mobile as they depend on the location of the observer on the surface of the Earth.

Verse 113

Nityānanda begins the section on the visible celestial sphere by describing the equinoctial and solstitial points of the ecliptic. These reference points of the ecliptic are determined by the intersection of the ecliptic (*bhavṛtta*) with the celestial equator (*viṣuvavṛtta*) and with the solstitial colure (*ayanavṛtta*) respectively.[28]

In the northern hemisphere, when considering the ecliptic as the sidereal zodiac (*nirāyana rāśimaṇḍala*) with no equinoctial displacement, the vernal and autumnal equinoctial points correspond to the intersection of the ecliptic with the celestial equator and are marked by the beginning of Aries (*Meṣādi* or 0° ♈) and the beginning of Libra (*Tulādi* or 180° ♎) respectively. Similarly, the summer and winter solstitial points correspond to the intersection of the ecliptic with the solstitial colure and are marked by the beginning of Cancer (*Karkādi* or 90° ♋) and the beginning of Capricorn (*Mṛgādi* or 270° ♑) respectively; see Figure 5.39. The northward journey (*uttarāyana*) of the Sun indicates the Sun's movement along the ecliptic from the beginning of Capricorn (*Mṛgādi* or 270° ♑) till the end of Gemini (90° ♊, identical to *Karkādi* or 90° ♋) passing through the vernal equinoctial point (*Meṣādi* or 0° ♈). Correspondingly, the southward journey (*dakṣiṇāyana*) of the Sun indicates the Sun's movement along the ecliptic from the beginning of Cancer (*Karkādi* or 90° ♋) till the end of Sagittarius (270° ♐, identical to *Mṛgādi* or 270° ♑) passing through the autumnal equinoctial point (*Tulādi* or 180° ♎).

Verse 114

The declination (*apama*) of the Sun is measured along the arc (*cāpa* or *dhanus*) of the direction circle of declination (*apamasūtra*) between the Sun and the celestial equator (*viṣuvavṛtta*); see Figure 5.39. At the end of its northward journey (*uttarāyana*), the Sun arrives at the beginning of Cancer (*Karkādi* or 90° ♋) to its summer solstitial position on the ecliptic (sidereal zodiac with no equinoctial displacement). In this position, the arc of the solstital colure (*ayanavṛtta*) between the Sun and celestial equator corresponds to the greatest declination (*para-apama*) of the Sun—in other words, the obliquity of the ecliptic. In verse 76, Nityānanda states the value of the ecliptic obliquity as 23° 30′, and hence accordingly, the northernmost maximum declination of the Sun (when it lies at the beginning of Cancer, *Karkādi* or 90° ♋) and the southernmost maximum declination of the Sun (when it lies at the beginning of Capricorn, *Mṛgādi* or 270° ♑) are equal and opposite in measure (i.e., alternatively $\pm 23°30'$).

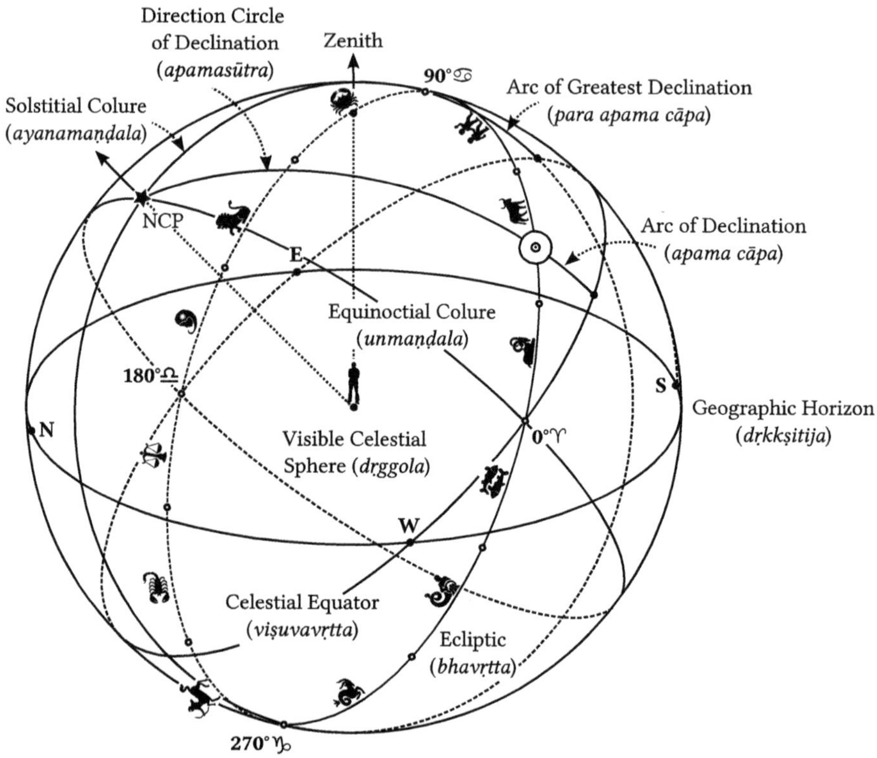

Figure 5.39 The visible celestial sphere depicting the relative positions of the equinoctial points (0° ♈ and 180° ♎) and solstitial points (90° ♋ and 270° ♑) of the ecliptic (the sidereal zodiac with no equinoctial displacement) in relation to the geographic horizon of a terrestrial observer in the northern hemisphere. The arc of the direction circle of solar declination (the secondary circle to the celestial equator that passes through the Sun, shown here to be in Taurus) that lies between the Sun and the celestial equator is arc of solar declination. When the Sun is at the beginning of Cancer (90° ♋), its declination lies along the arc of greatest declination and is equal to the obliquity of the ecliptic.

Verse 115

The declination (*apama*) of a planet, as already described in the notes to verse 79, is measured along the direction circle of declination (*apamasūtra*) as the arc (*śarāsana*) between the planet (*khecara*) and the celestial equator (*viṣuvamaṇḍala*). In essence, it is the measure of the planet's elevation above (to the north) and below (to the south) the plane of the celestial equator.

In verse 115, Nityānanda refers to this measure as the 'true declination' (*sphuṭatara-apama*) of the planet; see $\overset{\frown}{\mathrm{RA}}$ in Figure 5.40. Here, the position of the planet is identified with its *bimba*, lit. disk, orb, or sphere.[29] The use of the word *bimba* suggests that Nityānanda is referring to the true or visible position of the planet and not its mean position (*madhyagraha*)—in other words, the actual position of the planet (*sphuṭagraha*) in its orbit (*vikṣepavṛtta*) projected onto the sphere of asterism (i.e., the *bāṇavṛtta* or the latitudinal circle, see verse 78) at a given instant of time. The declination of the planet changes through the course of the day, and, at any given time of the day, its value can be determined using spherical trigonometry. See Misra (2021, 2022) for extensive discussions on true declination computations in Nityānanda's *Siddhāntasindhu* and *Sarvasiddhāntarāja*, and the history of their origins in Ulugh Beg's *Sūlṭānī Zīj* and Mullā Farīd's *Zīj-i Shāh Jahānī*.

Nityānanda's definition of the true declination in verse 115 mirrors Bhāskara II's statement on the 'correct declination' in his *Siddhāntaśiromaṇi* (II.6.16) where he says *sphuṭo nāḍikāvṛttākheṭāntarāle 'pamaḥ* 'the corrected declination [of any of the small planets and Moon] is the distance of the planet from the equinoctial in a circle of declination' (L. Wilkinson and B. D. Śāstrī 1861, p. 157).

Verse 116

In verse 116, Nityānanda defines the 'second declination' (*para-apama*, lit. other declination) as the arc (*dhanus*) of the great circle passing through the ecliptic pole and the planet that lies between the celestial equator (*viṣuvavṛtta*) and the ecliptic (*bhamaṇḍala*). The great circle passing through the ecliptic pole and the planet is the direction circle of ecliptic latitude (*viśikhasūtra* or *śilīmukhasūtraka*) as previously described in verse 79. In Figure 5.40, $\overset{\frown}{\mathrm{KC}}$ is the direction circle of ecliptic latitude, and $\overset{\frown}{\mathrm{DC}}$ represents the second declination. In the figure, the point D represents the ecliptic projection of the planet (situated at R) with respect to the north ecliptic pole K, making $\overset{\frown}{\Gamma\mathrm{D}}$ the ecliptic longitude (*bhoga*) of the planet and $\overset{\frown}{\mathrm{RD}}$ its ecliptic latitude (*vikṣepa*, *viśikha*, or *śara*), both measured in relation to the ecliptic pole.

Similar to the second declination, the arc of the great circle passing through the celestial pole P and point D (i.e., the arc of the secondary circle to the celestial equator passing through point D), and lying between the celestial equator and the ecliptic is called the 'first declination' (*prathama-krānti*, lit. first declination). In Figure 5.40, $\overset{\frown}{\mathrm{DB}}$ represents the arc of the first declination of the planet (situated at R).

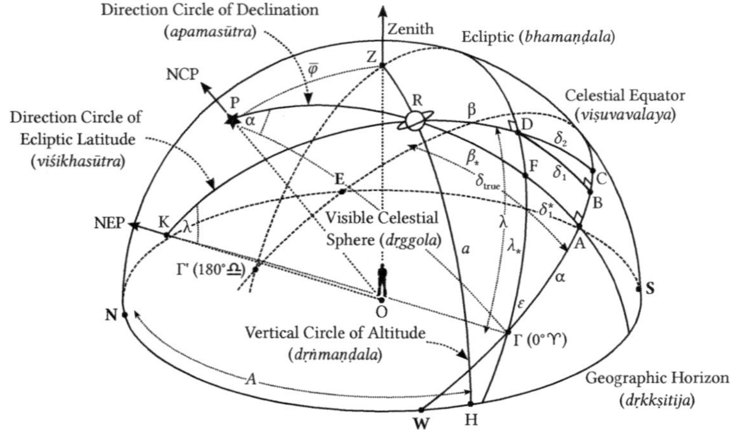

Ecliptic Longitude (*bhoga*, λ) ∢ ΓKD or $\widehat{\Gamma D}$	First Declination (*prathama apama*, δ_1) ∠DOB or $\widehat{DB}$
Ecliptic Latitude (*vikṣepa*, *viśikha*, or *śara*, β) ∠ROD or $\widehat{RD}$	Second Declination (*dvitīya*, *para* or *anya apama*, δ_2) ∠DOC or $\widehat{DC}$
Right Ascension (*kāla*, α) ∢ ΓPA or $\widehat{\Gamma A}$	Polar Longitude (*dhruvaka*, λ_*) ∢ ΓKF or $\widehat{\Gamma F}$
True Declination (*sphuṭatara apama*, δ_{true}) ∠ROA or $\widehat{RA}$	Polar Latitude (*vikṣepa*, β_*) ∠ROF or $\widehat{RF}$
Altitude (*utkrama*, *a*) ∠ROH or $\widehat{RH}$	Azimuth (*udagrā*, *A*) ∠NOH or $\widehat{NH}$
Obliquity of the Ecliptic (*paramakrānti*, ε) ∢ DΓC	Local Co-Latitude (*lamba*, $\overline{\varphi}$) ∠POZ or $\widehat{PZ}$
(First) Declination of Polar Longitude (*dhruvakakrānti*, δ_1^*) ∠FOA or $\widehat{FA}$	

Figure 5.40 The visible celestial sphere with the arcs of various secondary reference circles indicating the position of a planet. The position of the planet (at point R) is indicated with elliptic coordinates (β, λ), equatorial coordinates (δ_{true}, α), horizontal coordinates (a, A), and polar coordinates (λ_*, β_*). The first (δ_1) and second (δ_2) declinations of the planet are the arcs of the secondary circles to the celestial equator (direction circles of declination) and to the ecliptic (direction circle of ecliptic latitude) respectively, measuring the distance between the ecliptic projection of the planet (at point D) and the celestial equator. In the hybrid polar coordinate system, the first declination (δ_1^*) corresponding to the polar longitude (λ_*) is the arc of the direction circle of declination between the ecliptic and the celestial equator. As seen in the figure, the true declination of the planet δ_{true} is precisely equal to $\beta_* + \delta_1^*$, or approximately equal to $\beta + \delta_1$. This approximation is different to the 'curve of true declination', a quantity of

The measures of the first and the second declination of a planet are Islamicate concepts. In Arabic, the partial first and second declinations are called *al-mayl al-awwal al-juzʾī* and *al-mayl al-thānī al-juzʾī* respectively; see Kūshyār b. Labbān's *al-Zīj al-Jāmiᶜ*(ca. 1020–1025), chapter III (book 31): *Al-bāb al-mufrad fī jawāmiᶜ ᶜilm al-hayʾa* 'A special chapter on generalities of the science of cosmology', Bagheri (2006, p. 153). Mullā Farīd, in his *Zīj-i Shāh Jahānī*, follows the terminology of Ulugh Beg and refers to the first and second declinations as *mayl-i daraji-yi u* 'declination of its degree' and *mayl-i thānī-yi daraji-yi u* 'second declination of its degree' respectively. Correspondingly, in his *Siddhāntasindhu*, Nityānanda simply translates the first declination as *krānti* or *apama* 'declination' and the second as *anya-krānti* 'other declination', see Misra (2022).[30] It is worth noting that Nityānanda is among the earliest Sanskrit authors to explicitly (and accurately) describe the use of the first and second declinations in the computation of the true declination of a planet; see Misra (2022) for detailed discussions on how Nityānanda uses the first and second declinations of a planet in exact trigonometric expressions (in lieu of approximations) to compute the planet's true declination in his *Sarvasiddhāntarāja*.

In the last *pāda* of verse 116, Nityānanda explicitly defines the ecliptic latitude of a planet ($\overset{\frown}{RD}$ in Figure 5.40) as the difference (*vivara*) between the planet (*khaga*) and the ecliptic (*bhacakra*) measured along its direction circle (i.e., along the *viśikhasūtra*). Looking at Figure 5.40, it is easy to recognise how the arcs of the ecliptic latitude of a planet and its second declination are adjacent parts of the direction circle of ecliptic latitude (i.e., adjacent arcs of the *viśikhasūtra*).

Earlier Siddhāntic authors have also described the ecliptic latitude of a planet similarly For example, Bhāskara II, in his *Siddhāntaśiromaṇi* (II.6.16), says ...*krāntivṛttāccharaḥ kṣepavṛttāvadhis tiryag evam* 'celestial latitude is in like manner an arc of a great circle (which passes through the ecliptic poles) intercepted between the ecliptic and the *kshepa-vṛitta* [sic]' (L. Wilkinson and B. D. Śāstrī 1861, p. 157), where the *kṣepavṛtta* refers to the orbit of the planet.

Among the eight manuscripts of Nityānanda's *Sarvasiddhāntarāja* (*golādhyāya*) consulted in this study (see Section 2.1 for a description of the manuscripts, including Table 2.1 for the list of manuscript sigla cited below), only three manuscripts Bn.I, Np, and Sc attest verse 116 describing the second declination; the four manuscripts Bn.II, Br, Pm, and Rr do not include this verse. (MS. Pb in incomplete, ending with the middle of verse 91.) The omission of verse 116 (that describes the Islamicate idea of second declination) is an interesting aspect of the historical reception of Nityānanda's text. As Section 2.3 elaborates, the manuscript phylogeny suggests that the transmission of the *ur*-MS α splits into two different branches, MSS β and γ (see Figure 2.1), as its passed down. The four manuscripts that omit verse 116 derive from branch MS γ, while the three manuscripts that attest the verse originate from branch MS β. It is, perhaps, likely that at some point in the transmission of this text, scribe(s) excluded this verse on account of the foreignness in its origin and content.

Verse 117

In verse 117, Nityānanda defines the arc of the direction circle of ecliptic latitude (*viśikhasūtra*) between the planet (*khecara*) and the celestial equator (*viṣuvamaṇḍala*) as a quantity called the 'curve of true declination' (*sphuṭatara-apama-aṅka*). As Figure 5.40 shows, the curve of true declination ($\overset{\frown}{RC}$) of a planet (at R) is the sum of its second declination ($\overset{\frown}{DC}$) and its ecliptic latitude ($\overset{\frown}{RD}$).[31] In other words, $\overset{\frown}{RC} = \overset{\frown}{DC} + \overset{\frown}{RD} = \delta_2 + \beta$. In contrast, the true declination of the planet ($\overset{\frown}{RA}$) is the sum of the first declination ($\overset{\frown}{FA}$) of its polar longitude ($\overset{\frown}{\Gamma A}$) and its polar latitude ($\overset{\frown}{RF}$). In other words, $\overset{\frown}{RA} = \overset{\frown}{FA} + \overset{\frown}{RF}$ or $\delta_{\text{true}} = \delta_1^*(\lambda_*) + \beta_*$.[32] Comparing these two quantities, the curve of true declination $\overset{\frown}{RC}$ can be seen as the curved side (*aṅka*) of the right spherical triangle $\triangle$RAC with the other two sides being mutually perpendicular; in other words, the arcs of true declination $\overset{\frown}{RA}$ and ecliptic deviation $\overset{\frown}{AB}$ are orthogonal to one another.

The curve of true declination is also an Islamicate concept. In his *Siddhāntasindhu* (II.6) and *Sarvasiddhāntarāja* (I.*spaṣṭakrāntyādhikāra*), Nityānanda refers to the curve of true declination as *sphuṭa-apama-aṅka* (or *sphuṭatara-apama-aṅka*). However, in MS 4962 (from the Khasmohor Collection held at the City Palace Library of Jaipur) of the *Siddhāntasindhu* (II.6.1), on f. 20r, the curve of true declination is called the *sphuṭa-apama-aṃśa* 'share of true declination'. This expression mirrors Mullā Farīd's expression *ḥiṣṣi-yi-buᶜd* (in Persian) 'share of the distance' in his *Zīj-i Shāh Jahānī* (II.6.1)—an expression that is, in turn, borrowed from the equivalent Arabic expression *ḥiṣṣat al-buᶜd* from Ulugh Beg's *Sūlṭānī Zīj*. Al-Kāshī, in his *Khāqānī Zīj*, refers to this same quantity as *ḥiṣṣa-yi ᶜarḍ* 'share of the latitude' (Kennedy 1985, p. 9). Also, see Misra (2022) for discussions on the curve of true declination in Nityānanda's *Siddhāntasindhu* and *Sarvasiddhāntarāja*, and its role in one of the trigonometric methods to compute the true declination of a celestial object.

MSS Bn.II, Br, Pm, and Rr attest a different version of this verse:

> *viṣuvavṛttabhamaṇḍalamadhyagaṃ viśikhasūtradhanuryadihāsti saḥ* |
> *sphuṭatarāpamakāṅka udīrito gaṇitagolavicāravicakṣaṇaiḥ* ||

> What arc of the direction [circle] of [ecliptic] latitude is here between the celestial equator and the ecliptic, that [arc] is

declared as the curve of true declination by those who are wise in the investigations of computations and spheres.

This changes the meaning of the curve of true declination. According to these four manuscripts, the second declination $\widehat{DC}$ represents the curve of true declination (instead of the correct $\widehat{RC}$). This corruption is perhaps a haplography. As described in the notes of verse 116, MSS Bn.II, Br, Pm, and Rr omit verse 116. However, the first two *pāda*s of verse 116 are near-identical to the first two *pāda*s of verse 117 (cited above) in these manuscripts. The last two *pāda*s of verse 117 in these manuscripts and in the edited text are the same. It is, perhaps, likely that at some point in the transmission process, the scribe(s) of MS γ conflated the two verses being unfamiliar with their content, either intentionally or unintentionally.

Verse 118

For a terrestrial observer in the northern hemisphere, the local latitude can be determined by observing the elevation (*unnata*) of the north celestial pole (*dhruva* or polestar) above the northern geographic horizon (*kuja*). The elevation of the north celestial pole is commensurate with the depression (*nata*) of the south celestial pole (*anya-dhruvaka*, lit. the other pole) below (*adhas*) the southern geographic horizon. The amount of elevation or depression is equal to the local latitude φ°N; see Figure 5.41.

In verse 118, Nityānanda makes this observation, and then goes on to add that the (southwards) depression of the celestial equator (*viṣuva-vṛtta*) below the zenith (*ātma-śiras-nabha*, lit. one's own head in the sky) is equal in amount to the elevation of the north celestial pole above the northern geographic horizon, both of which are equal to the local latitude.

For an observer at the equator, the local latitude is zero, and accordingly, the north celestial pole (polestar) is coincident with the direction due north on the local geographic horizon and the celestial equator passes through the local zenith. Similarly, for an observer at the north pole, the local latitude is 90° and hence the celestial pole (polestar) is directly overhead, with the celestial equator appearing coincident with the local geographic horizon.

In the southern hemisphere, the elevation of the south celestial pole (typically, unmarked by a fixed star) above the southern horizon and the (northwards) depression of the celestial equator below the zenith are equal in measure to the southern terrestrial latitude. Thus, the variation in the terrestrial latitude of an observer has a direct impact on the orientation of their visible celestial [hemi-]sphere and the great circles that are associated with it.

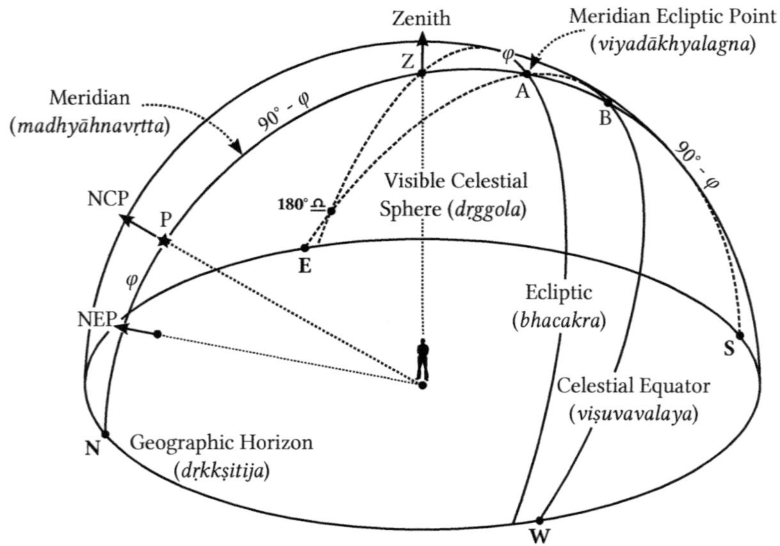

$\widehat{ZA}$ Zenith-Depression of Meridian Ecliptic Point (*madhyabhāga*)

$\widehat{ZB}$ Depression of the Celestial Equator below the Zenith, equal to the degrees of terrestrial latitude (*akṣalava*) φ° N

$\widehat{SB}$ Elevation of the Celestial Equator above the Horizon, equal to the degrees of terrestrial co-latitude (*lambabhāga*) $\overline{\varphi}^\circ$ N

$\widehat{ZP}$ Depression of the Polestar (*dhruva*) below the Zenith, equal to the degrees of terrestrial co-latitude (*lambabhāga*) $\overline{\varphi}^\circ$ N

$\widehat{NP}$ Elevation of the Polestar (*dhruva*) above the Horizon, equal to the degrees of terrestrial latitude (*akṣalava*) φ° N

Figure 5.41 The visible celestial sphere for a terrestrial observer in the northern hemisphere showing the local latitude and co-latitude in relation to the elevation and depression of the celestial pole and celestial equator, along with the zenith-depression of the meridian ecliptic point.

Earlier Siddhāntic authors have similarly described the change in the elevation (or depression) of the celestial pole with a change in an observer's terrestrial latitude; see, e.g.,Lalla's *Śiṣyadhīvṛddhidatantra* (XVIII.6–8), Śrīpati's *Siddhāntaśekhara* (XV.54–57), and Bhāskara II's *Siddhāntaśiromaṇi* (II.3.48–51).

Verse 119

Continuing from verse 118, Nityānanda now defines the degrees of latitude (*akṣalava*) and co-latitude (*lambabhāga*) at a terrestrial location equal to the depression (*nata*) of the celestial equator from the zenith and the elevation (*unnata*) of the celestial equator above the local geographic horizon respectively. The arcs of latitude and co-latitude (in relation to the position of the celestial equator) are represented by arc $\widehat{ZB}$ and $\widehat{SB}$ in Figure 5.41.

In the last two *pādas* of verse 119, Nityānanda calls the degrees of depression (*nata-aṃśaka*) of the meridian ecliptic point (*viyadākhya-lagna*) from the zenith (*khasvastika*) by its technical Sanskrit name *madhyabhāga*.[33] In Figure 5.41, $\widehat{ZA}$ is the zenith distance of the meridian-ecliptic point A—in other words, the *madhyabhāga*.

While defining the terrestrial latitude in the second *pāda* of verse 119, Nityānanda uses the adjective *prasiddha* 'well-established' or 'celebrated'. This is a semi-technical term used quite commonly in Sanskrit astronomy to refer to a canonical fact. For example, at the end of the preliminary section on definitions of technical terms (*paribhāṣā prakaraṇaṃ*) in his *Lilāvatī*, Bhāskara II's says, *śeṣā kālādiparibhāṣā lokataḥ prasiddhā jñeyā* 'The rest of definitions of time etc. are well established in the world [hence should be] understood [accordingly]'. By using the word *prasiddha*, Nityānanda emphasises that the definitions just stated are indeed common knowledge. This fact is easily verified by looking at near-identical statements on determining the terrestrial latitude (and co-latitude) in earlier *siddhāntas*, for example, in Lalla's *Śiṣyadhīvṛddhida-tantra* (XV.24), Śrīpati's *Siddhāntaśekhara* (XVI.44), and Bhāskara II's *Siddhāntaśiromaṇi* (II.7.33).

Verse 120

In verse 120, Nityānanda defines two measurements of a planet on a given day in relation to an observer's terrestrial location.

1 The degrees of maximum elevation (*para-unnata-aṃśa*) of a planet is the arc of the meridian (*madhyāhnavṛtta*) between the local geographic horizon (*kṣamāja*) and the disk of the planet (*graha-bimba*) as it reaches its upper culmination point in the sky.[34] In other words, it is the altitude of the planet at its local meridional transit

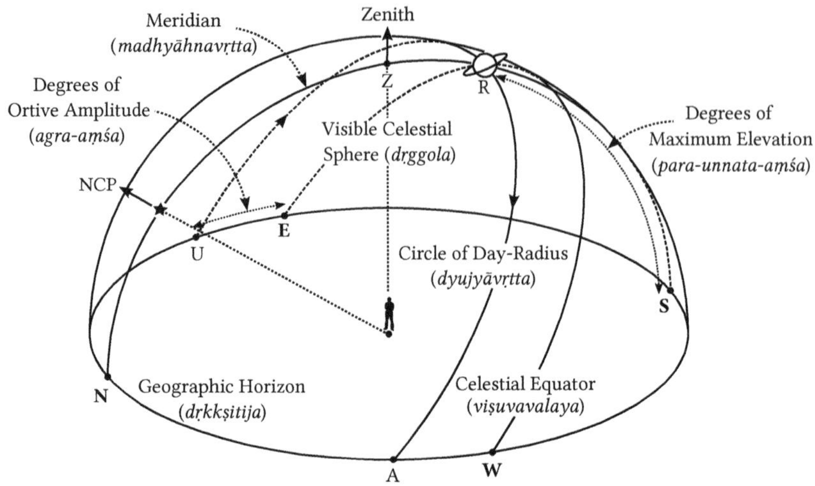

Figure 5.42 The visible celestial sphere with the degrees of maximum elevation and ortive amplitude of a planet.

on a given day. In verse 107, Nityānanda described the maximum elevation of the Sun (*para-unnata-aṃśa*) as its altitude measured on its vertical circle (*dṛṅmaṇḍala*) at the time of its upper culmination. Verse 120 generalises this statement to any celestial object. In essence, the vertical circle of altitude of a celestial object is coincident with the local meridian at the time of its meridional transit, and hence, the arc of its altitude above the local horizon is equal to its maximum elevation. In Figure 5.42, the degrees of maximum elevation of the planet (at R) is indicated by the arc $\widehat{\mathbf{RS}}$.

2 The degrees of rising or ortive amplitude (*agra-aṃśa*) of a planet is the arc of the local geographic horizon between the due east point and the point of rising of the planet on the eastern horizon, or alternatively, between the due west point and the point of setting of the planet on the western horizon. (This equivalence is due to the circle of day-radius of the planet being parallel to the celestial equator.) In Figure 5.42, the degrees of ortive amplitude of the planet (at R) is indicated by the arc $\widehat{\mathbf{UE}}$ on the eastern horizon or arc $\widehat{\mathbf{AW}}$ on the western horizon.

Like verse 115, the position of a planet is once again identified by its visible disk or orb (*bimba*) in the sky. This refers to the true position of the planet along its orbit (*vikṣepavṛtta*). On any given day, the day-circle of the planet, i.e., the circle of its day-radius (*dyujyāvṛtta*), is the small circle that depicts the westward movement of the planet (see verse 89). Nityānanda uses the word *graha-bimba* 'disk of the planet' to emphasise that the local altitude of the planet is measured with respect to the planet's visible body. The word *graha*, by itself, can sometimes imply the longitude of the planet (i.e., its theorised projected position on the ecliptic); for example, Bhāskara of Saudāmikā (fl. 1681) in his *Karaṇakesarī* (I.6a) says *yāmyottarediśi grahasya yutāyanāṃśāḥ* 'in the south [or] north direction, the precession-increased degrees of [the longitude of] the planet' (Montelle and Plofker 2014, p. 27).

Earlier Siddhāntic authors have also defined the altitude and amplitude of celestial bodies in their own texts; for example, Brahmagupta (in his *Brāhmasphuṭasiddhānta*, XXI.61), Lalla (in his *Śiṣyadhīvṛddhidatantra*, XV.18), Śrīpati (in his *Siddhāntaśekhara*, XVI.47), and Bhāskara II (in his *Siddhāntaśiromaṇi*, II.7.39) all describe the rising amplitude (or equivalently, its sine value) as the *agra* of the planet.

Verse 121

Compared to the degrees of ortive amplitude and maximum elevation of a planet (defined in verse 120), Nityānanda now describes the degrees of azimuth and altitude of a planet at a given instant of time. The azimuth and altitude are the horizontal coordinates of a celestial object from a terrestrial frame of reference (in the *dṛggola*).

1 The degrees of azimuth (*diś-aṃśaka*) is the angular separation along the local geographic horizon between the position of a planet (indicated by the eastern point of intersection of the planet's vertical circle of altitude (*dṛṅmaṇḍala*) and the local horizon) and the direction of due east (*pūrva*). Or equivalently, the position of a planet (indicated by the western point of intersection of the planet's vertical circle of altitude and the local horizon) and the direction of due west (*paścima*). In Figure 5.43, the equivalent arcs $\overset{\frown}{\mathrm{ME}}$ or $\overset{\frown}{\mathrm{QW}}$ represent the degrees of azimuth of a planet (at R) with respect to a terrestrial observer (at O) at a given time of the day, measured with respect to the due east or due west directions respectively. In essence, the degrees of azimuth is the angular separation between the planes of the prime vertical and vertical circle of altitude; in other words, $\overset{\frown}{\mathrm{ME}} = \angle\mathrm{EOM}$ and $\overset{\frown}{\mathrm{QW}} = \angle\mathrm{WOQ}$.[35]

   The ortive amplitude of the planet, arc $\overset{\frown}{\mathrm{UE}}$ in Figure 5.43, remains fixed even as its degrees of azimuth changes throughout the day. This is because the circle of day-radius $\overset{\frown}{\mathrm{URA}}$ is constant for any given day, whereas the azimuthal direction signified by the vertical circle of altitude $\overset{\frown}{\mathrm{MZRQ}}$ continually changes through the course of the day.

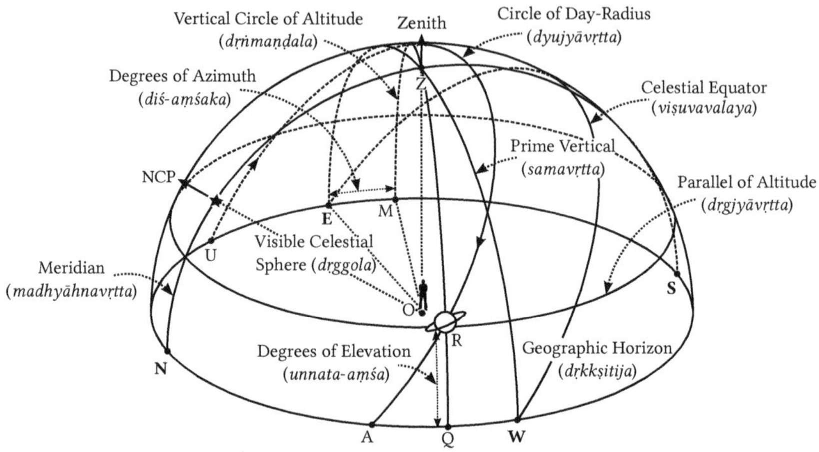

Figure 5.43 The visible celestial sphere with the degrees of azimuth and elevation of a planet along its vertical circle of altitude.

2 The degrees of elevation (*unnata-aṃśa*) is the altitude of the planet above the local geographic horizon measured along its vertical circle of altitude (*dṛṅmaṇḍala*). In Figure 5.43, $\widehat{RQ}$ represents the degrees of elevation of a planet (at R) towards the western horizon. If the altitude is negative—in other words, if the planet is below the horizon—the degrees of depressions (*adhas-aṃśa*) can be similarly determined.

As the vertical circle of altitude changes its orientation throughout the day with the movement of the planet, the elevation (or depression) of a planet also changes accordingly. At its zenithal meridian transit (upper culmination), the degrees of elevation are known as the degrees of maximum elevation (*para-unnata-aṃśa*) described in verse 120, and correspondingly, at its nadiral meridian transit (lower culmination), the degrees of depression are known as the degrees of maximum depression (*para-nata-aṃśa*). At any given instant of the day, the parallel of altitude (*dṛgjyāvṛtta*) indicates the altitude of the planet above the local horizon, or, equivalently, the complement of its zenith distance.

Other Siddhāntic authors have also described the azimuth and altitude of celestial objects in their own way. For example, Bhāskara II, in his *Siddhāntaśiromaṇi*, II.7.36), defines the altitude of a planet (or equivalently, the complement of its zenith distance) with its sine value, calling it the planet's *śaṅku* (or its *dṛgjyā*). In another section (I.3.45), he defines a planet's azimuth (*digāṃśa*) as the angular distance along the horizon between the prime vertical and the planet's vertical circle of altitude.

Verse 122

The degrees of zenith distance (*nata-aṃśaka*) of a planet at a particular time of day is measured on the vertical circle of altitude (*dṛṅmaṇḍala*) between the zenith (*kha-svastika*) and the planet (*kheṭa*). This is shown as arc $\widehat{ZR}$ in Figure 5.44 for a planet positioned at R. This distance is the complement of the degrees of elevation (*unnata-aṃśaka*) described in verse 121; in other words, $\widehat{RQ} = 90° - \widehat{ZR}$.

Also, the ascensional difference (*carārdha*, lit. half the *cara*) of a planet (on a particular day at a particular location) is the difference between the point of rising (or setting) of planet on the eastern (or western) horizon and the point of intersection of the equinoctial colure (*unmaṇḍala*) and the circle of day-radius (*dyujyāvṛtta*) of the planet, measured on the circle of day-radius on that day. In Figure 5.44, the arcs $\widehat{UC}$ and $\widehat{AD}$ (towards the eastern and western horizon) of the circle of day-radius (represented by the arc $\widehat{UCRDA}$) indicate the ascensional difference of the planet on the day. Essentially, the ascensional difference is the measure of the difference between the rising times of a celestial object at a certain terrestrial latitude when compared to its rising at the equator (zero degree latitude).

The difference in ascension times at different terrestrial locations is due to the changing obliquity of the celestial sphere (with a gain in terrestrial latitude) when compared to the right celestial sphere (at the equator).

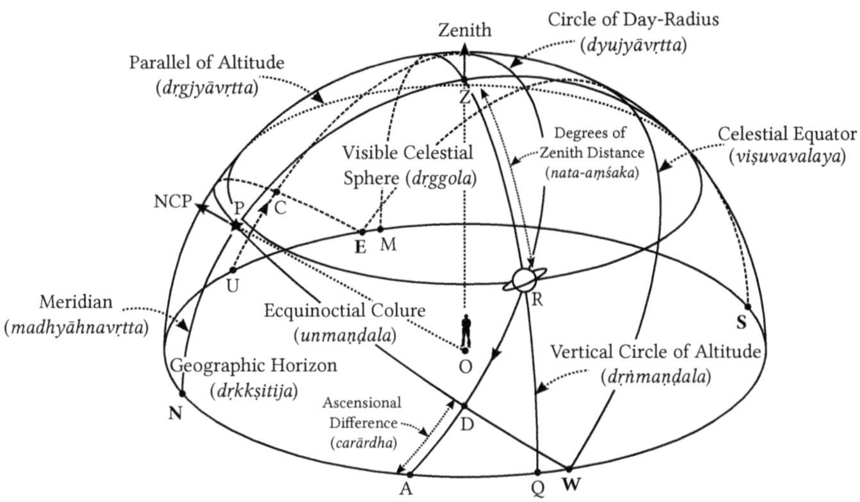

Figure 5.44 The visible celestial sphere with the degrees of zenith distance and the ascensional difference of a planet.

1 At the equator, there is no latitude (φ = 0°N), i.e., $\widehat{\mathbf{N}\mathrm{P}}$ = 0°, and hence the polestar (at P) is coincident with the due north point of the local geographic horizon. Here, the equinoctial colure $\widehat{\mathbf{EPW}}$ is coincident with the local horizon, and accordingly, the arcs of ascensional difference $\widehat{\mathrm{UC}}$ and $\widehat{\mathrm{AD}}$ are both zero. Effectively, the circles of day-radius of the planet are simply parallel to the celestial equator, with the planet rising simultaneously with the celestial equator.

2 At a terrestrial location of non-zero northern latitude (φ > 0°N), i.e., $\widehat{\mathbf{N}\mathrm{P}}$ > 0°, the equinoctial colure is inclined to the local horizon, at an angle equal to the latitude. Hence, so long as the circle of day-radius of the planet remains north of the celestial equator, the point C (or D) rises (or sets) simultaneously with the celestial equator, and accordingly the arcs $\widehat{\mathrm{UC}}$ and $\widehat{\mathrm{AD}}$ of the planet's circle of day-radius remain above the local horizon but below the equinoctial colure leading to a positive ascensional difference between the rising of the planet and the celestial equator.

3 At the north pole (φ = 90°N), i.e., $\widehat{\mathbf{N}\mathrm{P}}$ = 90°, the polestar (at P) is coincident with the zenith (at Z). Here, the equinoctial colure becomes perpendicular to the local horizon and the planet's day-circles are parallel to the local horizon. Thus, the arc of ascensional difference $\widehat{\mathrm{UC}}$ and $\widehat{\mathrm{AD}}$ become complete circle, effectively, implying the perpetual visibility of the planet above the local horizon. At the poles, the circle of day-radius of a celestial object forms its parallel of altitude (as the celestial equator is coincident with the local horizon); in other words, these objects are circumpolar.

Almost all Siddhāntic authors offer identical definitions of the zenith distance of a celestial object and its ascensional difference (at different terrestrial latitudes); see, for example, Āryabhaṭa I's*Āryabhaṭiya* (IV.26), Brahmagupta's *Brāhmasphuṭasiddhānta* (XXI.61), Jñānarāja's *Siddhāntasundara* (II.5.42), and Bhāskara II's *Siddhāntaśiromaṇi* (II.7.1).

Verse 123

Previously, in verse 73, Nityānanda defined the diurnal circle (*dyuvṛtta*) of luminaries (like the Sun) as spiralling circles that are loosely bound to the celestial equator, northwards and southwards of it. These circles trace the daily westward motion of the celestial objects, and on any particular day, they represent the arc of rising and setting of the luminary above and below the local geographic horizon. Accordingly, these circles are also called the *dinarātravṛtta*, lit. the day and night circle. Figure 5.45 depicts three diurnal circles $\odot U_1 M_1 A_1 I_1$, $\odot U_2 M_2 A_2 I_2$, and

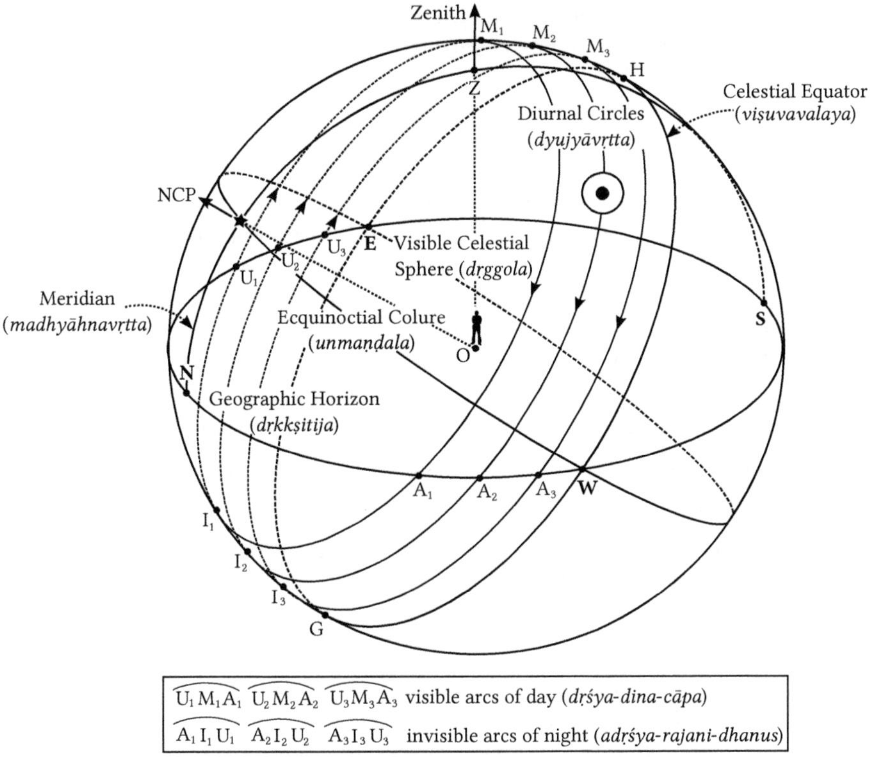

Figure 5.45 The visible celestial sphere with varying lengths of the visible arcs of day ($\overparen{U_1 M_1 A_1}$, $\overparen{U_2 M_2 A_2}$, and $\overparen{U_3 M_3 A_3}$) and the invisible arcs of night ($\overparen{A_1 I_1 U_1}$, $\overparen{A_2 I_2 U_2}$, and $\overparen{A_3 I_3 U_3}$) at a terrestrial location in the northern hemisphere when the Sun is north of the celestial equator.

$\odot U_3 M_3 A_3 I_3$ of the Sun at a terrestrial location in the northern hemisphere when the Sun is north of the celestial equator $\odot$EHWG.

In each of these diurnal circles, the arcs $\overparen{U_1 M_1 A_1}$, $\overparen{U_2 M_2 A_2}$, and $\overparen{U_3 M_3 A_3}$ represent the visible arcs (*dṛśya-cāpas*) above the local horizon (*kuja*), and are, accordingly, called the arc of day (*dina-cāpas*). Correspondingly, the arcs $\overparen{A_1 I_1 U_1}$, $\overparen{A_2 I_2 U_2}$, and $\overparen{A_3 I_3 U_3}$ are the invisible arcs (*adṛśya-cāpas*) below the local horizon and are hence called the arc of night (*rajani-dhanus*).

Among earlier Siddhāntic authors, Bhāskara II describes the arcs of day and night in his *Siddhāntaśiromaṇi* (II.7.5) in a manner identical to what Nityānanda has said in this verse. According to Bhāskara II, *dyurātravṛtte kṣitijād adhaḥsthe rātriryataḥ syād dinamānam ūrdhve* 'for, the length of the night is represented by that arc of the diurnal circle below the horizon, and the length of the day by that arc above the horizon' (L. Wilkinson and B. D. Śāstrī 1861, p. 162).

Verse 124

When the Sun is to the north of the celestial equator, the arc of day (*dina-cāpa*) is larger than the arc of night (*rajani-dhanus*) at any terrestrial location in the northern hemisphere. According to Nityānanda, this is due to the relatively larger parts of the diurnal circle (*dinarātra-vṛtta*) being visible above the local geographic horizon. Correspondingly, the length of day (*ahan-māna*) exceeds the length of night (*rajani-pramāṇa*) for observers in the northern hemisphere when the Sun remains north of the celestial equator (for half a solar year). This can be seen in Figure 5.45 where the visible arcs of day $\widehat{U_1\,M_1\,A_1}$, $\widehat{U_2\,M_2\,A_2}$, and $\widehat{U_3\,M_3\,A_3}$ are longer than their corresponding invisible arcs of night $\widehat{A_1\,I_1\,U_1}$, $\widehat{A_2\,I_2\,U_2}$, and $\widehat{A_3\,I_3\,U_3}$. In fact, the more northwards the diurnal circles get from the celestial equator (i.e., towards the north celestial pole or polestar), the larger is the proportion of day as compared to night in those northern locations. In other words, the arcs of day $\widehat{U_1\,M_1\,A_1} > \widehat{U_2\,M_2\,A_2} > \widehat{U_3\,M_3\,A_3}$, and by symmetry, the arcs of night $\widehat{A_1\,I_1\,U_1} < \widehat{A_2\,I_2\,U_2} < \widehat{A_3\,I_3\,U_3}$.

Conversely, when the Sun is to the south of the celestial equator, the arcs of day and night are reversed in their proportions. For half a solar year as the Sun remains southwards, northern observers experience longer nights than days. The more southwards the diurnal circles get from the celestial equator (i.e., towards the direction of the south celestial pole), the larger is the proportion of night as compared to day in those northern locations.

Verses 125–126

In verse 125, Nityānanda describes how the arc of the circle of day-radius (*dyujyāvṛtta*) of a celestial object (like the Sun) measures 180° (*cakra-dala-aṃśa*, lit. the degrees of a semicircle) between its eastern (*pūrva*) and western (*apara*) points of intersection with the equinoctial colure (*unmaṇḍala*) at any terrestrial location. The excess, i.e., the equal arcs of the circle of day-radius of the Sun between the equinoctial colure and the local horizon towards the east and west, is the ascensional difference (*cara*) (described in verse 122).

On a given day, the diurnal circle (*dyuvṛtta*) of the Sun corresponds to its circle of day-radius (*dyujyāvṛtta*) for that day. There, the arc of day

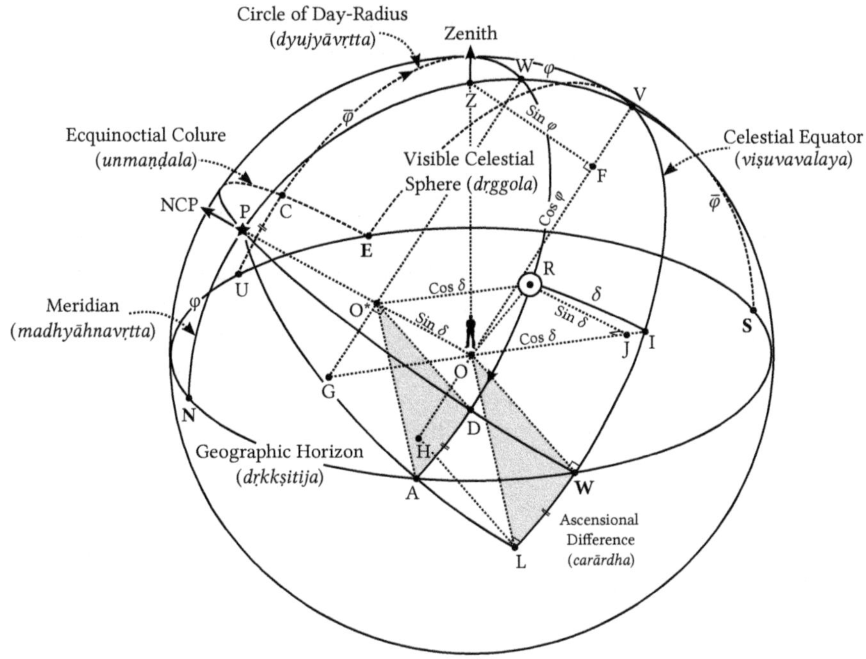

Figure 5.46 The visible celestial sphere depicting the arcs of ascensional difference on the day-circle of the Sun for a northern observer with Sun being situated north of the celestial equator.

(*dina-cāpa*), and accordingly the day-length (*dina-pramāṇa*), is then the visible part of its circle of day-radius (*dyujyāvṛtta*).[36] In Figure 5.46, arc $\widehat{\mathrm{UCWRDA}}$ represents this arc of day for the Sun positioned at R, at a terrestrial latitude of φ°N. Also, the arc $\widehat{\mathrm{CWRD}}$ forms a semicircle (i.e., 180°) and the arcs $\widehat{\mathrm{UC}} = \widehat{\mathrm{AD}}$ are the arcs of ascensional difference (i.e., the *carārdha*). Here, it can be seen that $\widehat{\mathrm{UCWRDA}} = \widehat{\mathrm{UC}} + \widehat{\mathrm{CWRD}} + \widehat{\mathrm{AD}}$ or effectively, *dina-cāpa* = 180°+ 2 *carārdha* = 180°+ *cara*.

As the notes to verse 122 describe, the ascensional difference varies according to an observer's terrestrial latitude. This variation, along with the relative position of the Sun with respect to the celestial equator on a particular day, implies that arc of day equals the semicircular arc (180°) increased (*āḍya*) or decreased (*hinā*) by the total ascensional difference (on that day).

1 At the equator (φ = o° N), the equinoctial colure is coincident with the local geographic horizon (i.e., with the *vyakṣa-kuja* 'equatorial horizon') and hence the arcs of the ascensional difference

(*carārdhas*) do not exist. Here, the arc of day is simply 180°, or in other words, the Sun remains visible in the sky (above the local horizon) for the same amount of time for which it remains invisible (below the horizon) creating equal days and nights. Nityānanda discusses this special case in verse 126.[37]

2 At a northern latitude ($\varphi > 0°$ N), when the Sun is north of the celestial equator (as seen in Figure 5.46), its arc of day is 180° increased by total ascensional difference. In other words, the Sun remains visible above the horizon for more than twelve hours as the total ascensional difference (in units of time) is added to it. Similarly, when the Sun is south of the celestial equator, its arc of day is 180° decreased by the total ascensional difference. Here, the Sun remains visible above the horizon for less than twelve hours as an amount of time equalling the total ascensional difference is reduced from it.

3 At the north pole ($\varphi = 90°$ N), the arc of ascensional difference is a complete circle, effectively implying that the arc of day is a complete circle. Here, if the Sun is north of the celestial equator (identified with the local horizon at the north pole), it remains perpetually visible above the horizon (i.e., the Sun is circumpolar). Similarly, if the Sun is south of the celestial equator, it is entirely invisible, remaining below the horizon at all times of the day.

The measure of the ascensional difference (*carārdha*) can be mathematically derived knowing the declination δ of the Sun and the local (northern) latitude φ. The enumerated lists below sketch this derivation (see Figure 5.46).

1 In ΔROJ, with $\widehat{RI} = \angle ROJ = \delta$, we have RJ = Sin δ and OJ = Cos δ. The radius of the visible celestial sphere is taken as $\mathscr{R}$ (the *sinus totus*). By symmetry, it can be seen that $OO^* = RJ = \text{Sin}\,\delta$ and $O^*R = OJ = \text{Cos}\,\delta$ where O^* is centre of the circle of day-radius of the Sun and O is the position of the terrestrial observer.

2 By comparing right-angled triangles ΔOO^*G and ΔOFZ we find

$$\frac{O^*G}{OO^*} = \frac{FZ}{OF} \Rightarrow \frac{O^*G}{\text{Sin}\,\delta} = \frac{\text{Sin}\,\varphi}{\text{Cos}\,\varphi}$$

$$\Rightarrow O^*G = \frac{\text{Sin}\,\delta \text{Sin}\,\varphi}{\text{Cos}\,\varphi} = \text{Sin}\,\delta \tan\varphi \qquad (5.2)$$

since $OO^* = \text{Sin}\,\delta$, and OF = Cos φ and FZ = Sin φ in right-angled triangle ΔZFO, where $\widehat{ZV} = \angle ZOF = \varphi$.

The segment O*G is the projection of the arc of ascensional difference $\widehat{\text{AD}}$ (or $\widehat{\text{UC}}$) along the radial arm of the day-circle. In Sanskrit astronomy, this measure is often called the Earth-sine (*bhūjyā* or *kṣitijyā*); for example, see Āryabhaṭa I's *Āryabhaṭiya* (IV.26) or Brahmagupta's *Brāhmasphuṭasiddhānta* (II.56–58).

As O*G constitutes a proportion of the radius O*R (or O*W) of the day-circle, it is proportional to the proportion OH of the radius OI (or OV) of the celestial equator. Thus,

$$\frac{\text{O}^*\text{G}}{\text{O}^*\text{R}} = \frac{\text{OH}}{\text{OI}} \Rightarrow \text{OH} = \frac{\text{OI} \times \text{O}^*\text{G}}{\text{O}^*\text{R}}$$

$$\text{OH} = \frac{\mathscr{R}\,\text{Sin}\,\delta \tan\varphi}{\text{Cos}\,\delta} = \mathscr{R} \tan\delta \tan\varphi \qquad (5.3)$$

since OI = $\mathscr{R}$, O*G = Sin δ tan φ and O*R = Cos δ.

3 The arc of ascensional difference $\widehat{\text{AD}}$ (= $\widehat{\text{UC}}$) on the circle of day-radius correspond to the arc $\widehat{\textbf{LW}}$ on the celestial equator. In other words, point A on the day-circle (at a northern latitude of φ) and point L on the celestial equator (at the equator) set simultaneously.

By treating the arcs $\widehat{\text{AD}}$ and $\widehat{\textbf{LW}}$ as segments, we find that AD ≈ O*G and L**W** ≈ OH. Thus, in planar triangles $\Delta\text{O}^*\text{AD}$ and $\Delta\text{OL}\textbf{W}$, we find

$$\text{AD}\,(\approx \text{O}^*\text{G}) = \text{O}^*\text{A}\,\sin(\angle\text{AO}^*\text{D}) \text{ and } \text{L}\textbf{W}\,(\approx \text{OH}) = \text{OL}\,\sin(\angle\text{LO}\textbf{W}).$$

And hence,

$$\text{AD} \approx \text{O}^*\text{G} = \text{Cos}\,\delta\,\sin(\angle\text{AO}^*\text{D}) \text{ and}$$
$$\text{L}\textbf{W} \approx \text{OH} = \mathscr{R}\,\sin(\angle\text{LO}\textbf{W}) = \text{Sin}\,(\angle\text{LO}\textbf{W}) \qquad (5.4)$$

since O*A = O*R = Cos δ (radius of the day-circle) and OL = OI = $\mathscr{R}$ (radius of the celestial equator).

4 Comparing the measures of O*G and OH from equations 5.2–5.4 we find

$$\text{O}^*\text{G} = \text{Cos}\,\delta\,\sin(\angle\text{AO}^*\text{D}) = \text{Sin}\,\delta \tan\varphi \text{ and}$$
$$\text{OH} = \text{Sin}\,(\angle\text{LO}\textbf{W}) = \mathscr{R} \tan\delta \tan\varphi.$$

In other words,

$$\sin(\angle\text{AO}^*\text{D}) = \frac{\text{Sin}\,\delta \tan\varphi}{\text{Cos}\,\delta} = \tan\delta \tan\varphi \text{ and}$$
$$\sin(\angle\text{LO}\textbf{W}) = \tan\delta \tan\varphi.$$

Hence,

$$\widehat{\text{AD}} = \angle\text{AO}^*\text{D} = \widehat{\text{L}\textbf{W}} = \angle\text{LO}\textbf{W} = \sin^{-1}\left(\tan\delta \tan\varphi\right). \qquad (5.5)$$

Thus, each arc of ascensional difference (i.e., the *carādhas*) can be computed with the expression $\sin^{-1}$ [tan(declination) × tan(latitude)] or

$$\textit{carārdha} = \sin^{-1}\left[\tan\delta\tan\varphi\right] \tag{5.6}$$

These *carārdhas* increase or decrease the hours of daylight at any particular place, depending on the solar declination on the day and the local latitude.

- The period between sunrise and midday, i.e., arc $\widehat{\mathrm{UW}}$, corresponds to $6^h \pm \widehat{\mathrm{UC}}^h$, with the maximum value of the *carārdha* $\widehat{\mathrm{UC}} = 6^h$ (15 *ghaṭīkās*) at the north pole.

- The period between midday and sunset, i.e., arc $\widehat{\mathrm{AD}}$, corresponds to $6^h \pm \widehat{\mathrm{AD}}^h$, with the maximum value of the *carārdha* $\widehat{\mathrm{AD}} = 6^h$ (15 *ghaṭīkās*) at the north pole.

The topic of ascensional difference, and its effect on the changing hours of daylight at a terrestrial location, is a topic discussed in almost all Siddhāntic texts; see, for example, the *Pañcasiddhāntikā* (IV.34) of Varāhamihira, the *Āryabhaṭiya* (IV.26) of Āryabhaṭa I, the *Brāhmasphuṭasiddhānta* (II.56–58) of Brahmagupta, the *Vaṭeśvarasiddhānta* (IV.8–10) of Vaṭeśvara, the *Śiṣyadhīvṛddhidatantra* (XVI.8–11) of Lalla, the *Siddhāntaśekhara* (XVI.25–28) of Śrīpati, and the *Siddhāntaśiromaṇi* (II.7.1–9) of Bhāskara II.

Verses 127–128

The local rising (or setting) of the zodiacal signs, i.e., the ecliptic (*apakramamaṇḍala*), at any terrestrial location is dependant on the inclination of the ecliptic to the local horizon (*kṣamāja*). In verse 127, Nityānanda observes that the more vertical (*daṇḍavat*, lit. like a straight line) the ecliptic is to the horizon, the longer is the amount of time (*kāla*) required for the signs to ascend (or descend). Whereas, the greater the inclination of the ecliptic to the local horizon, the faster the ascension (or descent) of the zodiacal signs.

This observation applies to most terrestrial locations, including the equator (*nirakṣa-deśa*, lit. a place with no latitude). While the circles of day-radius (or the diurnal circle) of a celestial object (like the Sun) may remain perpendicular to the local horizon at the equator (and hence the arcs of day and night at the equator are equal), the ecliptic, however, is inclined to the celestial horizon at a fixed obliquity (23° 30′). The diurnal westward motion of the rotating celestial sphere then causes a difference in the ascension times of the zodiacal signs at the equator.

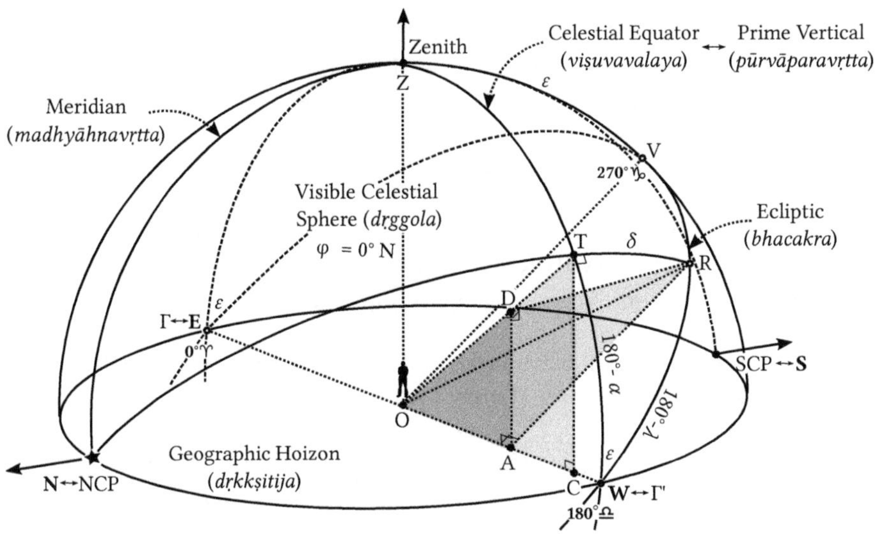

Figure 5.47 The visible celestial sphere with the right-angled triangles generated with respect to a central equatorial observer.

To understand the relation between the rising times of the zodiacal signs, i.e., their right ascension α, and their corresponding ecliptic longitude λ at the equator ($\varphi = 0°$ N), consider the position of the ecliptic $\widehat{\Gamma V \Gamma'}$ inclined to the celestial equator $\widehat{\Gamma Z \Gamma'}$ at an obliquity $\varepsilon = 23°30'$ in Figure 5.47.

1 For a point on the ecliptic (at R) with declination $\widehat{TR} = \delta$, right ascension $\widehat{\Gamma ZT} = \alpha$, and ecliptic longitude $\widehat{\Gamma VR} = \lambda$, we can construct the following right-angled triangles with respect to a central observer O at the equator ($\varphi = 0°$N).

a) Dropping a perpendicular from R onto the plane of the celestial equator, meeting the radial line OT at point D, we find $\triangle$DOR with $\angle$RDO = 90°, $\angle$DOR = $\widehat{TR} = \delta$, and OR = $\mathscr{R}$ (*sinus totus*), and hence DR = Sin δ and OD = Cos δ.

b) Dropping a perpendicular from T onto the plane of the geographic horizon, meeting the radial line OΓ' at point C, we find $\triangle$TOC with $\angle$TCO = 90°, $\angle$TOC (or 180°−$\angle\Gamma$OT) = $\widehat{T\Gamma'}$ = 180° − α, and OT = OΓ' = $\mathscr{R}$ (*sinus totus*), and hence TC = Sin (180° − α) = Sin α and OC = Cos (180° − α) = −Cos α.

c) Dropping a perpendicular from D onto the plane of the geographic horizon, meeting the radial line OΓ' at point A, we find $\triangle$DOA with $\angle$DAO = 90° and $\angle$DOA (or 180° − $\angle\Gamma$OT) =

$\widehat{\Upsilon\Gamma'} = 180° - \alpha$, and OD = Cos δ (from $\triangle$DOR), and hence DA = OD sin($\angle$DOA) = Cos δ sin(180° − α) = Cos δ sin α.

d) Joining the points R and A on the ecliptic plane, we find $\triangle$ROA with $\angle$RAO = 90°, $\angle$ROA (or 180° − $\angle\Gamma$OR) = $\widehat{R\Gamma'} = 180° - \lambda$, and OR = O$\Gamma'$ = $\mathscr{R}$ (*sinus totus*), and hence RA = Sin (180° − λ) = Sin λ and OA = Cos (180° − λ) = −Cos λ.

2 In the right-angled $\triangle$RDA, $\angle$RDA = 90°, and hence, DA = $\sqrt{RA^2 - DR^2}$. With RA = Sin λ (from $\triangle$ROA) and DR = Sin δ (from $\triangle$DOR), we have DA = $\sqrt{\text{Sin}^2\lambda - \text{Sin}^2\delta}$.

3 Comparing similar triangles $\triangle$TOC ~ $\triangle$DOA (where both $\angle$TOC and $\angle$DOA are 180° − α), we find

$$\frac{TC}{OT} = \frac{DA}{OD} \Rightarrow TC = \frac{OT \times DA}{OD} \Rightarrow TC = \frac{\mathscr{R}\sqrt{\text{Sin}^2\lambda - \text{Sin}^2\delta}}{\text{Cos}\,\delta}, \quad (5.7)$$

where the expression for DA is from paragraph 2 above, OT = $\mathscr{R}$ (from $\triangle$TOC), and OD = Cos δ (from $\triangle$DOR).

4 In spherical $\triangle$RTΓ' inscribed on the sphere by the arcs of declination $\widehat{TR} = \delta$, supplement of right ascension $\widehat{\Gamma' T} = 180° - \alpha$, and supplement of longitude $\widehat{\Gamma' R} = 180° - \lambda$, $\angle$RTΓ = 90° and $\sphericalangle$RΓ'T = ε (ecliptic obliquity). Thus, applying the spherical rule of sines (for non-unitary *sinus totus*) to $\triangle$RTΓ', we find

$$\frac{\text{Sin} \sphericalangle R\Gamma' T}{\text{Sin}\,\widehat{TR}} = \frac{\text{Sin}\,\angle RT\Gamma}{\text{Sin}\,\widehat{\Gamma' R}} \Rightarrow \frac{\text{Sin}\,\varepsilon}{\text{Sin}\,\delta} = \frac{\text{Sin}\,90°}{\text{Sin}\,(180° - \lambda)} \Rightarrow \text{Sin}\,\delta = \frac{\text{Sin}\,\varepsilon\ \text{Sin}\,\lambda}{\mathscr{R}}. \quad (5.8)$$

5 From equations 5.7 and 5.8, and noting that TC = Sin α (from $\triangle$TOC), we have

$$TC = \text{Sin}\,\alpha = \frac{\mathscr{R}\sqrt{\text{Sin}^2\lambda - \text{Sin}^2\delta}}{\text{Cos}\,\delta} = \frac{\text{Sin}\,\lambda\sqrt{\mathscr{R}^2 - \text{Sin}^2\varepsilon}}{\text{Cos}\,\delta}$$

$$\Rightarrow \text{Sin}\,\alpha = \frac{\text{Sin}\,\lambda\,\text{Cos}\,\varepsilon}{\text{Cos}\,\delta} \text{ or } \frac{\text{Sin}\,\lambda\,\text{Cos}\,\varepsilon}{\sqrt{\mathscr{R}^2 - \text{Sin}^2\delta}} = \frac{\mathscr{R}\,\text{Sin}\,\lambda\,\text{Cos}\,\varepsilon}{\sqrt{\mathscr{R}^4 - \text{Sin}^2\varepsilon\,\text{Sin}^2\lambda}}. \quad (5.9)$$

With ε = 23°30′ and $\mathscr{R}$ taken as 1, the right ascension α of a zodiacal sign with longitude λ can be approximately calculated with the expression

$$\sin\alpha \approx \frac{0.917\sin\lambda}{\sqrt{1 - 0.159\sin^2\lambda}}. \quad (5.10)$$

Table 5.4 (third column) lists the right ascensions α_j corresponding to the end of the twelve zodiacal signs (i.e., corresponding to the longitude λ_j of 30°, 60°, 90°, etc.) at the equator. These values are approximated using equation 5.10. The difference between the right ascensions of (the terminal cusps of) successive zodiacal signs, i.e., $\Delta\alpha_i = \alpha_j - \alpha_i$, for consecutive cusps with longitude λ_i and λ_j (such that $\lambda_j - \lambda_i = 30°$) corresponds to the times of rising of the entire sign at the equator. Table 5.5 lists these values in degrees and minutes. (The value in degrees is converted to units of time following the relation $1° = 4^{m}$.)

From Table 5.5, we can see that the first quarter (Aries to Gemini) and the third quarter (Libra to Sagittarius) of the ecliptic rise above the horizon with increasing amounts of time at the equator, whereas the second quarter (Cancer to Virgo) and the fourth quarter (Capricorn to Pisces) rise in decreasing times. The right ascension values of the zodiacal signs in Table 5.5 are near similar to Bhāskara II's estimates in his *Siddhāntaśiromaṇi* (II.7.16–23); see Table 5.3.

In verse 128, Nityānanda describes how the six zodiacal signs beginning from Capricorn (*nakra*), i.e., the part of the ecliptic between 270° ♑ and 90° ♊, rise above the local eastern horizon obliquely (*tiras*) going northwards (*saumya-dik*) from the equator, i.e., at a place with non-zero latitude. In other words, these six signs rise with shorter oblique ascending times. Conversely, the six zodiacal signs beginning from Cancer (*karka*), i.e., the part of the ecliptic between 90° ♋ and 270° ♐, rise above the local eastern horizon more vertically (*ṛjutva*, lit. [with] straightness); in other words, they rise with longer oblique ascending times.[38]

At northern non-zero latitudes, the oblique celestial sphere is inclined to the local horizon by an amount equal to the local latitude. The difference between the equatorial and oblique ascensions of the zodiacal signs is the ascensional difference. In essence, the oblique ascension ρ_j of the jth point on the ecliptic with longitude $\lambda_j \in [0°, 360°]$ and declination $\delta_j \in [-23°30', +23°30']$ at a terrestrial location (of latitude $\varphi°$ N) is related to its right ascension α_j by

$$\rho_j = \alpha_j - \omega_j = \alpha_j - \sin^{-1}\left[\tan\delta_j \tan\varphi\right], \tag{5.11}$$

where ω_j is the ascensional difference (*carārdha*) of the jth point on the ecliptic calculated with equation 5.6 (see notes to verses 122, 125, and 126). As Table 5.4 (fourth column) shows, the *carārdha* values corresponding to the end of the twelve zodiacal signs are positive when their declinations are positive (i.e., for signs between 0° ♈ and 180° ♍) and negative when their declinations are negative (i.e., signs between 180° ♎ and 360° ♓). This implies that the right and oblique ascensions (α_j and ρ_j respectively) of the jth ecliptic point are related to the ascensional difference ω_j as

$$\rho_j = \alpha_j - |\omega_j| \quad \forall\ \delta_j > 0 \quad \text{and} \quad \rho_j = \alpha_j + |\omega_j| \quad \forall\ \delta_j < 0$$

Table 5.4 The declination, right ascensions, ascensional differences, and oblique ascensions of the terminal cusps of the zodiacal signs at a terrestrial latitude $\varphi = 45°$ N for an ecliptic obliquity of 23° 30′.

End of the Zodiacal Sign	*Declination*	*Right Ascension*	*Ascensional Difference*	*Oblique Ascension*
Aries (30° ♈)	+11° 30′ 1″	27° 53′ 52″ E	+11° 44′ 21″	16° 9′ 31″ E
Taurus (60° ♉)	+20° 12′ 6″	57° 48′ 3″ E	+21° 35′ 24″	36° 12′ 39″ E
Gemini (90° ♊)	+23° 30′	90° E	+25° 46′ 24″	64° 13′ 36″ E
Cancer (120° ♋)	+20° 12′ 6″	122° 11′ 57″ E	+21° 35′ 24″	100° 36′ 33″ E
Leo (150° ♌)	+11° 30′ 1″	152° 6′ 8″ E	+11° 44′ 21″	140° 21′ 47″ E
Virgo (180° ♍)	0°	180° E	0°	180° E
Libra (210° ♎)	-11° 30′ 1″	207° 53′ 52″ E	-11° 44′ 21″	219° 38′ 13″ E
Scorpio (240° ♏)	-20° 12′ 6″	237° 48′ 3″ E	-21° 35′ 24″	259° 23′ 27″ E
Sagittarius (270° ♐)	-23° 30′	270° E	-25° 46′ 24″	295° 46′ 24″ E
Capricorn (300° ♑)	-20° 12′ 6″	302° 11′ 57″ E	-21° 35′ 24″	323° 47′ 21″ E
Aquarius (330° ♒)	-11° 30′ 1″	332° 6′ 8″ E	-11° 44′ 21″	343° 50′ 29″ E
Pisces (360° ♓)	0°	360° E	0°	360° E

Table 5.5 Table showing the equatorial and oblique rising times of the zodiacal signs in units of degrees of ascensional differences between the cusp of successive signs, and correspondingly, in units of time (where $1° = 4^{m}$) at a terrestrial latitude $\varphi = 45°$ N for an ecliptic obliquity of 23° 30′.

Zodiacal Sign	*Equatorial Rising Times*			*Oblique Rising Times (at 45° N)*		
Aries (0° to 30° ♈)	27 ° 53′ 52″	or	$1^{h}51^{m}35^{s}$	16° 9′ 31″	or	$1^{h}4^{m}38^{s}$
Taurus (30° to 60° ♉)	29° 54′ 11″	or	$1^{h}59^{m}37^{s}$	20° 3′ 8″	or	$1^{h}20^{m}13^{s}$
Gemini (60° to 90° ♊)	32° 11′ 57″	or	$2^{h}8^{m}46^{s}$	28° 0′ 57″	or	$1^{h}52^{m}4^{s}$
Cancer (90° to 120° ♋)	32° 11′ 57″	or	$2^{h}8^{m}46^{s}$	36° 22′ 57″	or	$2^{h}25^{m}32^{s}$
Leo (120° to 150° ♌)	29° 54′ 11″	or	$1^{h}59^{m}37^{s}$	39° 45′ 14″	or	$2^{h}39^{m}1^{s}$
Virgo (150° to 180° ♍)	27° 53′ 52″	or	$1^{h}51^{m}35^{s}$	39° 38′ 13″	or	$2^{h}38^{m}33^{s}$
Libra (180° to 210° ♎)	27° 53′ 59″	or	$1^{h}51^{m}35^{s}$	39° 38′ 13″	or	$2^{h}38^{m}33^{s}$
Scorpio (210° to 240° ♏)	29° 54′ 11″	or	$1^{h}59^{m}37^{s}$	39° 45′ 14″	or	$2^{h}39^{m}1^{s}$
Sagittarius (240° to 270° ♐)	32° 11′ 35″	or	$2^{h}8^{m}46^{s}$	36° 22′ 57″	or	$2^{h}25^{m}32^{s}$
Capricorn (270° to 300° ♑)	32° 11′ 57″	or	$2^{h}8^{m}46^{s}$	28° 0′ 57″	or	$1^{h}52^{m}4^{s}$
Aquarius (300° to 330° ♒)	29° 54′ 11″	or	$1^{h}59^{m}37^{s}$	20° 3′ 8″	or	$1^{h}20^{m}13^{s}$
Pisces (330° to 360° ♓)	27° 53′ 52″	or	$1^{h}51^{m}35^{s}$	16° 9′ 31″	or	$1^{h}4^{m}38^{s}$

Figure 5.48 depicts these two positions of the ecliptic at a latitude of $\varphi = 45°$ N. In the figure to the left, the beginning of Capricorn (270° ♑) is seen rising with oblique ascension greater than its right ascension, whereas, in the figure to the right, the beginning of Cancer (90° ♋) is seen rising with oblique ascension smaller than its right ascension. The oblique ascensions ρ_j of the end points of the twelve zodiacal signs (i.e., corresponding to the longitude λ_j of 30°, 60°, 90°, etc.) at a terrestrial latitude of $\varphi = 45°$N are tabulated in Table 5.4 (fifth column).

Like equatorial rising times, the difference between the oblique ascensions of (the terminal cusps of) successive zodiacal signs, i.e., $\Delta\rho_i = \rho_j - \rho_i$, for consecutive cusps with longitude λ_i and λ_j (such that $\lambda_j - \lambda_i = 30°$) corresponds to the times of oblique rising of the entire sign at a non-equatorial latitude. Table 5.5 lists these oblique rising times in degrees and minutes at a place with latitude 45° N. By comparing the equatorial and oblique rising times of the entire signs (in Table 5.5), we can observe how the equatorial rising times of the six ascending signs (i.e., 270° ♑ to 90° ♊ via 0° ♈ or the first and fourth quadrants in which the declination increases from −23.5°to +23.5°) are greater than their oblique rising at a northern latitude. Conversely, the oblique rising times of the six descending signs (i.e., 90° ♋ to 270° ♐ via 180° ♎ or the second and third quadrants in which the declination decreases from +23.5°to −23.5°) are greater than their equatorial rising.

This observation suggests that at a (northern) place with non-zero latitude (e.g., a place with latitude 45° N), the arc of the ecliptic beginning with the six zodiacal signs of Capricorn rises more obliquely to the local horizon, whereas the arc of the ecliptic beginning with the six zodiacal signs of Cancer rises more vertically. This fact is reflected in the measures of the oblique ascensions of the terminal cusps of these signs when compared to their right ascensions (see, e.g., Table 5.4). At the end of verse 128, Nityānanda briefly mentions how the inclination of the ascending and descending arms of the ecliptic are reversed in the southern hemisphere. In southern locations, the six zodiacal signs beginning with Capricorn rise more vertically (and hence take longer) compared to the six zodiacal signs beginning with Cancer that rise more obliquely (and hence quicker).[39]

I note that Nityānanda does provide any mathematical formulae in his *golādhyāya* to calculate the right or oblique ascensions of zodiacal signs (like, e.g., equation 5.9 or 5.11), or even the ascensional differences at various terrestrial latitudes (like equation 5.6). However, his statements on the differences in the rising times of the zodiacal signs at various terrestrial locations can be considered to be qualitatively equivalent. Most earlier Siddhāntic authors have made similar observations in their own texts; see, e.g., Brahmagupta's *Brāhmasphuṭasiddhānta* (XXI.59–60), Lalla's *Śiṣyadhīvṛddhidatantra* (XVI.13–16), Śrīpati's *Siddhāntaśekhara* (XVI.50–53, 58), and Bhāskara II's *Siddhāntaśiromaṇi* (II.7.16–23).

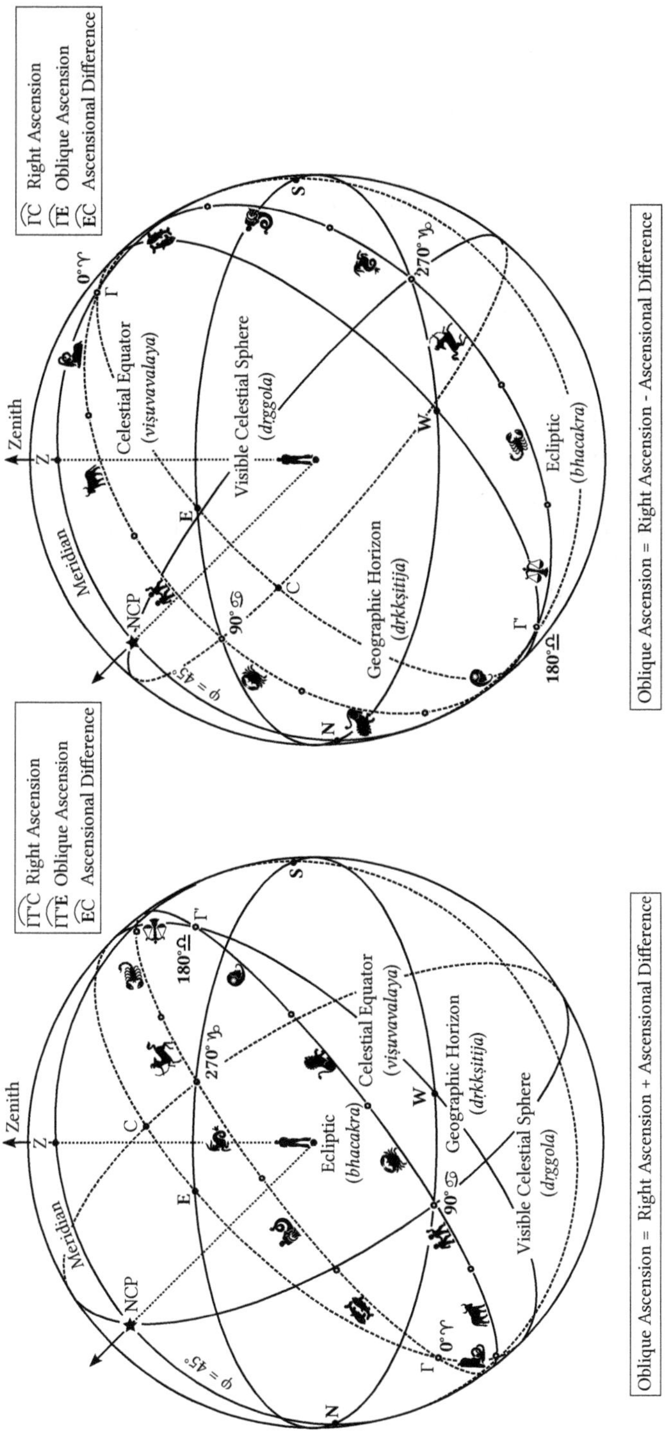

Figure 5.48 The two positions of the ecliptic at a latitude of $\varphi = 45°$ N with Capricorn (270° ♑) rising (to the left) and Cancer (90° ♋) rising (to the right). For the six signs beginning with Capricorn, the oblique ascension $\widehat{\Gamma\Gamma'\mathbf{E}}$ is the addition of the right ascension $\widehat{\Gamma\Gamma'C}$ and the ascensional difference $\widehat{\mathbf{E}C}$. Conversely, for the six signs beginning with Cancer, the oblique ascension $\widehat{\Gamma\mathbf{E}}$ is the difference between the right ascension $\widehat{\Gamma C}$ and the ascensional difference $\widehat{\mathbf{E}C}$.

Verses 129–130

Previously, in verse 88, Nityānanda defined the secondary circle to the ecliptic (*dṛkkṣepavṛtta*) as the great circle passing through the local zenith (*khasvastika*) and the ecliptic pole (*kadamba*). This secondary circle, although fixed to an observer's local zenith, is constantly moving in the sky due to the diurnal rotation of the ecliptic pole around the celestial pole. The point of intersection of this circle with the ecliptic is called the nonagesimal point (also called the central ecliptic point) that lies exactly 90° from the orient ecliptic point on the eastern horizon (*bhūja*). In Sanskrit astronomy, the nonagesimal point is variously called *tribhahīnalagna*, *vitribhalagna*, or *dṛkkṣepalagna*. Figure 5.49 shows the nonagesimal point V at the perpendicular intersection of the ecliptic $\odot\Gamma'\mathrm{CV}\Gamma\mathrm{D}$ and the secondary circle to the ecliptic $\odot$AKZB. The arcs of the ecliptic between the orient ecliptic point C and the nonagesimal point V, i.e., $\widehat{\mathrm{CV}}$, and the corresponding arc $\widehat{\mathrm{VD}}$ between the nonagesimal point and the setting ecliptic point D are both quadrants (90°). Also, the arc of the secondary circle to the ecliptic between the north ecliptic pole K and the nonagesimal point V, i.e., $\widehat{\mathrm{KV}}$, is also a quadrant.

In verses 129 and 130, Nityānanda defines two arcs of elevation on the secondary circle to the ecliptic, namely, the altitude and the zenith distance of the nonagesimal point and the ecliptic pole. These arcs are shown in Figure 5.49.

1 The zenith distance of the nonagesimal point (*dṛkkṣepacāpa*) is the arc $\widehat{\mathrm{ZV}}$ between the nonagesimal point V and the local zenith (at Z). This arc is equivalent in measure to the altitude of the ecliptic pole (*dṛkkṣepadhanus*), indicated as arc $\widehat{\mathrm{KA}}$, between the ecliptic pole K and the horizon (at A).

2 The altitude of the nonagesimal point (*dṛggati*) is the arc $\widehat{\mathrm{VB}}$ between the nonagesimal point V and the local horizon (at B). This arc is equivalent in measure to the zenith distance of the ecliptic pole (*dṛggati*), indicated as arc $\widehat{\mathrm{ZK}}$, between the ecliptic pole K and the local zenith (at Z).

The zenith distance and altitude of a celestial object (or of a conceived point on a great circle) are complementary to one another; in other words, the sum of their measures is 90°The altitudes (or zenith distances) of the ecliptic pole and nonagesimal point are related to the altitudes (or zenith distances) of the celestial pole and the celestial equator. At a given location, the elevation of north celestial pole above the local horizon to the due north, or correspondingly the depression of the celestial equator below the prime vertical, indicates the local latitude. Its complement, i.e., the co-latitude, is, accordingly, the zenithal depression

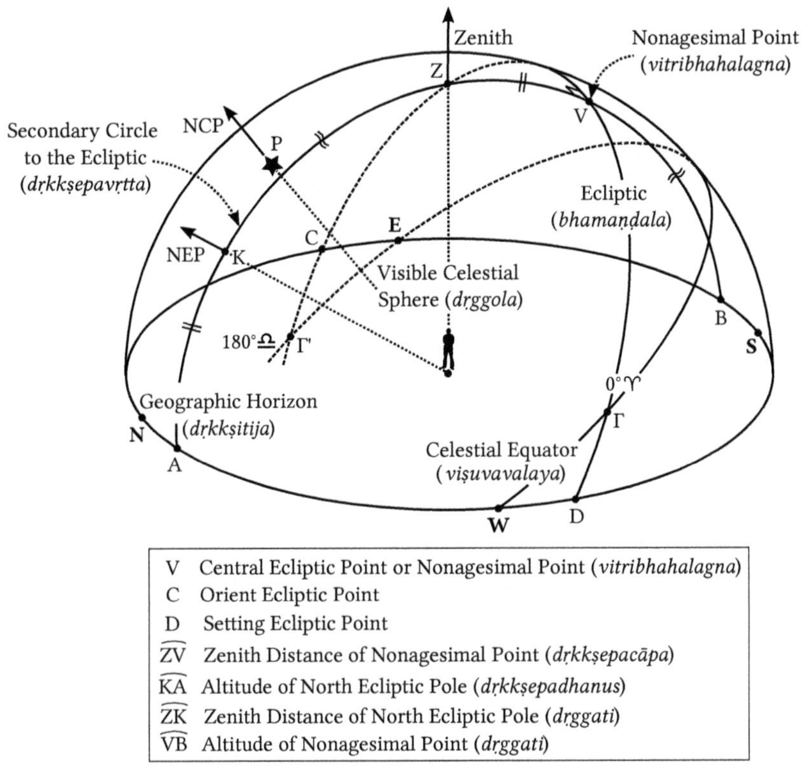

Figure 5.49 The visible celestial sphere with the arcs of the zenith distance and the altitude of the nonagesimal point and the north ecliptic pole.

of the celestial pole or the elevation of the celestial equator; see notes to verses 118 and 119. These quantities remain fixed at a particular terrestrial location independent of the time of the day. In contrast, the diurnal revolution of the ecliptic pole (around the celestial pole), and correspondingly the constant movement of the nonagesimal point, suggests that the inclinations of the nonagesimal point (on the ecliptic) and the ecliptic pole at a fixed terrestrial latitude changes throughout the day. (See Figure 5.27.)

Other Siddhāntic authors have also defined the altitude and the zenith distance of the nonagesimal point similarly; for example, Bhāskara II, in his *Siddhāntaśiromaṇi* (II.8.18), defines the *dṛkkṣepavṛtta* as the azimuth circles for the nonagesimal point and calls the sine of the zenith distance of the nonagesimal point (or the ecliptic latitude of the zenith) the *dṛkkṣepa*.

Verse 131

In the course of a day, as a planet (like the Sun) moves along its circle of day-radius (*dyujyāvṛtta*), its degrees of elevation (*unnata-aṃśa*) above the local horizon (*dṛkkṣtija*) as it crosses the intercardinal circle (*koṇamaṇḍala*) and the prime vertical (*samamaṇḍala*) are called its intercardinal altitude (*koṇaśaṅku*) and prime vertical altitude (*samanara*) respectively. In verses 86 and 87, Nityānanda defined the prime vertical and the intercardinal circles as the local vertical circles of altitude passing through certain specific directions on the local horizon. Here, in verse 129, he specifies the technical names of the altitude (*śaṅku*) of the points on these circles that intersect the circle of day-radius of a planet.[40]

Figure 5.50 shows the instantaneous positions R_1 and R_2 of a planet as it transits the intercardinal circle ⊙**NE**-Z-**SW** and the prime vertical ⊙**EZW** respectively, and correspondingly, the shorter arcs $\widehat{R_1 - \mathbf{SW}}$ and $\widehat{R_2\mathbf{W}}$ are the planet's intercardinal altitude and prime vertical altitude respectively.

Other Siddhāntic authors have also described the different types of solar altitudes corresponding to the Sun's transit across various vertical directional circles in the course of a day; see, e.g., Brahmagupta's *Brāhmasphuṭasiddhānta* (XXI.62–63), Lalla's *Śiṣyadhīvṛddhidatantra* (XV.22), Śrīpati's *Siddhāntaśekhara* (XVI.41), and Bhāskara II's *Siddhāntaśiromaṇi* (II.7.37).

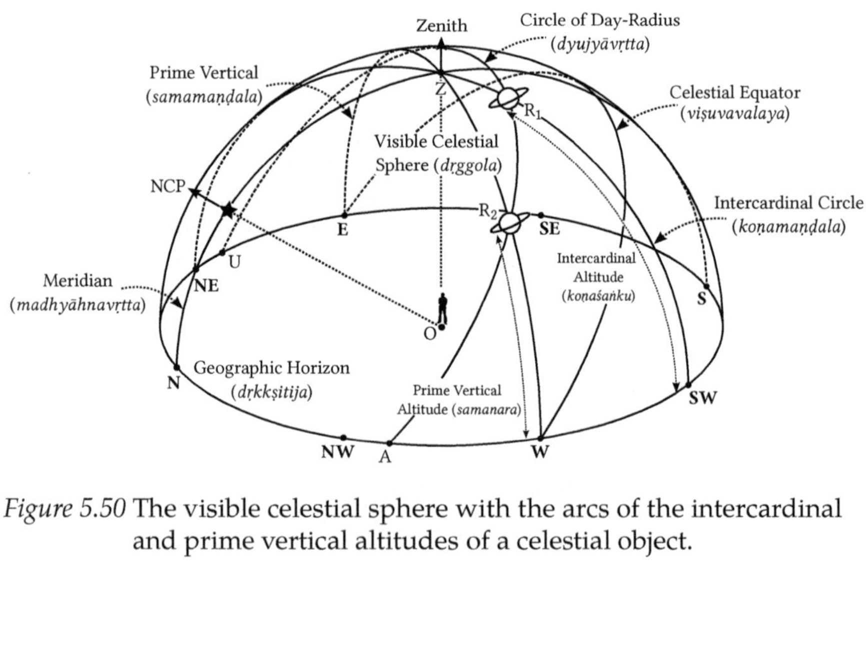

Figure 5.50 The visible celestial sphere with the arcs of the intercardinal and prime vertical altitudes of a celestial object.

Verse 132

When the Sun (or more generally, any celestial object) transits the local equinoctial colure (*unmaṇḍala*), its degrees of elevation (*unmiti*) above the local horizon on its vertical circle of altitude (*dṛṅmaṇḍala*) is called its equinoctial colure altitude (*udvṛttaśaṅku*). In verses 87, Nityānanda defined the vertical circle of altitude of a celestial object as the direction circle passing through the local zenith and a desired direction: here, the desired direction is the one that points towards the Sun, i.e., $\odot$MRZT in Figure 5.51 where the circle of day-radius $\odot$URA of the Sun intersects the equinoctial colure $\odot$EP**W** at R. The equinoctial colure altitude of the Sun is the shorter arc $\widehat{\mathrm{RM}}$ (or $\angle$ROM), and correspondingly, the complement of its altitude, i.e., the zenith distance of the Sun on the equinoctial colure, is $\widehat{\mathrm{ZR}}$ (or $\angle$ZOR). It can be easily seen that $\widehat{\mathrm{ZR}} + \widehat{\mathrm{RM}} = 90^\circ$ (i.e., $\angle\mathrm{ZOM} = \angle\mathrm{ZOR} + \angle\mathrm{ROM} = 90^\circ$.

In the last *pāda* of verse 132, Nityānanda calls the Sine of the zenith distance of the Sun (*dṛgjyā*) measured on the vertical circle of altitude intersecting through the equinoctial colure, i.e., Sin $\widehat{\mathrm{ZR}}$, as the degrees of the perpendicular (*bhujā-aṃśa*) in one's own direction (*svadiś*). This corresponds to the distance QR in Figure 5.51. By dropping the perpendicular RL to the horizontal (radial) direction OM towards the Sun (with respect to an observer at O), we observe that the right-angled triangles $\triangle$ORL $\sim$ $\triangle$OQR since $\angle\mathrm{RLO} = \angle\mathrm{RQO} = 90^\circ$. Thus,

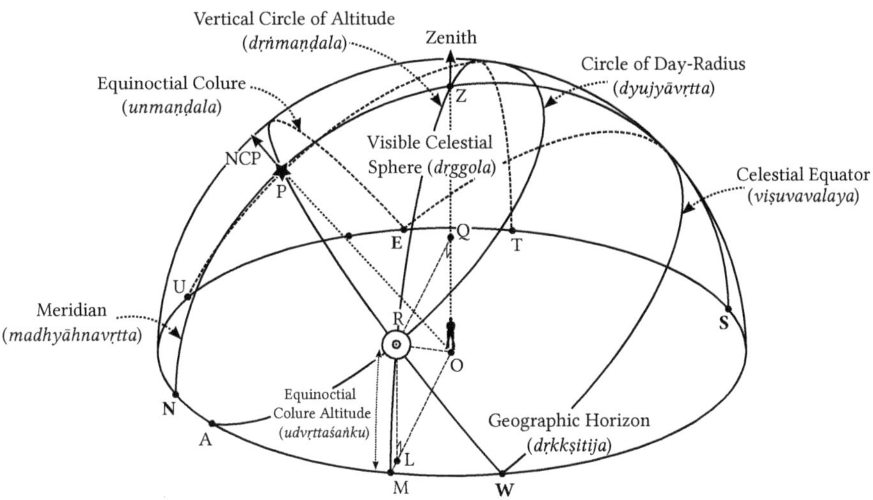

Figure 5.51 The visible celestial sphere with the complementary arcs of the equinoctial colure altitude and zenith distance of the Sun on its vertical circle of altitude.

1. the *koṭis* (vertical perpendiculars) RL = Sin $\widehat{RM}$ = Sin ∠ROM and QO = Cos $\widehat{ZR}$ = Cos ∠ZOR are equivalent. In other words, the Sine of the equinoctial colure altitude of the Sun is equal to the Cosine of its zenith distance on its vertical circle of altitude traversing the local equinoctial colure.

2. And the *bhujās* (horizontal bases) OL = Cos $\widehat{RM}$ = Cos ∠ROM and QR = Sin $\widehat{ZR}$ = Sin ∠ZOR are equivalent. In other words, the Cosine of the equinoctial colure altitude of the Sun is equal to the Sine of its zenith distance on its vertical circle of altitude traversing the local equinoctial colure.

The distance RL or QO (corresponding to $\widehat{RM}$) is the vertical altitude (synonymously called *śaṅku*, *nṛ*, or *nara*) of the Sun on the equinoctial colure, and QR or OL (corresponding to $\widehat{ZR}$) indicates the horizontal distance (*dṛgjyā*) of the Sun on its vertical circle of altitude at the moment it transits the equinoctial colure.

Earlier Siddhāntic authors have also used the arcs of altitude and zenith distance of a celestial object (along a particular vertical circle of reference) to define the *śaṅku* (elevation) and *dṛgjyā* (horizontal distance). For example, Vaṭeśvara, in his *Vaṭeśvarasiddhānta* (II.3.17), says, *sarvatraiva narāḥ samunnataguṇā dṛgjyā natajyāḥ kramāt* 'in all positions of the heavenly body, the Sine of the altitude is called *nara* and Sine of the zenith distance is called *dṛgjyā*'. Also, see endnote 40 on page 329 for Bhāskara II's definitions of *śaṅku* and *dṛgjyā*.

Verse 133

In Sanskrit mathematics, the complementary arcs in a quadrant of a circle are called *bhujā* and *koṭi*. Bhāskara I, in his seventh-century work *Laghubhāskarīya* (II.1-2), defined the *bhujā* (or *bāhu*) as the 'arc traversed' and *koṭi* as the 'arc to be traversed' in odd quadrants, and the reverse in even quadrants.[41] In each quadrant, the Sines and Cosines of the *bhujā* and *koṭi* are mutually related. Based on the appropriate *bhujās* and *koṭis*, other trigonometric ratios, e.g., the versed sines or sagittas (*viśikha*), can be accordingly calculated.

Figure 5.52 shows the four quadrants of a circle with the radial segments OE, OG, OF, and OH all measuring $\mathscr{R}$, the *sinus totus*. In each of these quadrants, the Sines and Cosines of corresponding complementary *bhujās* and *koṭis* are related as follows:

1. In the first quadrant, $\widehat{AE}$ is the *bhujā* and $\widehat{EC}$ is the *koṭi* with

$$\text{Sin}\,\widehat{AE} = EI = \text{Cos}\,\widehat{EC} = OK \text{ and } \text{Cos}\,\widehat{AE} = OI = \text{Sin}\,\widehat{EC} = EK.$$

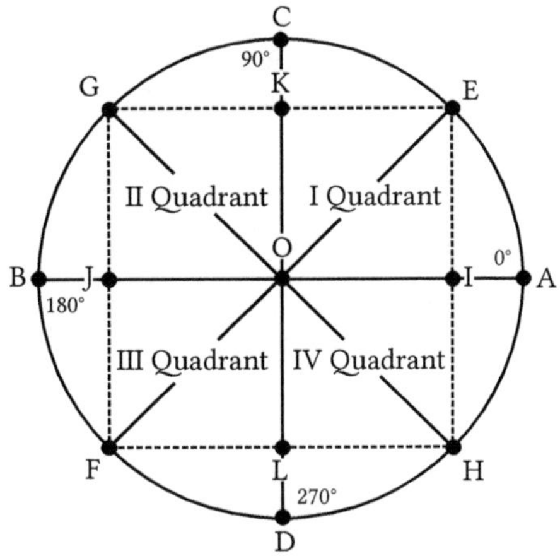

Figure 5.52 The complementary *bhujā* and *koṭi* arc in the four quadrants of a circle.

2 In the second quadrant, $\overset{\frown}{GB}$ is the *bhujā* and $\overset{\frown}{CG}$ is the *koṭi* with

$$\text{Sin}\,\overset{\frown}{GB} = GJ = \text{Cos}\,\overset{\frown}{CG} = OK \text{ and } \text{Cos}\,\overset{\frown}{GB} = OJ = \text{Sin}\,\overset{\frown}{CG} = GK.$$

3 In the third quadrant, $\overset{\frown}{BF}$ is the *bhujā* and $\overset{\frown}{FD}$ is the *koṭi* with

$$\text{Sin}\,\overset{\frown}{BF} = FJ = \text{Cos}\,\overset{\frown}{FD} = OL \text{ and } \text{Cos}\,\overset{\frown}{BF} = OJ = \text{Sin}\,\overset{\frown}{FD} = FL.$$

4 In the fourth quadrant, $\overset{\frown}{HA}$ is the *bhujā* and $\overset{\frown}{DH}$ is the *koṭi* with

$$\text{Sin}\,\overset{\frown}{HA} = HI = \text{Cos}\,\overset{\frown}{DH} = OL \text{ and } \text{Cos}\,\overset{\frown}{HA} = OI = \text{Sin}\,\overset{\frown}{DH} = HL.$$

Corresponding to the Sine (*jyā* or, more specifically, the *jyārdha*), the versed Sines (*viśikha*s) in the different quadrants are

1 Vers $\overset{\frown}{AE}$ = Vers $\overset{\frown}{HA}$ = IA, where IA = OA – OI = $\mathscr{R}$ – Cos $\overset{\frown}{AE}$ or $\mathscr{R}$ – Cos $\overset{\frown}{HA}$.

2 Vers $\overset{\frown}{EC}$ = Vers $\overset{\frown}{CG}$ = KC, where KC = OC – OK = $\mathscr{R}$ – Cos $\overset{\frown}{EC}$ or $\mathscr{R}$ – Cos $\overset{\frown}{CG}$.

3 Vers $\overset{\frown}{GB}$ = Vers $\overset{\frown}{BF}$ = JB, where JB = OB – OJ = $\mathscr{R}$ – Cos $\overset{\frown}{GB}$ or $\mathscr{R}$ – Cos $\overset{\frown}{BF}$.

4 Vers $\overset{\frown}{FD}$ = Vers $\overset{\frown}{DH}$ = LD, where LD = OD – OL = $\mathscr{R}$ – Cos $\overset{\frown}{FD}$ or $\mathscr{R}$ – Cos $\overset{\frown}{DH}$.

In Sanskrit astronomy, trigonometric ratios (defined in the spherical triangles made by intercepting small and great circles on the celestial sphere) are integral to the computations of various astronomical quantities. These computations often rely on using the rule of three (*trairāśika*) to compute a desired quantity (*icchā-phala*) where the known ratios are trigonometric functions; see, e.g., Varāhamihira's *Pañcasiddhāntikā* (IV.5), Brahmagupta's *Brāhmasphuṭasiddhānta* (XXI.23), or Bhāskara II's *Siddhāntaśiromaṇi* (II.5.5). Also, see Datta, Singh, and Shukla (1983, pp. 50–56)for a more detailed discussion on the various trigonometric functions used in Sanskrit astronomy.

Verse 134

An armillary sphere (*gola*, *golayantra*, or *golabandha*) is an assembly of movable and immovable rings that represent the different circles in the celestial sphere and is pivoted on a central axis. The rings are calibrated to indicate various measurements (or at least, demonstrate the effect of the movement of the celestial circles) and are fabricated (*racanā*) from different materials. In verse 134, Nityānanda provides a brief list of the materials typically used to construct the different parts of the armillary sphere. According to him, the spheres of the planets are made using clay (*mṛd*), while the circular rings (representing the different celestial circles) are made from metal (*dhātu*), wood (*kāṣṭhā*), straight pieces of bamboo (*śarala-vaṃśa-śalākikā*), animal horns (*viṣāṇa*), or ivory (*kuñjara-danta*).

In Siddhāntic literature, most authors have emphasised the importance of the armillary sphere as a tool to understand the geometry of the heavens; for example, Lalla, in his *Śiṣyadhīvṛddhidatantra* (XIV.4), says, *saṅkhyāya śāstravid eva vadanti golaṃ golāgame puṭadhiyo grahaceṣṭitajñāḥ* 'the astronomer stresses that the armillary sphere is solely (an object) of mathematical calculations; and that those who wish to study the planets must be experts in the subject of the sphere' (Chatterjee 1981, p. 227). Siddhāntic authors often dedicated separate chapters (or sections) on the construction and composition of the armillary spheres wherein they discussed the different types of materials used in constructing its different parts. For example, Āryabhaṭa, in his *Āryabhaṭiya* (IV.22), recommends

> *kāṣṭhamayaṃ samavṛttaṃ samantataḥ samaguruṃ laghuṃ golam |*
> *pāradatailajalaistam bhramayet svadhiyā ca kālasamam ||*
>
> The Sphere (*gola-yantra*) which is made of wood, perfectly spherical, uniformly dense all around but light (in weight) should be made to rotate keeping pace with time with the help of mercury, oil and water by the application of one's own intellect.
>
> (Shukla and Sarma 1976, p. 129)

Sūryadeva Yajvan, commenting on this verse, suggested that bamboo strips (*venu-śalākikā*) should be used to demarcate the different reference circles on the wooden sphere (K. V. Sarma 1976, p. 143). Bhāskara I added that *śrīparṇi* wood (*Gmelina arborea* or *gamhar*), a type of native Indian beech, should be used to construct the wooden armillary sphere (or parts thereof) (Shukla 1976, p. 268). In fact, in an earlier discussion on the best type of wood for constructing the armillary sphere, Bhāskara I suggests using wood of the *vañcula* tree (in addition to the *śrīparṇi* wood) as an alternative (Ibid, p. 240).[42] A little later, he says that bamboo strips (*vaṃśa-śalākikā*), or perhaps hollow reeds, can also be used to construct the armillary sphere (Ibid, p. 243). These different suggestions are indicative of the different methods and materials with which Siddhāntic authors fabricated the armillary sphere. See Ôhashi (1994, pp. 259–273), K. V. Sarma (2008), S. R. Sarma (2019, pp. 3383–3408), and Devadevan (2016) for fuller discussions on the history of armillary spheres in Sanskrit astronomy.

Notes

1 Śaṅkara Vāriyar, a student of Nīlakaṇṭha Somayājī in a long line of Sanskrit mathematicians from the Kerala (Niḷā) school of astronomers and mathematicians, remarks on the importance of a *maṅgalācaraṇa* in his *Laghuvivṛti* (ca. 1550) by saying *maṅgalācārayuktānāṃ vinipāto na vidyate* 'those who follow the practice of starting a work with invocation do not suffer a fall' (Ramasubramanian and Sriram 2011, p. 2).

2 Following the *Śarvadarśanasaṅgraha* of Mādhava Vidyāranya (ca. 1331), a medieval compendium on Indian philosophical thought, the Sāṃkhya, Śaiva, and Vedānta doctrines can be briefly summarised as follows:

1 Sāṃkhya is the philosophical doctrine of dualistic realism where the ultimate reality has a dual basis in the dynamic (active) state of Nature and static (passive) state of the Self. The primitive cause (Prakṛti or Pradhāna), devoid of conscience, is the active agency that forms the material substratum upon which all things (gross and subtle) base their existence, while the supreme self of pure consciousness (Puruṣa) is without any corporeality and forms the static transcendental subject of all experience. According to the Sāṃkhyas, the evolution of the world is brought about by the contact (*saṃyoga*) between these two separate entities.

2 Śaiva (or Śaivism) is the religious doctrine based on different philosophical schools of dualism (*bheda*), non-dualism (*abheda*), and both dualism and non-dualism (*bhedābheda*). Despite a lack of unity in their philosophical, esoteric, or ritualistic practices, the adherents of these schools still consider Śiva, the auspicious one, as the Supreme Being.

3 Vedānta (lit. the end of the Vedas) is the philosophical view characterised by the Vedic (Upaniṣadic) concept of Brahman—the self-existing consciousness of bliss (*sat-cit-ānanda*) that is the eternal and absolute nature of reality. Over the course of its history, varying opinions on (i) the nature of the supreme Brahman in relation to the individual Ātman (inner self or soul), (ii) the ultimate ontological status of the world (*jagat*) as

real (*satya*) or unreal (*mithyā*), and (iii) soteriological and theophilosophical questions have led to different schools of Vedāntic thought, namely, non-dualism (*advaita*), dualism (*dvaita*), qualified non-dualism (*viśiṣṭa-advaita*), inconceivable dualism and non-dualism (*acintya-bheda-abheda*), etc.

3 As briefly described in endnote 2, the Sāṃkhya cosmology speaks of Prakṛti as the unconscious primordial materiality and Puruṣa as the transcendental self of pure consciousness; the contact of these two principle realities causes the coalescence of the unmanifested Prakṛti (as *jaḍa* or insentient matter) into the gross elements of the *pañca-mahābhūta* by dynamically and energetically affecting the three substantive and innate tendencies or *guṇas*—lucidity (*sattva*), activity (*rajas*), and lethargy (*tamas*)—that constitute all things in the universe. This material cosmogony is what makes the classical elements have an inherent qualia (e.g., the heat of fire or fluidity of water) and, by extension, the absence of any motion of the *terra firma* of the Earth. The Advaita Vedāntic idea that *māyā* is the inherent and inseparable part of Brahman (*īśvara*) and the manifestation of all material existence through its energy (*śakti*) hints at a similar theme of a quiescent energy inherent in all things including the Earth.

4 A *yojana* is an Indian measure of linear distance approximately equivalent to a distance of about five to eight miles (i.e., on the order of around ten kilometres). Historically, different Sanskrit mathematicians have understood this measure differently in different time periods, and hence, an exact value of a *yojana* in modern units is indeterminate.

5 al-Bīrūnī expressed a similar doubt on the city of Siddhapurī being inhabited when he said, 'How the Hindus came to suppose the existence of Siddhapura I do not know, for they believe, like ourselves, that behind the inhabited half-circle there is nothing but unnavigable seas' (Sachau [1910] 2013, p. 304).

6 In Vaiṣṇava Vedāntic theosophy, the movement of the celestial sphere is seen as the eternal movement of the wheel of time, the wheel of the universe, and the wheel of the *yugas*—a movement that has the divine will of Keśava (Kṛṣṇa) as its primal cause; see, e.g., the *Mahābhārata* (V.66.12).

7 A descriptive essay with digital images of one of Ṣādiq Iṣfahānī's maps (from MS Egerton 1016, f. 335r, held at the British Library in London) can be seen at http://www.myoldmaps.com/early-medieval-monographs/204-map-of-the-inhabited/204-sadiq-isfahani.pdf.

8 The word *ghaṭī* (synonymous with *ghaṭīkā*, *nāḍī*, *nāḍikā*) is a unit of time equal to twenty-four minutes, or equivalently, one-sixtieth of a twenty-four-hour day. This measure is extensively used in Sanskrit astronomy to indicate day-lengths, e.g., Bhāskara II, in his auto-commentary *Vāsanābhāṣya* on his *Siddhāntaśiromaṇi* (I.1.16–18), states *ghaṭīnāṃ ṣaṣṭyā dinam* 'the nychthemeron (*dina*) is sixty *ghaṭīkās*'. A *ghaṭīkā* is further divided into the smaller units of 60 *vighaṭis* or 3600 *liptās*.

9 Ptolemy's *Almagest* (Μαθηματικὴ Σύνταξις) and *Geography* (Γεωγραφικὴ Ὑφήγησις), both written in the second century, describe the division of the surface of the Earth (oecumene) into different parallels of latitude, segregated from one another by an isochronous distribution of daylight hours across each parallel circle. Going from the equator to the polar regions, the hours of daylight increase asymmetrically with the corresponding (uniform) increase in latitude; see Ptolemy's *Almagest* (II.6) (Toomer 1998, pp. 82–89).

Historically, beginning from the days of Eratosthenes (ca. second century BCE) all the way up until Ptolemy's predecessor Marinus of Tyre (ca. 70–130

BCE), the angle of inclination of the axis of the celestial sphere and the plane of the horizon (κλίμα), or effectively the elevation of the polestar above the northern horizon, was considered as the measure of latitude expressed in terms of the length of the longest solstitial day. This helped provide a traditional division of the Earth's surface into latitudinal strips or *klimata* where the solstitial day-length was more or less uniform in each *klima*. Classical authors from various schools wrote different lists of *klimata* covering a range of latitudes. By Ptolemy's time, seven *klimata* had been accepted as convention. Ptolemy himself chose his seven climes ranging from around 17° N (13 hours of daylight) to 48° N (16 hours of daylight) (Berggren and Jones 2000, p. 10).

In his *Almagest* and *Geography*, Ptolemy provides two different reasons to visualise the Earth's surface as zones. In the *Almagest*, the division of the Earth's latitudes helps determine the rising and setting times of celestial objects for different terrestrial observers, whereas, in *Geography*, the latitudinal *climes* of the Earth are based on daylight hours to describe the Earth's physical geography (Pedersen 2011, p. 109).

10 An identical expression for calculating the ascensional difference (*cara*) of the Sun can be found in most Siddhāntic texts, e.g., Brahmagupta's *Brāhmasphuṭasiddhānta* (II.55–59), Lalla's *Śiṣyadhīvṛddhidatantra* (III.17–19), Bhāskara's *Siddhāntaśiromaṇi* (II.8.47–49, 53), Nīlakaṇṭha Somayājī's *Tantrasaṅgraha* (II.23–27), or Jñānarāja's *Siddhāntasundara* (II.1.34–35). The *cara* is the arc of the celestial equator lying between the six o'clock hour circle (arc PE in Figure 5.12) and the hour circle of the Sun at rising (arc PB)—in other words, ∢ **E**PB (measured in degrees of arc) equal to the hour angle ∢ ZPS* (also expressed in degrees of arc).

11 Pingree (1996b, p. 186) describes MS α 424 (of anonymous authorship, ca. seventeenth century) from the collection of the Wellcome Institute in London that contains a miscellaneous assortment of astrological and astronomical material. F. 59 of the manuscript describes the distances (in *yojanas*) between the different cities on the Indian prime meridian in a Sanskrit verse (in the *sragdharā* metre). The verse declares Gargarāta to be at distance of nine *yojanas* from Avanti (Ujjain).

12 The word *śara*, lit. arrow or shaft, often implies the 'height of an arc or segment of a circle' (Sarasvati Amma [1979] 2007, p. 265). In the present case, it refers to the height of the clime that forms the spherical segment/zone on the sphere. In other words, the latitudinal extent measures in *yojanas*. The word *śara* is also used to refer to the ecliptic latitude of a celestial body in verse 116.

13 MSS Br, Np, Pm, and Rr contain an alternative reading of the first *pāda* of this verse where they explicitly refer to the periphery of the heavens (*vyomakakṣāvadhisthaḥ*) as the region as far as the [outermost] orbit of the stars (*ṛkṣākhyakakṣā*).

14 As noted in endnote 64 on page 56, I am currently preparing a critical edition (with an annotated English translation) on these twenty-one verses on the formation of the universe from Nityānanda's *Sarvasiddhāntarāja* (I.3.180–200) in association with Christopher Minkowski.

15 Gerd Graßhoff, in his book *The History of Ptolemy's Star Catalogue*, describes the developments to Ptolemaic star catalogue as the 'history of [its] interpretations by its readers', a history where discussions on the astronomy of fixed stars fuelled a long succession of revisions and emendations to the original catalogue by Arabic and European scholars (Graßhoff 2013, p. 5).

16 For example, see f. 1v (on p. 16) of MS NAK_NGMCP_A_1068_36_55+10ff (bundled collation, incomplete), https://catalogue.ngmcp.uni-hamburg.de/receive/aaingmcp_ngmcpdocument_00098177.

17 See Pingree (1978a, pp. 564–565) for the list of junction stars (from the *Paitāmahasiddhānta*) along with their polar coordinates, equivalent ecliptic coordinates, and an approximate identification with the stars from Ptolemy's catalogue.

18 The word *ananta*, lit. without an end, is more commonly understood as endless or infinite. However, in the notes to verse 84, we learn that the outermost orbit of constellations (*nakṣatrakakṣa*) is 150,086,000 *yojana*s which limits the physical (spatial) dimensions of the celestial sphere. However, understanding the celestial sphere as an eternally existent structure allows for the temporal meaning of *ananta* to be preserved.

19 The vertical circle of altitude (*dṛṅmaṇḍala*) is sometimes referred to as the *iṣṭa-digvṛtta* 'desired direction circle', and its secondary circle (perpendicular to it) is called the *viparita-digvṛtta* 'reversed direction circle' (see, e.g., Ramasubramanian and Sriram 2011, p. 469). Nityānanda does not specify a secondary to the vertical circle of altitude (in verse 87); he does, however, identify the vertical circle as passing through the *iṣṭa-dik* 'desired direction' and then *tad-viparita-kāṣṭhā* 'its reversed direction'.

20 The word *pūrva* is polysemous—it can be variously translated as 'first', 'previous', 'aforesaid', 'eastern', or even 'ancient' and 'traditional'. This makes it difficult to establish a fixed meaning of the expressions *pūrvābhidaṃ*. In the present context, the geometrical description of the small circle (in the last four *pāda*s of verse 90) describes the parallel of meridian. However, without a historical precedent of the Sanskrit equivalent, I understand the name *pūrva* (*pūrvābhidaṃ*) as a reference to the name *madhya* (*madhyābhidaṃ*) in the first *pāda* of verse 90. In both contexts, the word *madhya* refers to the meridian, with *sūtra* being the great circle and *vṛtta* as the small circle. Understood differently, the parallel of the meridian identified as *pūrvavṛtta* (instead of *madhyavṛtta*) could suggest the 'first/ancient/oriental circle'.

21 The Arabic word for the parallel circles of altitude is *dāʾira al-irtifāᶜ* or *al-muqanṭarah*, which in Latin came to called *almucantar*. Celestial objects on the almucantars are considered to have equal elevation.

22 In Indian astrology, the astrological houses (*bhāva*s), and the degrees of the zodiac that occupy the houses at a given place and time, marked ominous occasions in the course of one's life and reveal the characteristics of an individual through divination. The Indian genethlialogy practices of drawing natal charts (*jātaka*) and casting horoscopes (*janmapatrikā*) relied extensively on these fixed house divisions. The prognostication of events, actions, benefits and losses in one's life depended on the conjunction (*yoga*) of the heavenly bodies in these fixed houses.

23 In Islamicate (Arabic) texts, these four cardinal houses-cusps (*al-watad*, pl. *al-autād*) are commonly called (i) the first house of the Ascendant (*al-ṭāliᶜ*), (ii) the fourth house of Imum Coeli or lower culmination (*al-rābiᶜ*), (iii) the seventh house of the Descendant (*al-ghārīb* or *al-sābiᶜ*), and (iv) the tenth house of the Medium Coeli or upper culmination (*al-ᶜāshir* or *wasaṭ al-samāᶜ*).

24 In verse 94, Nityānanda refers to the cardinal house cusps corresponding to the ecliptic positions by different names: the first house cusp or Ascendant is called *lagna*, the fourth house cusp or Imum Coeli is called the house of happiness (*sukha*) or simply the fourth (*caturtha*), the seventh house cusp or

Descendant is called the house of women (*strī*), and the tenth house cusp or Medium Coeli is called the sky (*kha* or *viyat*).

25 In the Indian system of natal astrology based on dividing the ecliptic into houses that are then identified with the zodiacal signs, the Pārāśarī system (derived from Parāśara's *Bṛhatpārāśarahorā*, ca. seventh century) regards the first house cusp with the exact degrees of the ascending sign. Subsequently, each house then extends 30° along the ecliptic, with the cusp of the tenth house being the nonagesimal point (*vitribhalagna* or central ecliptic point). In this Pārāśarī system of houses of equal extent, two zodiacal signs can occupy the same house.

Alternatively, according to the Jaiminīya system (derived from Jaimini's *Jaiminisūtra*, sometimes known as *Upadeśasūtra*, ca. fourth century), the first house corresponds to the zero degree of the ascending sign and not the actual degree of the mounting point (*ārūḍha*). Each house is then simply equal to an entire zodiacal sign, with the cusp of the tenth house located at beginning of the zodiacal sign at the nonagesimal point. This sign-wise equal house division of the zodiac is among the most common system of house divisions in Indian natal astrology.

A third system of Śrīpati (described in his *Śrīpatipaddhati*, ca. eleventh century) regards the cusps of the houses as the middle of the house (*bhāvamadhya*). The cusp of the angular houses (first, fourth, seventh, and tenth) corresponds to the cardinal positions of the ecliptic (Ascendant, Imum Coeli Descendant, and Medium Coeli), and trisecting the intercardinal sectors of the ecliptic produces the boundary between the houses (*bhāvasaṃdhi*). The start of the first house is 15° below the Ascendant and the second house begins, accordingly, 15° above the Ascendant. Directly opposite these houses, the start of the seventh house is 15° below the Descendant while the beginning of the eight house is 15° above the Descendant. Here, the triads of houses in opposing quadrants are equal in length even as all the houses are not equal.

26 At the Arctic Circle, the momentary horizontal position of the ecliptic is a regular diurnal phenomenon independent of the position of Sun along the ecliptic. Although Nityānanda makes the case for the summer solstitial position of the Sun, the simultaneous ascension (and descension) of opposing sets of six zodiacal signs immediately following the movement of the ecliptic from the horizon is a regular phenomenon every day of the year.

27 The twenty-seven *nakṣatra*s together constitute the entire ecliptic and hence each *nakṣatra* extends across 13° 20′ (the 27th part of 360°) of ecliptic longitude. Accordingly, in regular order, the fifth *nakṣatra* called *indubha* or *mṛgaśirṣa* extends from 53° 20′ to 66° 40′ of ecliptic longitude.

28 The solstitial colure (*ayanavṛtta*) is a great circle of the celestial sphere that passes through the celestial poles and the solstitial positions of the Sun on the ecliptic. Along with the equinoctial colure (*unmaṇḍala*), see verse 88, it forms the two main secondary circles to the celestial equator in the celestial sphere.

29 In Islamicate (Arabic) astronomy, the orb of a planet (*falak*, pl. *aflāk*) and sphere of a planet (*kurah*, pl. *kurāt* or *ukar*) carried different meanings. The orb of a planet often referred to the concentric or eccentric solid structure into which the sphere of the planet was embedded. And even though the distinction between these two terms has never been consistently applied, they have, however, been used at times to distinguish the spherical bodies of the planets (sing. *jism kurī*) from the orbs of the heavens in which they were embedded. See Janos (2012, Section II.1.1 on pp. 115–119) for a wider discussion on the

planetary spheres and orbs in the context of Islamicate astronomy of the late tenth and early eleventh centuries.

30 In the tables of the *Siddhāntasindhu*, Nityānanda often uses the expression *prathama-krānti* 'first declination' (in the table titles) to differentiate it from the second declination; e.g., the title *meṣatulayoḥ prathamakrāntikoṣṭhakāḥ* 'table of first declinations of [the zodiacal signs] Aries and Libra' seen on f. 47v of MS 4962 from the Khasmohor Collection held at the City Palace Library of Jaipur. Also, in the chapter on computations (*gaṇitādhyāya*) of his *Sarvasiddhāntarāja* (I.4.49–50), Nityānanda explicitly refers to the second declination as *dvitīya* 'second' even as he calls it *para* or *anya* (both meaning 'other') elsewhere in the text.

31 Figure 5.40 depicts the position of the planet with its ecliptic latitude and second declination similarly oriented; in other words, the ecliptic latitude and second declination of the planet are in the same direction north of the celestial equator. Hence, here, the curve of true declination is the sum of the two quantities. When the ecliptic latitude and second declination are oppositely oriented, the curve of true declination is the difference between its second declination and ecliptic latitude; see Misra (2022).

32 Typically, in Siddhāntic astronomy, the ecliptic and polar coordinate systems are not differentiated; in other words, the ecliptic coordinates (β, λ) of a planet are considered equivalent to its polar coordinates (β_*, λ_*). Hence, the true declination $\widehat{RA} = \widehat{FA} + \widehat{RF} \approx \widehat{DB} + \widehat{RD}$, or $\delta_{\text{true}} = \delta_1^*(\lambda_*) + \beta_* \approx \delta_1(\lambda) + \beta$; see Plofker (2000, pp. 39–41).

33 In verse 93, Nityānanda called the meridian ecliptic point the *madhyalagna*, lit. the central point of contact. Again, in verse 94, he refers to the meridian ecliptic point as the Medium Coeli (mid-heaven) when speaking about the astrological divisions on the sky. In the present verse, he uses the expression *viyadākhyalagna*, lit. the point of contact in the sky, as a synonym of the word *madhyalagna*.

34 Nityānanda does not explicitly mention the meridian; he simply says that the distance measured between the horizon and the disk of the planet (*kṣamāja-graha-bimba-madhye*) should be considered as the degrees of maximum elevation (*paronnatāṃśa*) by wise men (*vibudha*). The use of the word *madhye* 'in the middle' or 'in between' is an oblique reference to *madhyāhnavṛtta*, a technical expression for the meridian or midday circle.

35 In Sanskrit astronomy, the azimuth is typically measured with respect to the due east (or due west) direction of the local horizon, whereas in western positional astronomy, it is measured from the due north direction. If the degrees of azimuth (*diś-aṃśaka*) are measured from the due north, its complement is sometimes called the amplitude (*diś-agra* or *āśā-agra*); see, e.g., the *Tantrasaṅgraha* of Nīlakaṇṭha Somayājī (Ramasubramanian and Sriram 2011, pp. 229 and 468).

36 To convert the arc of day (*dina-cāpa*) in degrees to the measure of day-length (*dina-pramāṇa*) in hours, minutes, etc., we note that $360^\circ \leftrightarrow 24^h$, and hence, $1^\circ = 4^m$.

37 In his discussions here, Nityānanda does not mention the equation of time corrections due to the eccentricity of the Earth's orbit (the *bhūjāntara* correction) and the ecliptic obliquity (the *udayāntara* correction) that affect the equality in the lengths of day and night at the equator.

38 In verse 128, Nityānanda uses the terms 'ascending' and 'descending' to refer to the six zodiacal signs beginning with Capricorn and Cancer respectively. These terms relate to the change in the solar declination as the Sun moves through the zodiac. In going from the beginning of Capricorn to the end of

Gemini, the solar declination increases towards the north and hence these signs are considered ascending. Conversely, in the Sun's southern journey from the beginning of Cancer to the end of Sagittarius the solar declination decreases and hence the signs are termed descending.

39 In verses 98–100, Nityānanda has discussed the oblique ascension of the zodiacal sign and the sidereal day-lengths at the Arctic Circle (66° 30′) at different times of the year. Therein, in verse 100, he mentions that on the day of the summer solstice in the northern hemisphere, the oblique ascension of the zodiacal signs at the Arctic Circle is twice their right ascension (at the equator). In verses 127 and 128, we find his general observations on the right and oblique ascensions of the zodiacal signs (i.e., the ecliptic) at equatorial and non-equatorial regions respectively.

40 Technically, the words *śaṅku* and *nara* refer to the height of a gnomon; however, in Siddhāntic literature, they are often used to refer to the degrees of altitude (or its Sine). For example, Bhāskara II, in his *Siddhāntaśiromaṇi* (II.7.36), says *śaṅkur unnatalavajyakā bhaved dṛgguṇaś ca natabhāgaśiñjinī* 'the *śaṅku* is the sine of the degrees of altitude: and the *dṛigjyā* [sic] is the sine of the zenith distance' (L. Wilkinson and B. D. Śāstrī 1861, p. 171), where *dṛgguṇa* and *dṛgjyā* are synonyms.

41 In Sanskrit mathematics, this identification of the arcs of the *bhujā* and *koṭi* in the four quadrants of a circle (with the *sinus totus* as its radius) was conventionally established (see, e.g., Subbarayappa and Sarma 1985, pp. 299–300).

42 I suspect the *vañcula* tree is the *bakula* tree (*Mimusops elengi*), the bullet-wood tree famous throughout the Indian subcontinent.

Appendix A Nityānanda's Geodetic Method vis-à-vis al-Bīrūnī's Method to Calculate the Earth's Radius

Nityānanda discusses two geodetic methods to determine the elevation and distance of high mountain-summit in verses 8–12 (see notes on pages 5). In the first of these two methods, Nityānanda offers a mathematical algorithm (without any derivation or justification) to compute the height of a mountain given its distance from an observer; see equation 5.1. Nityānanda's method closely resembles al-Bīrūnī's method to compute the radius of the Earth (knowing the height of a mountain) in his *al-Qānūn al-Masᶜūdī* (V.7) (see the edited text of the fifth *maqāla*, chapter seven, in Baranī 1954).[1]

A.1 al-Bīrūnī's method of calculating the Earth's radius

A.1.1 To determine the height of a mountain

To determine the height h of a mountain, al-Bīrūnī uses three measurements, namely, the level distance d between two points A and B, and the two angles θ_1 and θ_2 subtended between the horizontal surface at those points and the mountain top F as shown in Figure A.1. With these measurements, it can be seen that in the right-angled triangles $\triangle$BFD and $\triangle$AFD,

$$\tan\theta_2 = \frac{\mathrm{FD}}{\mathrm{BD}} = \frac{h}{\mathrm{BD}} \Rightarrow \mathrm{BD} = \frac{h}{\tan\theta_2} \text{ and}$$

$$\tan\theta_1 = \frac{\mathrm{FD}}{\mathrm{AD}} = \frac{\mathrm{FD}}{\mathrm{AB}+\mathrm{BD}} = \frac{h}{d+\mathrm{BD}} = \frac{h}{d+\dfrac{h}{\tan\theta_2}} = \frac{h\tan\theta_2}{d\tan\theta_2 + h},$$

and hence

$$h = \frac{d\tan\theta_1\tan\theta_2}{\tan\theta_2 - \tan\theta_1}. \tag{A.1}$$

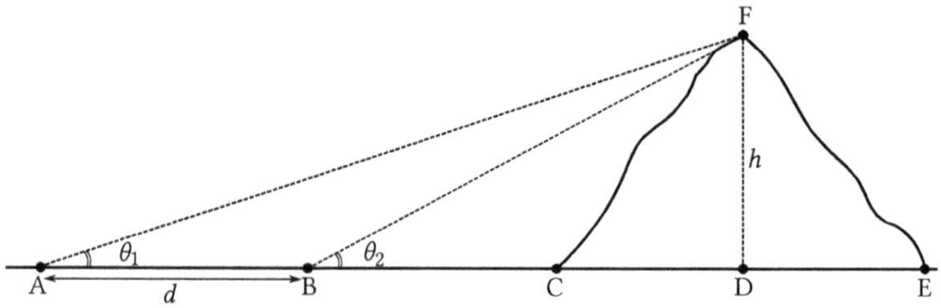

Figure A.1 al-Bīrūnī's method to determine the height of the mountain.

A.1.2 *To calculate the radius of the Earth*

Looking at Figure A.2, we can see that the angle subtended by radial lines OS and OF at the centre of the Earth O, i.e., $\delta\phi$ is equal to the dip of the line FS to the horizontal from the mountain top F. Also, the right-angled triangle $\triangle$FDQ implies that $\angle$FQD $= \delta\phi$ and $\angle$QFD $= 90° - \delta\phi$ (since $\triangle$FSO is a right-angled triangle with $\angle$FOS $= \delta\phi$). Accordingly, by applying the law of sines to the right-angled triangle $\triangle$FDQ, we find

$$\frac{\mathrm{QD}}{\sin\angle\mathrm{QFD}} = \frac{\mathrm{FD}}{\sin\angle\mathrm{FQD}} \Rightarrow \frac{\mathrm{QD}}{\sin(90° - \delta\phi)} = \frac{\mathrm{FD}}{\sin\delta\phi}$$

$$\Rightarrow \mathrm{QD} = \frac{h\sin(90° - \delta\phi)}{\sin\delta\phi} = \frac{h\cos\delta\phi}{\sin\delta\phi},$$

$$\text{and hence QF} = \sqrt{\mathrm{FD}^2 + \mathrm{QD}^2} = \sqrt{h^2 + \frac{h^2\cos^2\delta\phi}{\sin^2\delta\phi}} = \frac{h}{\sin\delta\phi}.$$

Thus,

$$\mathrm{SF} = \mathrm{SQ} + \mathrm{QF} = \mathrm{QD} + \mathrm{QF} = \frac{h}{\sin\delta\phi}(1 + \cos\delta\phi) \qquad \text{(A.2)}$$

as SQ and QD are equal by symmetry.

And again, applying the law of sines to the right-angled triangle $\triangle$FSO, we find

$$\frac{\mathrm{SF}}{\sin\angle\mathrm{FOS}} = \frac{\mathrm{OF}}{\sin\angle\mathrm{FSO}} \Rightarrow \frac{\mathrm{SF}}{\sin\delta\phi} = \frac{\mathrm{FD} + \mathrm{OD}}{\sin 90°} \Rightarrow \mathrm{SF} = \sin\delta\phi\,(h + R_e) \qquad \text{(A.3)}$$

where OD $= R_e$ is the radius of the Earth.

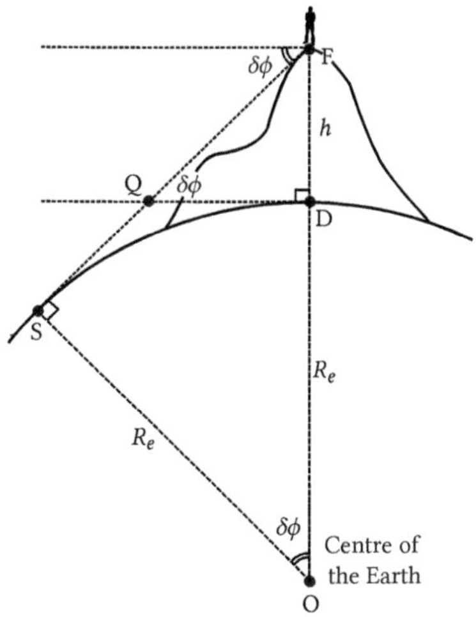

Figure A.2 The equality of the angle between the line of sight from the mountain top F to the horizon (at point S) and the angle subtended at the centre O of the Earth by the perpendicular of the mountain's height (radial line OF) and the horizontal point S.

From equations A.2 and A.3, we conclude

$$\frac{h}{\sin \delta\phi}(1 + \cos \delta\phi) = \sin \delta\phi \, (h + R_e)$$

$$\Rightarrow R_e = h - \frac{h(1 + \cos \delta\phi)}{\sin^2 \delta\phi} = h - \frac{h}{1 - \cos \delta\phi},$$

$$\text{or } R_e = \frac{h \cos \delta\phi}{1 - \cos \delta\phi}. \qquad \text{(A.4)}$$

A.2 Comparing with Nityānanda's expression for the elevation of a mountain-summit

In equation 5.1, Nityānanda describes the height h of a mountain at a distance of d (equalling 30 *yojana*s) with $\delta\phi = \frac{30^{yojanas} \times 360^\circ}{6000^{yojanas}} = 1.8^\circ$ (see step 1 on page 197). Nityānanda takes the value of the Earth's circumference as 6000 (somewhat arbitrarily) and uses a *sinus totus* $\mathscr{R}$ of 60 (see Sections 1.3.9.1 and 1.3.9.2). Nityānanda's expression in equation 5.1 can be rewritten as

$$h = \frac{R_e\,(\mathscr{R} - \mathrm{Cos}\,\delta\phi)}{\mathrm{Cos}\,\delta\phi} \Rightarrow R_e = \frac{h \times \mathrm{Cos}\,\delta\phi}{\mathscr{R} - \mathrm{Cos}\,\delta\phi} = \frac{h \cos \delta\phi}{1 - \cos \delta\phi} \qquad \text{(A.5)}$$

which is identical to equation A.4.

Note

1 Also, see Baranī (1951) for a succinct historical review of the Islamicate methods of geodesy, in particular, al-Bīrūnī's method to compute the dimensions of the Earth.

Appendix B The Cosmography of the Purāṇas

In Hinduism, the Purāṇas (lit. ancient accounts) are considered as sacred encyclopaedic records of cosmogenic history, divine mythology, heroic chronicles, and moral antiquity.[1] The authoritativeness of these texts meant that, historically, many aspects of Purāṇic cosmology and geography were important to Siddhāntic authors. In the *golādhyāya* of his *Sarvasiddhāntarāja*, Nityānanda follows in the tradition of earlier Siddhāntic authors by repeating Purāṇic descriptions of the physical geography of the Earth (see Table 1.5). In the course of its development from the early centuries of the common era, Purāṇic literature included material that originated from (or was influenced by) Greco-Babylonian ideas of earlier times (see, e.g., Pingree 1990). A detailed description of Purāṇic cosmology is beyond the scope of this book; instead, I provide here a comprehensive overview of Purāṇic cosmography without dwelling on the aspects of cosmogony (*viśvaśṛṣṭi*).

I note that while there are differences in the names, numbers, and ordering of the various geographical and astronomical objects described in the different sectarian Purāṇas, the overall description of these objects is similar in most of these texts. This overview of Purāṇic cosmography is based on the following sources: *Śrīmad Bhāgavatam* or *Bhāgavata Purāṇa*, Canto V (Prabhupāda 1975); *Viṣṇu Purāṇa*, Book II (Wilson 1864–77); *Vāyu Purāṇa, Anuṣaṅgapāda*: VIII–XLIV (Tagare [1960] 1987–88); *Mārkaṇḍeya Purāṇa*, Chapters LI–LVIII (Pargiter 1904 and Banerjea [1862] 2004); *Mahābhārata, Jambūkhaṇḍa-nirmāṇa-parva* and *Bhūmī-parva* of Book VI: *Bhīṣma Parva* (Cherniak 2009); *Yogadarśana* (Jhā [1934] 2003); and the secondary literature in Kirfel (1920), Pingree (1978a), Sircar (1990) and Wujastyk (2004).

B.1 Structure of the universe

According to the Purāṇas, the material universe (*brahmāṇḍa*), one among many universes, is enclosed in an egg-shaped, cosmic shell suspended in an infinite causal ocean (Kāraṇodaka). It is brought into existence by the absolute, eternal, and primal causer, *ens entium* Mahā Viṣṇu. It consist of a central, horizontal Earth-disk called the Bhūmaṇḍala that separates the seven empyrean realms of the celestial beings above from the subterranean regions below, while remaining suspended in an ocean filled with amniotic water.[2] The firmament (Bhūmaṇḍala) is a thick circular disk that contains the earthly oceans and continents on its surface, and the seven netherworlds below it inhabited by different races of demon-gods, serpents, and natural spirits.[3] The region under the Bhūmaṇḍala is believed to be filled with the waters of the Garbhodaka ocean, a wave-less continuum of primordial matter from which all things emerge upon creation and into which all things return upon dissolution. The realm of the forefathers (Pitṛloka) and the various kinds of hells (Naraka or Yamaloka) are located in the region between the Bhūmaṇḍala and the Garbhodaka ocean. The entire universe is thought to be encircled by the endless coils of the cosmic-serpent Ananta Śeṣa who rests at the bottom of the Garbhodaka ocean (at a distance of 100,000 *yojana*s from the Bhūmaṇḍala). Ananta Śeṣa carries the weight of all creation as a demiurgic being; the *Śrīmad Bhāgavatam* considered Ananta Śeṣa to be an incarnation of Mahā Viṣṇu himself. Figure B.1 sketches the relative arrangement of the different realms in the universe.

B.2 Topography of the Bhūloka

The surface of the Bhūmaṇḍala, i.e., the Bhūloka, contains alternating concentric rings of the seven island-continents (*sapta-dvīpa*) and the seven oceans (*sapta-samudra*), and three larger annular regions extending further all the way to the edge of the Bhūmaṇḍala.[4] While the names and order of the seven island-continents are different in different Purāṇas, their general arrangement is depicted in Figure B.2. All of the Purāṇas agree that a central circular continent called Jambūdvīpa (land of the black plum trees) supports the world-mountain, the *axis mundi* Mt Sumeru. This central island-continent is surrounded by successive annular oceans and continents in the (approximate) order of Lavaṇa (ocean of salt water), Plakṣa (land of holy fig trees), Ikṣu (ocean of sugar-cane juice), Śālamali (land of silk-cotton trees), Surā (ocean of wine), Kuśa (land of salt reed-grass), Sarpis (ocean of clarified butter), Krauñca (land of Mt Krauñca), Dadhi or Dadhimaṇḍoda (ocean of emulsified yoghurt), Śāka (land of teak trees), Dugdha or Kṣīroda (ocean of milk), Puṣkara (land of blue lotuses), and Jala or Svādūdaka (ocean of sweet water).

The *Śrīmad Bhāgavatam* (V.12.30) places the circular mountain range Mānasottara in the outermost island-continent Puṣkaradvīpa, occupying half the width of the island-continent. This mountain is believed to be the circular track along which the Sun's one-wheeled chariot continually turns, with the axle supported by Mt Sumeru at the other end. Beyond the seven island-continents, in successive order, are the annular regions of the inhabited lands (Lokavarṣa), the golden land (Kāñcanī Bhūmi), the Lokāloka Mountain, and the uninhabited lands (Alokavarṣa) extending all the way to the shell of the universe

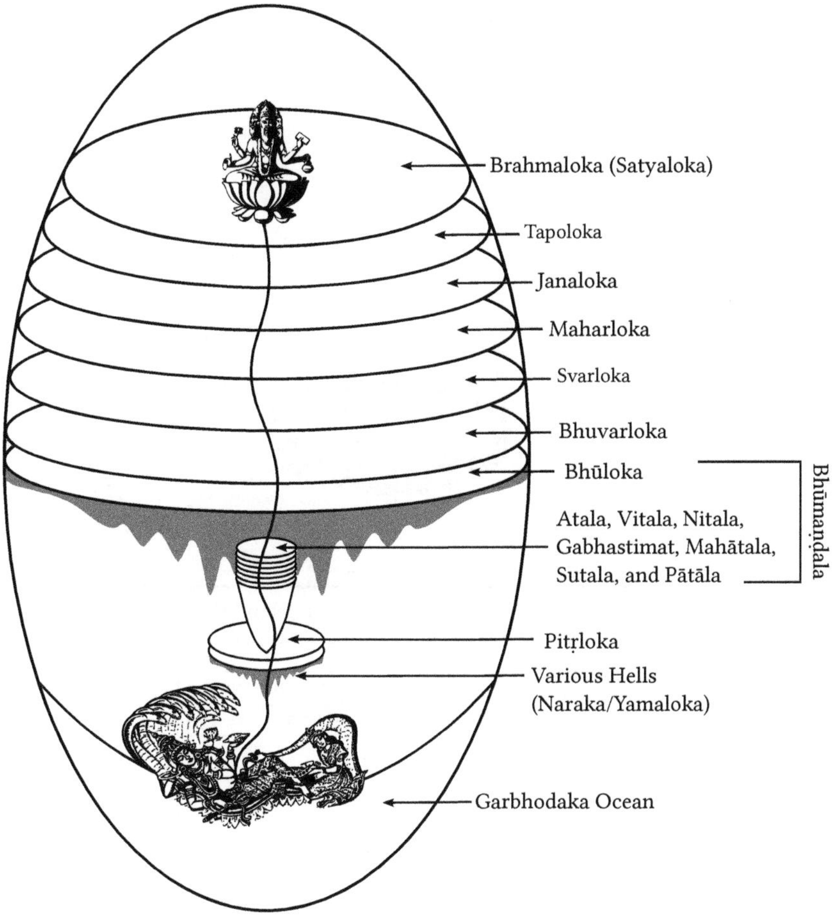

Figure B.1 A schematic representation of Purāṇic cosmography.

(*brahmāṇḍa-paridhi*).[5] The *Śrīmad Bhāgavatam* (V.12.38) and the Viṣṇu Purāṇa (II.4.65) estimate the diameter of the central circular disk of Bhūmaṇḍala, with its immensely large island-continents and oceans, to be five hundred million *yojanas*—roughly equivalent to the perihelion-distance of Neptune's orbit in our solar system.

B.3 Geography of Jambūdvīpa

The Purāṇas also describe the nine countries or regions (*varṣas*) separated by eight mountain ranges on the Jambūdvīpa. This central island-continent measures 100,000 *yojanas* and Mt Sumeru is at its centre. Some

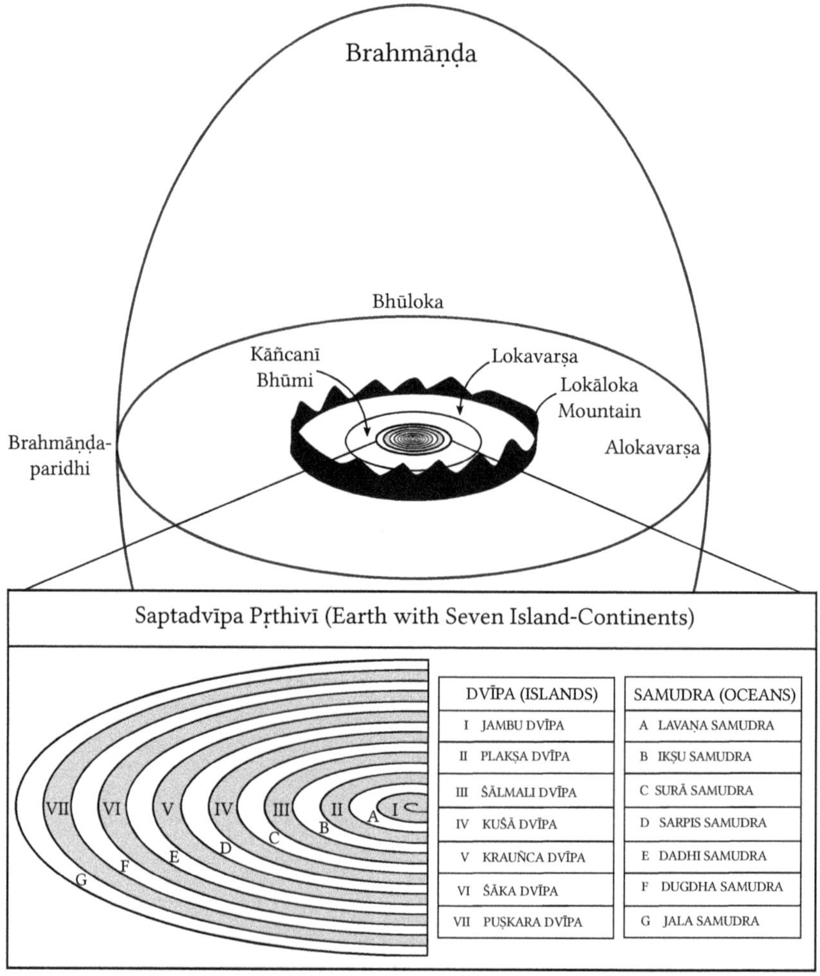

Figure B.2 A schematic representation of the terrestrial world according to the Purāṇas.

of the salient geographical features of this island-continent are listed below.[6]

1 Uttarakuru *varṣa,* the northernmost frontier extending 9000 *yojanas* north of the mountain range Śṛṅgavat (whose altitude is 2000 *yojanas* and length 80,000 *yojanas*) and having the Lavaṇa ocean of salt water to the north.

2 Hiraṇmaya *varṣa,* the mid-northern region extending 9000 *yojanas* north of the mountain range Śvetavat (whose altitude is 2000 *yojanas* and length 90,000 *yojanas*) and south of Śṛṅgavat.

3 Ramyaka *varṣa*, the lower-northern region extending 9000 *yojana*s north of the mountain range Nīlavat (whose altitude is 2000 *yojana*s and length 100,000 *yojana*s) and south of Śvetavat.

4 Going from the west to the east of Jambūdvīpa, there are the three central regions of Ketumāla, Ilāvṛtta, and Bhadrāśva situated north of the mountain range Niṣadha (whose altitude is 2000 *yojana*s and length 100,000 *yojana*s) by 18,000 *yojana*s. The western region of Ketumāla is separated from the central Ilāvṛtta by the mountain range Mālyavat (whose altitude is 2000 *yojana*s and length 18,000 *yojana*s) whereas the eastern region of Bhadrāśva is separated from the central Ilāvṛtta by the mountain range Gandhamādana (whose altitude is 2000 *yojana*s and length 18,000 *yojana*s) . Both these mountain ranges extend in the north-south direction going from the northern Nilāvata mountain range to the southern Niṣadha mountain range.

5 Mt Sumeru stands in the central region of Ilāvṛtta that extends 9000 *yojana*s radially outwards from the edge of this world-mountain. Mt Sumeru itself has enormous dimensions: its height is 84,000 *yojana*s , its depth under the Earth's surface is 16,000 *yojana*s , its basal diameter is 16,000 *yojana*s , and its apical-diameter is 32,000 *yojana*s (Viṣṇu Purāṇa, II.2.8–10).[7] Along its four cardinal directions, Mt Sumeru is supported by four buttress mountains with altitudes and lengths of 10,000 *yojana*s each, namely, Mt Mandara (to the east), Mt Merumadana (to the south), Mt Vipula (to the west), and Mt Suparśva (to the north).

6 Hari *varṣa*, the lower-southern region extending 9000 *yojana*s north of the mountain range Mahāhimavat or Hemakuṭa (whose altitude is 2000 *yojana*s and length 100,000 *yojana*s) and south of the mountain range Niṣadha.

7 Kiṃpuruṣa *varṣa*, the mid-southern region extending 9000 *yojana*s north of the mountain range Himavat or Himālaya (whose altitude is 2000 *yojana*s and length 100,000 *yojana*s) and south of Mahāhimavat.

8 Bhārata *varṣa*, the southernmost frontier extending 9000 *yojana*s south of the mountain range Himālaya and reaching up to the Lavaṇa ocean of salt water to the south.

B.3.1 On the world-mountain Mt Sumeru

The *Śrīmad Bhāgavatam* (V.12.26) lists several other mountains that encircle the world-mountain Mt Sumeru like 'filaments around the whorl of a lotus'. These are the mountains of Kuraṇga, Kurara, Kusumbha,

Vaikaṇka, Trikūṭa, Śiśira, Pataṅga, Rucaka, Niṣadha, Sinīvāsa, Kapila, Śaṅkha, Vaidūrya, Jārudhi, Haṃsa, Ṛṣabha, Nāga, Kālañjara, and Nārada.

Additionally, four pairs of mountain ranges are said to surround Mt Sumeru in the four cardinal directions, each having a length of 18,000 *yojanas*, a height of 2000 *yojanas*, and a width of 2000 *yojanas*. The four pairs of mountain ranges are Jaṭhara and Devakūṭa to the east of Mt Sumeru along the north-south direction, Pavana and Pāriyātra to the west of Mt Sumeru and also along the north-south direction, Kailāśa and Karavīra to the south of Mt Sumeru along the east-west direction, and Triśṛṅga and Makara to the north south of Mt Sumeru along the east-west direction (Ibid, V.12.27).

Brahmapurī or Manovatī, the abode of Brahmā, is thought to be situated on the summit of Mt Sumeru and is encircled by the celestial river Gaṅgā. This city is in turn surrounded by the cities of the eight regents of the world (*aṣṭalokapālas*; sometimes called *aṣṭadikpatis*, lit. eight lords of the directions); namely, Amarāvatī of Indra (in the east), Yaśovatī of Īśāna (in the north-east), Mahodayā of Kuvera (in the north), Gandhavatī of Vāyu (in the north-west), Śraddhāvatī of Varuṇa (in the west), Kṛṣṇāġanā of Nirṛti (in the south-west), Samyāmanī of Yama (to the south), and Tejovatī of Agni (in the south-east) (Ibid, V.12.29).

B.3.2 On the river Gaṅgā

The river Gaṅgā is believed to emanate from the waters of the causal ocean (Kāraṇodaka) that surrounds the universe. After having descended from the empyrean realms, Gaṅgā falls on Mt Sumeru and thereafter splits into four streams flowing in the four cardinal directions. These are the distributary rivers of Gaṅgā, namely,

1 Sītā to the east that falls on Mt Gandhamādana and then flows through the region of Bhadrāśva towards the eastern side of the Lavaṇa ocean of salt water,

2 Alakanandā to the south that falls on Mt Niṣadha, Mt Hemakuṭa, and Mt Himavat and then flows through the region of Bhārata before dividing into the seven arterial rivers that pour into the Lavaṇa ocean of salt water to the south,

3 Cakṣu to the west that falls on Mt Mālyavat and then flows through the region of Ketumāla towards the western side of the Lavaṇa ocean of salt water, and

4 Bhadrā to the north that falls on Mt Nīlavat, Mt Śvetavat, and Mt Śṛṅgavat and then flows through the region of Uttarakuru to pour into the Lavaṇa ocean of salt water to the north.

B.4 Astronomy in the Purāṇas

In Purāṇic astronomy, the motions of the planetary bodies are restricted to the regions of Bhuvarloka and Svarloka, a space collectively called *antarikṣa*. The *Śrīmad Bhāgavatam* (V.21.2) compares *antarikṣa* to the space between the two halves (cotyledons) of a lablab bean. The Bhuvarloka, a region extending from the surface of the Earth up to the orbit of the Sun, is of the same dimension as the Bhūmaṇḍala (or Bhūloka), i.e., it has a diameter of 500 million *yojanas*. The orb of the Sun is at a distance of 100,000 *yojanas* above the Bhūloka. The orbits of the other celestial bodies are in the Svarloka beyond the Bhuvarloka in the following order: Moon (100,000 *yojanas* above the Sun), Lunar Mansions (*nakṣatras*) (100,000 *yojanas* above the Moon), Mercury (200,000 *yojanas* above the Lunar Mansions), Venus (200,000 *yojanas* above Mercury), Mars (200,000 *yojanas* above Venus), Jupiter (200,000 *yojanas* above Mars), Saturn (250,000 *yojanas* above Jupiter), Ursa Major (Saptaṛṣi) (100,000 *yojanas* above Saturn), and (North) polestar (100,000 *yojanas* above Ursa Major).[8]

The orbits of the celestial bodies in the *antarikṣa* are believed to be parallel to the plane of the Bhūmaṇḍala. These bodies revolve in their respective orbits resembling the motion of a wheel with the axle being Mt Sumeru. The north polestar is situated on the top of this exoteric cosmic axle passing through Mt Sumeru, and all orbiting bodies are yoked to the polestar with aerial cords of varying lengths. The motion of these bodies is brought about by the flowing of the Pravaha wind, the cosmic wind issued forth by Brahmā himself. This in turn causes the rotation of the cosmic axle-like axis.[9]

The visibility (after rising) and invisibility (after setting) of the planets (like the Sun) is on account of them being obstructed by Mt Sumeru in the course of their daily rotation. According to the Purāṇas, the mountain range Lokāloka divides the visible world from the invisible; its height was believed to be large enough to block the light of the polestar in the invisible (uninhabited) lands beyond it. On the top of the Lokāloka mountain, Brahmā placed the celestial elephants Ṛṣabha, Puṣkaracūḍa, Vāmana, and Aparājita, each pointing in one of the four cardinal directions, to maintain the planetary systems of the universe (*Śrīmad Bhāgavatam*, V.20.37–39). Beyond the Svarloka, the higher empyrean realms (Mahar, Jana, Tapas, and Satya) extended out over vast distances till the edge of the universe (*brahmāṇḍa-paridhi*) at 250 million *yojanas*.

The time scales in the Purāṇas are equally immense; see, e.g., *Śrīmad Bhāgavatam* (Canto III.11) for the names and sizes of the different time units. Purāṇic time is cyclical, with the cumulatively increasing units (i.e., days, months, years, etc.) understood differently in human and divine experiences. For example, a nychthemeron (one day and night period) for humans lasts twenty-four hours, for the forefather (*pitṛ*) fifteen days, for the gods and demons (*deva-daitya*) six months, and for Brahmā 8.64 billion years. Beyond this, the moment of time in the twinkling of an eye (*nimeṣa*) of Mahā Viṣṇu is equal to 311.04 trillion years, the very lifespan of Brahmā (*Śrīmad Bhāgavatam*, III.11.38).

Notes

1 The Purāṇic literary tradition is not unique to Hinduism. In fact, Jainism also contains similar texts, although the Jain texts are often written with a different sociohistorical focus. The Jaina Purāṇas integrate their own legends of liberated beings (*tīrthaṃkaras*) in the standard repertoire of heroic chronicles found in most Indian epics (*itihāsa*) (see, e.g., Jaini 2000). This difference notwithstanding, the cosmography in Jaina and Hindu Purāṇas remains intimately and intricately connected to one another (see, e.g., Schubring [1962] 2000, Chapters IV and V). Similarly, the Buddhist commentaries in the Mahāyāna and Theravāda *Abhidharma* traditions describe an esoteric cosmogony and an exoteric cosmology that are also influenced by Hindu Purāṇas (see, e.g., Gethin 1998, Chapter V).

2 The seven realms, going in an ascending order from the Bhūmaṇḍala, are:

- i the Bhūloka, the terrestrial realm of men, animals, forests, demons, and spirits;
- ii the Bhuvarloka, the realm of the sky between the Earth and the Sun inhabited by seers and semi-divine beings;
- iii the Svarloka, the realm of the heaven between the Sun and the polestar inhabited by various celestial beings like choristers, minstrels, nymphs, and magicians, as well as the gods lead by Indra;
- iv the Maharloka, the realm of the sages beyond the polestar inhabited by lords of progeny and other celestial tribes;
- v the Janaloka, the realm of productivity inhabited by the four adopted sons (*manas-putras*, lit. mind-born sons) of Brahmā and other divine beings;
- vi the Tapoloka, the realm of austerities inhabited by semi-divine manes, mendicants, and ascetics; and
- vii the Brahmaloka (or Satyaloka), the highest realm of absolute truth and abode of the creator god Brahmā.

3 The seven netherworlds begin 7000 *yojanas* below the surface of the Earth, and measure up to 10,000 *yojanas* each in extent (*Viṣṇu Purāṇa*, II.5.1–2). Their names are Atala, Vitala, Nitala, Gabhastimat, Mahātala, Sutala, and Pātāla. Different demon races, semi-divine beings, and serpent-gods live in these nether regions enjoying resplendent lives in abundance (Ibid, II.5.4–12).

4 According to the *Vāyu Purāṇa* (XXXII.4), *pṛthivī sarvā saptadvīpasamanvitā* 'the Earth is fully constituted by the seven island-continents'.

5 By most Purāṇic accounts, the Lokāloka mountain is thought to be the boundary between the illuminated regions of the Bhūmaṇḍala, i.e., the

Lokavarṣa, and the dark uninhabited lands further afar in the Alokavarṣa; see, e.g., *Śrīmad Bhāgavatam* (V.12.34).

6 Nityānanda provides a near-identical description of the Jambūdvīpa in his *golādhyāya*, verses 46–58. See Figures 5.13 and 5.14 for a schematic depiction of the physical geography of Jambūdvīpa.

7 Although these measures suggest that Mt Sumeru is shaped like an inverted-cone, its shape has been variously described in different Purāṇas (see, e.g., Wilson 1864–77).

8 Early Purāṇic texts simply order the Sun, the Moon, the constellations (*nakṣatras*), and Saptarṣis; later texts include and order the planets (*graha*) in different ways (Pingree 1978a).

9 The *Viṣṇu Purāṇa* (II.12.27–29) explains the orbital motion of the planets by an analogy:

> *tailapīḍā yathā cakraṃ bhramanto bhrāmayanti vai* |
> *tathā bhramanti jyotīṃṣi vātaviddhāni sarvaśaḥ* ||
>
> *alātacakravadyānti vātacakreritāni tu* |
> *yasmājjyotīṃṣi vahati pravahastena sa smṛtaḥ* ||
>
> In the same manner as the oil-man himself, going round, causes the spindle to revolve, so the planets travel round, suspended by cords of air, which are circling round a (whirling) centre.
>
> The air, which is called Pravaha, is so termed because it bears along the planets, which turn round, like a disc of fire, driven by the aerial wheel. (Wilson 1864–77, pp. 305–306 in Volume II).

Appendix C Numbering of verses in the Critical Edition vis-à-vis the Eight Manuscripts of the *golādhyāya* in Nityānanda's *Sarvasiddhāntarāja*

Table C.1 represents the numbering of the verses in the critical edition of the *golādhyāya* in Nityānanda's *Sarvasiddhāntarāja* (in Chapter 3) and the eight manuscript witnesses of the text; namely, Bn.I_B (ff. 67v–75v), Bn.II_B (ff. 68v–76v), Br_B (ff. 37v–47v), Np_γ (ff. 1v–9v), Pb_α (ff. 1v–4r), Pm_B (ff. 44r–48v), Rr_B (ff. 49v–55r), and Sc_B (ff. 155v–174r).

In the table, an unmarked verse in a manuscript is represented by '-' while an absent verse is indicated by a blank cell. Occasionally, verses in a manuscript are irregularly combined with half-verses from the critical edition. In such cases, the verses from the critical edition are split with affixes 'ab' and 'cd' representing the first and second hemistiches (and for the tristich-hypermetric verse 90, an additional 'ef'). Verse 73, with its two alternative readings found in different manuscript groups, is labelled with numeral affixes '1' and '2' (as is also done in Chapters 3 and 4). MS Pb is incomplete and terminates in first quarter of verse 91; this is abbreviated in table as '*term*'.

Table C.1 Numbering of the verses in the critical edition (in Chapter 3) vis-à-vis the eight manuscripts of the *golādhyāya* in Nityānanda's *Sarvasiddhāntarāja*

<table>
<tr><th></th><th colspan="8">Verse Numbering</th></tr>
<tr><th>Critical Edition</th><th>Bn.I$_B$</th><th>Bn.II$_B$</th><th>Br$_B$</th><th>Np$_\gamma$</th><th>Pb$_\alpha$</th><th>Pm$_B$</th><th>Rr$_B$</th><th>Sc$_B$</th></tr>
<tr><td>1</td><td>1</td><td>1</td><td>1</td><td>1</td><td>1</td><td>1</td><td>1</td><td>1</td></tr>
<tr><td>2</td><td>2</td><td>1</td><td>1</td><td>2</td><td>2</td><td>1</td><td>1</td><td>-</td></tr>
<tr><td>3</td><td>3</td><td>2</td><td>2</td><td>3</td><td>3</td><td>2</td><td>2</td><td>3</td></tr>
<tr><td>4</td><td>4</td><td>3</td><td>3</td><td>4</td><td>4</td><td>3</td><td>4</td><td>4</td></tr>
<tr><td>5</td><td>5</td><td>4</td><td>4</td><td>5</td><td>5</td><td>4</td><td>-</td><td>5</td></tr>
<tr><td>6</td><td>6</td><td>5</td><td>5</td><td>6</td><td>6</td><td>5</td><td>5</td><td>6</td></tr>
<tr><td>7</td><td>7</td><td>6</td><td>6</td><td>7</td><td>7</td><td>6</td><td>6</td><td>-</td></tr>
<tr><td>8</td><td>8</td><td>7</td><td>7</td><td>8</td><td>8</td><td>7</td><td>7</td><td>2</td></tr>
<tr><td>9</td><td>-</td><td>8</td><td>8</td><td>9</td><td>9</td><td>8</td><td>8</td><td>3</td></tr>
<tr><td>10</td><td>10</td><td>9</td><td>9</td><td>10</td><td>10</td><td>9</td><td>9</td><td>4</td></tr>
<tr><td>11</td><td>11</td><td>10</td><td>10</td><td>11</td><td>11</td><td>10</td><td>10</td><td>5</td></tr>
<tr><td>12</td><td>12</td><td>11</td><td>11</td><td>12</td><td>12</td><td>11</td><td>11</td><td>6</td></tr>
<tr><td>13</td><td>13</td><td>12</td><td>12</td><td>13</td><td>13</td><td>12</td><td>12</td><td>-</td></tr>
<tr><td>14</td><td>14</td><td rowspan="2">13</td><td rowspan="2">13</td><td>14</td><td>14</td><td rowspan="2">13</td><td rowspan="2">13</td><td>-</td></tr>
<tr><td>15ab</td><td rowspan="2">15</td><td rowspan="2">15</td><td rowspan="2">15</td><td rowspan="2">-</td></tr>
<tr><td>15cd</td><td rowspan="2">14</td><td rowspan="2">14</td><td rowspan="2">14</td><td rowspan="2">14</td></tr>
<tr><td>16ab</td><td rowspan="2">16</td><td rowspan="2">16</td><td rowspan="2">16</td><td rowspan="2">-</td></tr>
<tr><td>16cd</td><td rowspan="2">15</td><td rowspan="2">15</td><td rowspan="2">15</td><td rowspan="2">15</td></tr>
<tr><td>17ab</td><td rowspan="2">17</td><td rowspan="2">17</td><td rowspan="2">17</td><td rowspan="2">-</td></tr>
<tr><td>17cd</td><td rowspan="2">16</td><td rowspan="2">16</td><td rowspan="2">16</td><td rowspan="2">16</td></tr>
<tr><td>18ab</td><td rowspan="2">18</td><td rowspan="2">18</td><td rowspan="2">18</td><td rowspan="2">-</td></tr>
<tr><td>18cd</td><td rowspan="2">17</td><td rowspan="2">17</td><td rowspan="2">17</td><td rowspan="2">17</td></tr>
<tr><td>19ab</td><td rowspan="2">19</td><td rowspan="2">19</td><td rowspan="2">19</td><td rowspan="2">-</td></tr>
<tr><td>19cd</td><td>17</td><td>17</td><td>17</td><td rowspan="2">18</td></tr>
<tr><td>20</td><td>20</td><td>18</td><td>18</td><td>20</td><td>20</td><td>18</td><td>-</td></tr>
<tr><td>21</td><td>21</td><td>19</td><td>19</td><td>21</td><td>21</td><td>19</td><td>19</td><td>-</td></tr>
<tr><td>22</td><td>22</td><td>20</td><td>20</td><td>22</td><td>22</td><td>20</td><td>20</td><td>-</td></tr>
<tr><td>23</td><td>23</td><td>21</td><td>21</td><td>23</td><td>23</td><td>21</td><td>21</td><td>-</td></tr>
<tr><td>24</td><td>24</td><td>22</td><td>22</td><td>24</td><td>24</td><td>22</td><td>22</td><td>-</td></tr>
<tr><td>25</td><td>25</td><td>23</td><td>23</td><td>25</td><td>25</td><td>23</td><td>23</td><td>-</td></tr>
</table>

Table C.1 (Continued)

				Verse Numbering				
Critical Edition	$Bn.I_B$	$Bn.II_B$	Br_B	Np_γ	Pb_α	Pm_B	Rr_B	Sc_B
26	26	24	24	26	26	24	24	-
27	27	25	25	27	27	25	25	-
28	28	26	26	28	28	26	26	-
29	29	-	27	29	29	27	-	-
30	30	28	28	30	30	28	28	-
31	31	28	29	-	31	29	29	-
32	32	-	30	31	32	20	-	-
33	33	31	31	32	33	21	31	-
34	34	32	32	33	34	22	32	-
35	35	33	33	34	35	23	-	1
36	36	34	34	35	36	24	34	2
37	37	35	35	36	37	25	35	3
38	38	36	36	37	38	26	36	4
39	39	37	37	38	39	27	37	5
40	40	38	38	39	40	28	38	6
41	41	39	39	40	41	39	39	7
42	42	40	40	41	42	40	40	8
43	43	41	41	42	43	41	41	9
44	44	42	42	43	44	42	42	10
45	45	43	43	44	45	43	43	11
46	46	44	44	45	46	44	44	11
47	47	45	45	46	47	45	45	12
48	48	46	46	47	48	46	46	13
49	49	47	47	48	49	47	47	-
50	50	48	48	49	50	48	48	15
51	51	49	49	50	51	49	49	16
52	52	50	50	51	52	50	50	-
53	53	51	51	52	53	51	51	18

(*Continued*)

Table C.1 (Continued)

	Verse Numbering							
Critical Edition	Bn.I$_B$	Bn.II$_B$	Br$_B$	Np$_\gamma$	Pb$_\alpha$	Pm$_B$	Rr$_B$	Sc$_B$
54	54	52	52	53	54	52	52	19
55	55	53	53	54	55	53	53	20
56	56	54	54	55	56	54	54	21
57	57	55	55	56	57	55	55	22
58	58	56	56	57	58	56	56	23
59	59	57	57	58	59	57	57	24
60	60	58	58	59	60	58	58	25
61	61	59	59	50	61	59	59	26
62	62	60	60	61	62	60	60	27
63	63	61	61	62	63	61	61	-
64	64	62	62	63	64	62	62	29
65	65	63	63	64	65	63	63	30
66	66	64	64	65	66	64	64	31
67	67	65	65	66	67	65	65	32
68	68	66	66	67	68	66	66	33
69	69	67	67	68	69	67	67	34
70	70	68	68	69	70	68	68	35
71	71	69	69	70	71	69	69	36
72	72	70	70	71	72	70	70	1
73.1	73			72	73			2
73.2		71	71			71	71	
74	74	72	72	73	74	72	72	3
75	75	73	73	74	75	73	73	4
76	76	74	74	75	76	74	74	5
77	77	75	75	76	77	75	75	6
78	78	76	76	77	78	76	76	7
79	79	77	77	78	79	77	77	9
80	80	78	78	79	80	78	78	9

Table C.1 (Continued)

<table>
<tr><th colspan="9">Verse Numbering</th></tr>
<tr><th>Critical Edition</th><th>Bn.I$_B$</th><th>Bn.II$_B$</th><th>Br$_B$</th><th>Np$_\gamma$</th><th>Pb$_\alpha$</th><th>Pm$_B$</th><th>Rr$_B$</th><th>Sc$_B$</th></tr>
<tr><td>81</td><td>81</td><td>79</td><td>79</td><td>80</td><td>81</td><td>79</td><td>79</td><td>910</td></tr>
<tr><td>82</td><td>82</td><td>80</td><td>80</td><td>81</td><td>82</td><td>80</td><td>80</td><td>11</td></tr>
<tr><td>83</td><td>82</td><td>81</td><td>81</td><td>82</td><td>83</td><td>81</td><td>81</td><td>12</td></tr>
<tr><td>84</td><td>83</td><td>82</td><td>82</td><td>83</td><td>84</td><td>82</td><td>82</td><td>13</td></tr>
<tr><td>85</td><td>84</td><td>83</td><td>83</td><td>84</td><td>85</td><td>83</td><td>83</td><td>1</td></tr>
<tr><td>86</td><td>85</td><td>84</td><td>84</td><td>85</td><td>86</td><td>84</td><td>84</td><td>2</td></tr>
<tr><td>87</td><td>86</td><td>85</td><td>85</td><td>86</td><td>87</td><td>85</td><td>85</td><td>3</td></tr>
<tr><td>88</td><td>87</td><td>86</td><td>86</td><td>87</td><td>88</td><td>86</td><td>86</td><td>4</td></tr>
<tr><td>89</td><td>88</td><td>87</td><td>87</td><td>88</td><td>89</td><td>87</td><td>87</td><td>5</td></tr>
<tr><td>90ab</td><td rowspan="3">89</td><td rowspan="2">88</td><td rowspan="2">88</td><td rowspan="3">89</td><td rowspan="2">90</td><td rowspan="2">88</td><td rowspan="2">88</td><td>5</td></tr>
<tr><td>90cd</td><td rowspan="2">89</td></tr>
<tr><td>90ef</td><td rowspan="2">89</td><td rowspan="2">89</td><td rowspan="7">term.</td><td rowspan="2">89</td><td rowspan="2">89</td></tr>
<tr><td>91ab</td><td rowspan="2">90</td><td rowspan="2">90</td><td rowspan="2">7</td></tr>
<tr><td>91cd</td><td rowspan="2">90</td><td rowspan="2">90</td><td rowspan="2">90</td><td rowspan="2">90</td></tr>
<tr><td>92ab</td><td rowspan="2">91</td><td rowspan="2">91</td><td rowspan="2">-</td></tr>
<tr><td>92cd</td><td rowspan="2">91</td><td rowspan="2">91</td><td rowspan="2">91</td><td rowspan="2">91</td></tr>
<tr><td>93</td><td>92</td><td>92</td><td>-</td></tr>
<tr><td>94</td><td>93</td><td>92</td><td>92</td><td>93</td><td>92</td><td>92</td><td>-</td></tr>
<tr><td>95</td><td>94</td><td>30</td><td>30</td><td>94</td><td></td><td>93</td><td>30</td><td>11</td></tr>
<tr><td>96</td><td>95</td><td>31</td><td>31</td><td>95</td><td></td><td>31</td><td>31</td><td>13</td></tr>
<tr><td>97</td><td>96</td><td>32</td><td>32</td><td>96</td><td></td><td>32</td><td>32</td><td>13</td></tr>
<tr><td>98</td><td>97</td><td>33</td><td>33</td><td>97</td><td></td><td>33</td><td>33</td><td>14</td></tr>
<tr><td>99</td><td>98</td><td>34</td><td>34</td><td>98</td><td></td><td>34</td><td>34</td><td>15</td></tr>
<tr><td>100</td><td>99</td><td>35</td><td>35</td><td>99</td><td></td><td>35</td><td>35</td><td>-</td></tr>
<tr><td>101</td><td>100</td><td>36</td><td>36</td><td>100</td><td></td><td>36</td><td>36</td><td>-</td></tr>
<tr><td>102ab</td><td rowspan="2">101</td><td rowspan="2">37</td><td rowspan="2">37</td><td rowspan="2">101</td><td rowspan="2"></td><td rowspan="2">37</td><td rowspan="2">37</td><td>17</td></tr>
<tr><td>102cd</td><td>18</td></tr>
<tr><td>103</td><td>102</td><td>38</td><td>38</td><td>102</td><td></td><td>38</td><td>38</td><td>19</td></tr>
<tr><td>104</td><td>103</td><td>39</td><td>39</td><td>103</td><td></td><td>39</td><td>39</td><td>20</td></tr>
<tr><td>105</td><td>104</td><td>40</td><td>40</td><td>104</td><td></td><td>40</td><td>40</td><td>-</td></tr>
</table>

(*Continued*)

Table C.1 (Continued)

	Verse Numbering							
Critical Edition	Bn.I_B	Bn.II_B	Br_B	Np_γ	Pb_α	Pm_B	Rr_B	Sc_B
106	105	105	105	105		105	105	21
107	106	106	106	106		106	106	-
108	107	7	17	7		107	107	-
109ab	108	18	108	8		108	108	23
109cd								25
110	109	9	19	9		9	9	
111	110	110	110	110		110	110	26
112ab	111			11				27
112cd		111	111			110	111	
113	112			12				28
114	113	12	112	113		12	12	29
115	114	13	113	114		13	13	30
116	115			115				31
117	116	14	114	16		14	14	32
118	117	15	115	17		115	15	33
119	118	16	116	18		116	116	-
120	119	117	117	19		117	117	35
121	120	18	118	20		18	18	36
122	121	19	119	121		119	19	37
123	122	120	120	122		120	120	38
124ab	123	21	121	123		120	21	39
124cd		122	122			122	122	
125ab	124			124				40
125cd		123	123			123	123	
126	125			125				41
127ab		124	124			124	124	
127cd	126			126				42
128ab		125	125			125	125	
128cd	127			127				43
129ab		126	126			126	126	
129cd	128			128				44
130ab		127	127			127	127	
130cd	29							

Table C.1 (Continued)

<table>
<tr><th colspan="9">Verse Numbering</th></tr>
<tr><th>Critical Edition</th><th>$Bn.I_B$</th><th>$Bn.II_B$</th><th>Br_B</th><th>Np_γ</th><th>Pb_α</th><th>Pm_B</th><th>Rr_B</th><th>Sc_B</th></tr>
<tr><td>131ab</td><td></td><td rowspan="2">-</td><td rowspan="2">128</td><td rowspan="2">129</td><td rowspan="2"></td><td rowspan="2">-</td><td rowspan="2">28</td><td rowspan="2">-</td></tr>
<tr><td>131cd</td><td rowspan="2">130</td></tr>
<tr><td>132ab</td><td rowspan="2">29</td><td rowspan="2">129</td><td rowspan="2">130</td><td rowspan="2"></td><td rowspan="2">29</td><td rowspan="2">29</td><td rowspan="2">46</td></tr>
<tr><td>132cd</td><td rowspan="2">131</td></tr>
<tr><td>133ab</td><td rowspan="2">130</td><td rowspan="2">130</td><td rowspan="2">131</td><td rowspan="2"></td><td rowspan="2">130</td><td rowspan="2">130</td><td rowspan="2">47</td></tr>
<tr><td>133cd</td><td>132</td></tr>
<tr><td>134</td><td>133</td><td>131</td><td>131</td><td>132</td><td></td><td>131</td><td>131</td><td>48</td></tr>
<tr><td>135</td><td>-</td><td>132</td><td>132</td><td>-</td><td></td><td>132</td><td>132</td><td>-</td></tr>
</table>

Bibliography

Manuscript Catalogues

1941. **Catalogue of Sanskrit Manuscripts in the Punjab University Library, Lahore (SSS).**
SSS (Shanti Saroop Saith). *The Shanti Saroop Saith, Catalogue of Sanskrit Manuscripts in the Punjab University Library, Lahore.* Vol. 2. Lahore: University of Punjab, 1941. https://pulibrary.edu.pk/hindi.php.

1968. **Sanskrit Astronomical Tables in the United States (SATIUS).**
David Pingree. 'Sanskrit Astronomical Tables in the United States'. *Transactions of the American Philosophical Society Held at Philadelphia for Promoting Useful Knowledge* 58, no. 3 (1968): 1–77. https://doi.org/10.2307/1006018.

1970–94. **Census of the Exact Sciences in Sanskrit (CESS).**
CESS A1 and A2 (David Pingree). *Census of the Exact Sciences in Sanskrit.* Vol. 1 and 2. Memoirs of the American Philosophical Society: Series A. Philadelphia: The American Philosophical Society, 1970.
CESS A3 (David Pingree). *Census of the Exact Sciences in Sanskrit.* Vol. 3. Memoirs of the American Philosophical Society: Series A. Philadelphia: The American Philosophical Society, 1976.
CESS A4 (David Pingree). *Census of the Exact Sciences in Sanskrit.* Vol. 4. Memoirs of the American Philosophical Society: Series A. Philadelphia: The American Philosophical Society, 1981.
CESS A5 (David Pingree). *Census of the Exact Sciences in Sanskrit.* Vol. 5. Memoirs of the American Philosophical Society: Series A. Unfinished at the time of the author's death in 2005. Philadelphia: The American Philosophical Society, 1994.

1973. **Sanskrit Astronomical Tables in England (SATE).**
SATE (David Pingree). Sanskrit Astronomical Tables in England. Madras: Kuppuswami Sastri Research Institute, 1973.

1983–85. **Catalogue of Sanskrit Manuscripts in the Scindia Oriental Institute, Ujjain (SOI).**
SOI (Ramesh Chandra Purohit and V. Venkatachalam). *A Descriptive Catalogue of Manuscripts in the Scindia Oriental Institute, Vikram University*, Ujjain.

2 vols. Sindhiyā Prācya Granthamālā Nos. 5 and 6. Ujjain: Sindhiyā Prācya Saṃsthāna, Vikrama Viśvavidyālaya, 1983–85.

2002–16. **The Nepalese-German Manuscript Cataloguing Project (NGMCP).**
NGMCP. *The Nepalese-German Manuscript Cataloguing Project, Asia-Africa Institute, Universität Hamburg: Catalogue of Indic manuscripts,* 2002–16. https://catalogue.ngmcp.uni-hamburg.de/content/index.xml.

2003. **Catalogue of the Sanskrit Astronomical Manuscripts Preserved at the Maharaja Man Singh II Museum, Jaipur (MMSM).**
MMSM (David Pingree). *A Descriptive Catalogue of the Sanskrit Astronomical Manuscripts Preserved at the Maharaja Man Singh II Museum in Jaipur, India.* Philadelphia: The American Philosophical Society, 2003. ISBN: 9780871692504.

Primary Sources

ca. second century. **The *Almagest* of Ptolemy.**
Ptolemy's Almagest. Revised edition. Translated by Gerald James Toomer. Princeton: Princeton University Press, 1998. ISBN: 9780691002606.

499. ***Āryabhaṭīya* of Āryabhaṭa I.**
Āryabhaṭīya of Āryabhaṭa: Critically edited with Introduction, English Translation, Notes, Comments, and Indexes. Edited and translated by Kripa Shankar Shukla and Krishna Venkateswara Sarma. New Delhi: Indian National Science Academy, 1976.

ca. sixth century. **Bṛhajjātaka of Varāhamihira.**
The Brihat Jataka of Varaha Mihira. Translated into English. 2nd ed. Translated by N. Chidambaram Aiyar. Aryan Miscellany Astrological Series. Madras: Theosophical Publishing House, 1905.

ca. 575. ***Pañcasiddhāntikā* of Varāhamihira.**
The Pañcasiddhāntikā: The Astronomical Work of Varāhamihira: The Text Edited with an Original Commentary in Sanskrit and an English translation and introduction. 2nd ed. Edited and translated by George Thibaut and Sudhākar Dvivedī. With a commentary by Sudhākar Dvivedī. The Chowkhamba Sanskrit Series 68. Varanasi: The Chowkhamba Sanskrit Series Office, 1968.
The Pañcasiddhāntikā of Varāhamihira. Edited and translated by Otto Neugebauer and David Pingree. 2 vols. Det Kongelige Danske Videnskabernes Selskab Historisk-Filosofiske Skrifter, 6.1 and 6.2. Copenhagen: Munksgaard, 1970–71.

ca. seventh century. ***Bṛhatpārāśarahorā* of Maharṣi Parāśara.**
Brihat Parasara Hora Sastra of Maharshi Parasara: With English Translation, Commentary, Annotation, and Editing. Edited and translated by Rangachari Santhanam. 2 vols. New Delhi: Ranjan Publications, 1984–88.

ca. seventh century. ***Laghubhāskarīya* of Bhāskara I.**
Bhāskara I and His Works, Part III: Laghu-Bhāskarīya, Edited and translated into English, with Explanatory and Critical Notes and Comments, etc. Edited and

translated by Kripa Shankar Shukla. Hindu Astronomical and Mathematical Text Series 4. Lucknow: Lucknow University Press, 1963.

628. ***Brāhmasphuṭasiddhānta* of Brahmagupta.**
Shri Brahmagupta viracita Brāhma-sphuṭa Siddhānta, with Vāsanā, Vijñāna and Hindi Commentaries. Edited by a board of editors headed by Acharyavara Ram Swarup Sharma (chief editor). 4 vols. New Delhi: Indian Institute of Astronomical and Sanskrit Research, 1966.

ante 629. ***Mahābhāskarīya* of Bhāskara I.**
Bhāskara I and His Works, Part II: Mahā-Bhāskarīya. Edited and translated by Kripa Shankar Shukla. Hindu Astronomical and Mathematical Text Series 3. Lucknow: University of Lucknow, 1960.

ca. late eighth century. **The *Sūryasiddhānta*.**
Translation of the Sūrya Siddhānta: A Textbook on Hindu Astronomy with Notes and an Appendix. Reprinted from the edition of 1860. Edited by Phanindralal Gangooly. Translated by Ebenezer Burgess. Calcutta: University of Calcutta, 1935.

ca. ninth century. ***Śiṣyadhīvṛddhidatantra* of Lalla.**
Śiṣyadhīvṛddhida Tantra of Lalla, with the commentary of Mallikārjuna Sūri: Critical Edition with Introduction, English Translation, Mathematical Notes and Indices. Edited and translated by Bina Chatterjee. 2 vols. New Delhi: Indian National Science Academy, 1981.

904. ***Vaṭeśvarasiddhānta* of Vaṭeśvara.**
Vaṭeśvara-Siddhānta and Gola of Vaṭeśvara: Critical Edited with English Translation and Commentary. Edited and translated by Kripa Shankar Shukla. 2 vols. New Delhi: Indian National Science Academy, 1985–86.

ca. mid eleventh century. ***Siddhāntaśekhara* of Śrīpati.**
The Siddhānta-Śekhara of Śrīpati: A Sanskrit Astronomical Work of the 11th Century: Edited, with the Commentary of Makkibhaṭṭa (Chaps. I–IV) and an Original Commentary (Chaps. IV–X). Part I: Chapters I–X and Part II: Chapters XIII–XX. Edited and commented by Babuajī (Śrīkṛṣṇa) Miśra Maithila. 2 vols. Calcutta: Calcutta University Press, 1932–47.

ca. mid eleventh century. ***Śrīpatipaddhatiḥ* of Śrīpati.**
Śrīpatipaddhatiḥ: Translated into English with Notes and a Sample Horoscope Worked Out. Edited and translated by V. Subrahmanya Sastri. Bangalore: V. B. Soobbiah and Sons, 1937.

1150. ***Līlāvatī* of Bhāskara II.**
Līlāvatī of Bhāskaracārya with the Commentary Kriyākramakarī of Śaṅkara and Nārāyaṇa. Edited by Krishna Venkateswara Sarma. Vishveshvaranand Indological Series 66. Hoshiarpur: Vishveshvaranand Vedic Research Institute, 1975.

1150. ***Siddhāntaśiromaṇi* (*Golādhyāya*) of Bhāskara II.**
Translation of the Sūrya-siddhānta by Pundit Bāpu Deva Śāstri and of the Siddhāntaśiromaṇi by Late Lancelot Wilkinson, Esq. C. S. Revised by Pundit Bāpū

Deva Śāstri from the Sanskrit. Bibliotheca Indica 32. Calcutta: The Asiatic Society of Bengal, 1861.

1150. ***Siddhāntaśiromaṇi* of Bhāskara II, with His Auto-commentary Vāsanābhāṣya and Nṛsiṃha Daivajña's *Vāsanāvārttika*.**
Siddhāntaśiromaṇi of Bhāskarācārya with His Autocommentary Vāsanābhāṣya and Vārttika of Nṛsiṃha Daivajña. Edited by Muralidhara Chaturvedi. Library Rare Text Publication Series 5. Varanasi: Sampurnanand Sanskrit University, 1981.

1432. ***Siddhāntadīpikā* of Parameśvara.**
Mahābhāskarīya of Bhāskarācārya with the bhāṣya of Govindasvāmin and the super-commentary Siddhāntadīpikā of Parameśvara. Edited by T. S. Kuppanna Shastri. Madras: Government Oriental Manuscripts Library, 1957.

1443. ***Goladīpikā* I of Parameśvara.**
The Goladīpikā by Parameśvara: Edited with Introduction, Translation and Notes. Edited by Krishna Venkateswara Sarma. Adyar Library Pamphlet Series 32. Madras: The Adyar Library and Research Centre, 1957.

ca. 1450. ***Goladīpikā* II of Parameśvara.**
The Goladīpikā by Śrī Parameśvara. Edited by Taruvāgrahāram Gaṇapati Śastrī. Trivandrum Sanskrit Series XLIX. Trivandrum: Oriental Research Institute and Manuscripts Library, 1916.

ca. 1500. ***Golasāra* of Nilakaṇṭha Somayājī.**
Golasāra of Gārya-Kerala Nīlakaṇṭha Somayājī: Critically Edited with Introduction. Edited by Krishna Venkateswara Sarma. Vishveshvaranand Indological Series 47. Hoshiarpur: Vishveshvaranand Research Institute, 1970.

ca. early nineteenth century. ***Siddhāntasundara* of Jñānarāja.**
The Siddhāntasundara of Jñānarāja: An English Translation with Commentary. Translated by Toke Lindegaard Knudsen. Baltimore: John Hopkins University Press, 2014. ISBN: 9781421414423.

1501. ***Tantrasaṅgraha* of Nilakaṇṭha Somayājī.**
Tantrasaṅgraha of Nilakaṇṭha Somayājī with Yuktidīpikā and Laghuvivṛti of Śaṅkara (An Elaborate Exposition of the Rationale of Hindu Astronomy): Critically Edited with Introduction and Appendices. Edited by Krishna Venkateswara Sarma. Panjab University Indological Series 10. Hoshiarpur: Vishveshvaranand Vishva Bandhu Institute of Sanskrit and Indological Studies, 1977.
Tantrasaṅgraha of Nilakaṇṭha Somayājī. 1st ed. Translated by Krishnamurthi Ramasubramanian and M. S. Sriram. Sources and Studies in the History of Mathematics and Physical Sciences. New Delhi: Springer-Verlag / Hindustan Book Agency, 2011. https://doi.org/10.1007/978-0-85729-036-6.

ca. seventeenth century. ***Karaṇakesarī* of Bhāskara (fl. ca. 1681).**
Clemency Montelle and Kim Plofker. 'The Karaṇakesari of Bhāskara: A 17th-century Table Text for Computing Eclipses'. *History of Science in South Asia* 2 (2014): 1–62. https://doi.org/10.18732/H2CC7F.

1646. ***Siddhāntasārvabhauma* of Munīśvara.**
The Siddhāntasārvabhauma by Śrī Munīśvara: Edited with Introduction etc. Edited by Gopī Nātha Kavirāja. 2 vols. The Princess of Wales Sarasvatī Bhavana Texts 41. Varanasi: Sampurnanand Sanskrit University, 1932–35.
Siddhāntasārvabhauma by Munīśvara. Edited by Mīṭhālāla Ojhā. Vol. 3. Saravatī Bhavana Granthamāla 41. Varanasi: Sampurnanand Sanskrit University, 1978.

1658. ***Siddhāntatattvaviveka* of Kamalākara.**
Siddhāntatattvaviveka of Śrī Kamalākara Bhaṭṭa with the Commentary Vasanābhāṣya of Śrī Gaṅgādhara Śarmā. Edited by Kṛṣna Candra Dvivedī. 3 vols. M. M. Sudhākaradvivedī Granthamālā 3. Varanasi: Sampurnanand Sanskrit University, 1993–98.
Siddhāntatattvaviveka of Śrī Kamalākara Bhaṭṭa with Hindi Commentary. Edited by Satyadeva Śarmā. 3 vols. Chaukhambā Surabhāratī Granthamālā 590. Varanasi: Chaukhambā Surabhāratī Prakāśana, 2015. ISBN: 9789385005190.

1732. ***Siddhāntasamrāṭ* of Jagannātha.**
Samrād Jagannāth-Viracita Samrāṭ Siddhānta (Siddhānta-sāra-kaustubha): With Various Readings on the Basis of Different Mss. Edited by a board of editors headed by Acharyavara Ram Swarup Sharma (chief editor). 2 vols. New Delhi: Indian Institute of Astronomical and Sanskrit Research, 1967.

Secondary Sources

Alam, Muzaffar. 2003. 'The Culture and Politics of Persian in Precolonial Hindustan'. Chap. 2 in *Literary Cultures in History: Reconstructions from South Asia*, edited by Sheldon Pollock, 131–198. Berkeley: University of California Press. https://doi.org/10.1525/9780520926738-007.

Ali, Syed Muzafer. 1983. *The Geography of the Purāṇas.* 3rd ed. New Delhi: People's Publication House.

Ansari, Shaikh Mohammad Razaullah. 1985. 'Introduction of modern western astronomy in India during 18–19 centuries'. In *History of Astronomy in India*, 363–402. New Delhi: Institute of History of Medicine / Medical Research.

———. 1995. 'On the Transmission of Arabic-Islamic Astronomy to Medieval India'. *Archives Internationales d'Histoire des Sciences* 45 (135): 273–297.

———. 2005. 'Hindu's Scientific Contributions in Indo-Persian'. *Indian Journal of History of Science* 40 (2): 205–221.

———. 2015. 'Survey of *Zījes* Written in the Subcontinent'. *Indian Journal of History of Science* 50 (4): 573–601.

———. 2016. 'Astronomy in Medieval India'. In *Encyclopaedia of the History of Science, Technology, and Medicine in NonWestern Cultures*, edited by Helaine Selin, 717–726. Dordrech: Springer Netherlands. https://doi.org/10.1007/978-94-007-7747-7_10114.

Apte, Vaman Shivaram. (1890) 1965. *The Practical Sanskrit-English Dictionary: Containing Appendices on Sanskrit Prosody and Important Literary and Geographical Names of Ancient India*. 4th revised and enlarged edition. Benaras: Motilal Banarsidass Publishers. ISBN: 9780895811714.

Ashfaque, Syed Mohammad. 1977. 'Astronomy in the Indus Valley Civilization: A Survey of the Problems and Possibilities of the Ancient Indian Astronomy and Cosmology in the Light of Indus Script Decipherment by the Finnish Scholars'. *Centaurus* 21 (2): 149–193. https://doi.org/10.1111/j.1600-0498.1977.tb00351.x.

Aussant, Emilie. 2014. 'Sanskrit Theories on Homonymy and Polysemy'. *Bulletin d'Études Indiennes*, Les études sur les langues indiennes. Leur contribution à l'histoire des idées linguistiques et à la linguistique contemporaine, 32. https://halshs.archives-ouvertes.fr/halshs-01502381.

Bagheri, Mohammad. 2006. 'Kūshyār ibn Labbān's Glossary of Astronomy'. SCIAMVS 7:145–174.

Banerjea, Krishna Mohan, ed. (1862) 2004. *The Mārcaṇḍeya Purāna: In the Original Sanskrit*. Reprint edition. 2 vols. Bibliotheca Indica 29. New Delhi: Cosmo Publications.

Baranī, Sayyid Ḥasan. 1951. 'Muslim Researches in Geodesy'. In *Al-Bīrūnī Commemoration Volume*, 1–52. Calcutta: Iran Society.

———, ed. 1954. *Kitāb al-Qānūn al-Masᶜūdī fiᵓ l-hayᵓ ā waᵓ l-nojūm (Canon Masudicus): An Encyclopaedia of Astronomical Sciences*. Vol. III. Hyderabad Arabic Publications 103. Hyderabad: Osmania Oriental Publication Bureau.

Begley, Wayne Edison, and Ziauddin Abdul Hayy Desai, eds. 1990. *The Shah Jahan Nama of ᶜInayat Khan: An Abridged History of the Mughal Emperor Shah Jahan, Compiled by His Royal Librarian*. The nineteenth-century manuscript translation of A. R. Fuller (British Library, Add. 30, 777). New Delhi: Oxford University Press. ISBN: 9780195624892.

Berggren, John Lennart. 1996. 'Al-Kūhī's "Filling a Lacuna in Book II of Archimedes" in the Version of Naṣīr al-Dīn al-Tuṣī'. *Centaurus* 38 (2–3): 140–207. https://doi.org/10.1111/j.1600-0498.1996.tb00609.x.

———. 2016. *Episodes in the Mathematics of Medieval Islam*. 2nd ed. New York: Springer. https://doi.org/10.1007/978-1-4939-3780-6.

Berggren, John Lennart, and Alexander Jones. 2000. *Ptolemy's Geography: An Annotated Translation of the Theoretical Chapters*. Illustrated reprint edition. Princeton: Princeton University Press. ISBN: 9780691092591.

Bhandarkar, Ramkrishna Gopal. 1905. *Mālatī-Mādhava by Bhavabhūti with the commentary of Jagaddhara: Edited with Notes, Critical and Explanatory*. 2nd ed. Bombay: Government Central Book Depot.

Bhaṭṭācārya, Tārānātha Tarkavācaspati. 1962. *Vācaspatyam: Bṛhat Saṃskṛtābhidhānam (A Comprehensive Sanskrit Dictionary)*. Vol. VI. The Chowkhamba Sanskrit Series 94. Varanasi: The Chowkhamba Sanskrit Series Office.

Bhaṭṭācārya, Vibhūti Bhūṣaṇa, and Baladeva Upadhyaya, eds. 1967. *Hayata.* Sarasvatī Bhavana Granthamālā 96. Varanasi: Varanaseya Sanskrit Vishvavidyalaya.

Bilimoria, Purushottama. 2017. 'Pramāṇa Epistemology: Origins and Developments'. Chap. 3 in *History of Indian Philosophy*, 1st ed., edited by Purushottama Bilimoria, J. N. Mohanty, Amy Rayner, John Powers, Stephen Phillips, Richard King and Christopher Key Chapple, 27–39. Routledge History of World Philosophies. Abingdon-on-Thames: Routledge. ISBN: 9780415309769.

Blake, Stephen P. 2013. *Time in Early Modern Islam: Calendar, Ceremony, and Chronology in the Safavid, Mughal and Ottoman Empires.* New York: Cambridge University Press. https://doi.org/10.1017/CBO9781139343305.

Bowen, Alan C., and Francesca Rochberg, eds. 2020. *Hellenistic Astronomy: The Science in Its Contexts.* Brill's Companions to Classical Studies. Leiden: Brill. https://doi.org/10.1163/9789004400566.

Brummelen, Glen Van. 2009. *The Mathematics of the Heavens and the Earth: The Early History of Trigonometry.* Princeton: Princeton University Press. ISBN: 9780691129730.

Chann, Naindeep Singh. 2009. 'Lord of the Auspicious Conjunction: Origins of the Ṣāḥib-Qirān'. *Iran and the Caucasus* 13 (1): 93–110. https://doi.org/10.1163/160984909X12476379007927.

Cherniak, Alex. 2009. *Mahabharata: Book Six Bhishma.* 2 vols. Clay Sanskrit Library 53. New York: New York University Press and JJC Foundation. ISBN: 9780814717059.

Datta, Bibhutibhusan, Avadhesh Narayan Singh and Kripa Shankar Shukla. 1983. 'Hindu Trigonometry'. *Indian Journal of History of Science* 18 (1): 39–108.

Delire, Jean Michel, ed. 2012. *Astronomy and Mathematics in Ancient India (Astronomie Et Mathmathiques de L'Inde Ancienne).* Bilingual edition. Lettres Orientales Series 17. Actes de la journée d'études organisée le 24 avril 2009 à l'Université Libre de Bruxelles. Leuven: Peeters Publishers and Booksellers. ISBN: 9789042926141.

Devadevan, Manu V. 2016. 'Armillary Spheres in India'. In *Encyclopaedia of the History of Science, Technology, and Medicine in Non-Western Cultures*, edited by Helaine Selin, 587–589. Dordrecht: Springer Netherlands. https://doi.org/10.1007/978-94-007-7747-7_10284.

Dicks, David Reginald. 1985. *Early Greek Astronomy to Aristotle.* Illustrated reprint edition. Aspects of Greek and Roman Life. Ithaca, New York: Cornell University Press. ISBN: 9780801493102.

Dikshit, R. K. 1976. *The Candellas of Jejākabhukti.* New Delhi: Abhinav Publications. ISBN: 9788170170464.

Dikshit, Sankar Balkrishna. 1969. *English Translation of Bharatiya Jyotish Sastra (History of Indian Astronomy): History of Astronomy during Vedic and Vedāṅga Periods.* Translated by R. V. Vaidya. Vol. I. Original text *Bhāratīya Jyotiṣa*

Śāstrācā Prācīna Āṇi Arvācīna Itihāsa in Marathi (Poona, 1896). New Delhi: Government of India Press.

———. 1981. *English Translation of Bharatiya Jyotish Sastra (History of Indian Astronomy): History of Astronomy during Siddhantic and Modern Periods.* Translated by R. V. Vaidya. Vol. II. Original text *Bhāratīya Jyotiṣa Śāstrācā Prācīna Āṇi Arvācīna Itihāsa* in Marathi (Poona, 1896). New Delhi: Government of India Press.

Divakaran, P. P. 2018. *The Mathematics of India: Concepts, Methods, Connections.* Sources and Studies in the History of Mathematics and Physical Sciences. Singapore: Springer-Verlag and Hindustan Book Agency. https://doi.org/10.1007/978-981-13-1774-3.

Dressler, Markus, Armando Salvatore and Monika Wohlrab-Sahr. 2019. 'Islamicate Secularities: New Perspectives on a Contested Concept'. Historical Social Research 44 (3): 7–34. https://doi.org/10.12759/hsr.44.2019.3.7-34.

Dreyer, John Louis Emil, and William Harris Stahl. (1906) 2019. *A History of Astronomy from Thales to Kepler.* 2nd revised edition. Classics of Science. Originally titled 'History of the planetary systems from Thales to Kepler'. New York: Dover Publications. ISBN: 9780486600796.

Dvivedī, Sudhākara. 1933. *Gaṇakataraṅgiṇī: Lives of Hindu astronomers.* Varanasi: Jyotish Prakash Press.

Evans, James. 1998. *The History and Practice of Ancient Astronomy.* Illustrated edition. Oxford: Oxford University Press. ISBN: 9780195095395.

Fleet, Kate, Gudrun Krämer, Denis Matringe, John Nawas and Everett Rowson, eds. 2007–20. 'Encyclopaedia of Islam, THREE'. Brill Online reference. https://referenceworks.brillonline.com/pages/help/transliteration-islam.

Gansten, Martin. 2019. 'Samarasiṃha and the Early Transmission of Tājika Astrology'. *Journal of South Asian Intellectual History* 1 (1): 79–132. https://doi.org/10.1163/25425552-12340005.

———. 2020. *The Jewel of Annual Astrology: A Parallel Sanskrit-English Critical Edition of Balabhadra's* Hāyanaratna. Sir Henry Wellcome Asian Series 19. Leiden: Brill. https://doi.org/10.1163/9789004433717.

Gethin, Rupert. 1998. The Foundations of Buddhism. Oxford: Oxford University Press. ISBN: 9780192892232.

Ghori, Shabbir Ahmad Khan. 2008. 'Development of Zīj literature in India'. In *The Tradition of Astronomy in India* Jyotiḥśāstra, edited by Bidare Venkatasubbaiah Subbarayappa, vol. IV, bk. 4, 383–415. History of Science, Philosophy and Culture in Indian Civilization. New Delhi: Project of History of Indian Science, Philosophy and Culture.

Graßhoff, Gerd. 2013. *The History of Ptolemy's Star Catalogue.* 1st ed. Studies in the History of Mathematics and Physical Sciences 14. New York: Springer. https://doi.org/10.1007/978-1-4612-4468-4.

Gupta, Radha Charan. 1977. 'Indian Values of the Sinus Totus'. *Indian Journal of History of Science* 13 (2): 125–143. Paper presented at the XVth International Congress of History of Science, Edinburgh, 1977.

Gutas, Dimitri. 1998. *Greek Thought, Arabic Culture: The Graeco-Arabic Translation Movement in Baghdād and Early ʿAbbāsid Society (2nd–4th/5th–10th centuries)*. 1st ed. London: Routledge. https://doi.org/10.4324/9780203017432.

Hafez, Ihsan. 2010. 'ʿAbd al-Raḥmān al-Ṣūfī and His Book of the Fixed Stars: a Journey of Re-discovery'. PhD diss., James Cook University. https://researchonline.jcu.edu.au/28854.

Haider, Najaf. 2011. 'Translating Texts and Straddling Worlds: Intercultural Communication in Mughal India'. In *The Varied Facets of History: Essays in Honour of Aniruddha Ray*, edited by Ishrat Alam and Syed Ejaz Hussain, 115–124. New Delhi: Primus Books. ISBN: 9789380607160.

Houtsma, Martijn Theodoor, ed. (1913–36) 1993. *E. J. Brill's First Encyclopaedia of Islam, 1913–1936*. Reprint edition. Vol. VII. Leiden: Brill.

Huff, Toby E. 2010. *Intellectual Curiosity and the Scientific Revolution: A Global Perspective*. Cambridge: Cambridge University Press. https://doi.org/10.1017/CBO9780511782206.

Hunger, Hermann, and David Pingree. 1999. *Astral Sciences in Mesopotamia*. Handbook of Oriental Studies: Section 1 The Near and Middle East 44. Leiden: Brill. https://doi.org/10.1163/9789004294134.

Huxley, George Leonard. 1964. *The Interaction of Greek and Babylonian Astronomy*. New Lecture Series 16. An Inaugural Lecture delivered before The Queen's University of Belfast on 29 January 1964. Belfast: Queen's University.

Ingalls, Daniel Henry Holmes, trans. 1965. *An Anthology of Sanskrit Court Poetry: Vidyākara's 'Subhāṣitaratnakoṣa'*. Harvard Oriental Series 44. Cambridge Massachusetts: Harvard University Press.

Jaini, Padmanabh Shrivarma. 2000. 'Jaina Purāṇa: A Purāṇic Counter Tradition'. In *Collected Papers on Jaina Studies*, edited by Padmanabh Shrivarma Jaini, 375–428. New Delhi: Motilal Banarsidass Publishers. ISBN: 9788120816916.

Janos, Damien. 2012. *Method, Structure, and Development in al-Fārābī's Cosmology*. Islamic Philosophy, Theology and Science. Texts and Studies 85. Leiden: Brill. https://doi.org/10.1163/9789004217324.

Jhā, Gaṅgānātha, trans. 1939. *Gautama's Nyāyasutras with Vātsyāyana Bhāṣya: Translated into English with His Own Revised Notes*. Poona Oriental Series 59. Poona: Oriental Book Agency.

———, ed. and trans. (1925) 1967. *Kāvyaprakāsha of Mammaṭa with English Translation (Chapters I to IX) (With Indices and Appendices)*. Reprint edition. Varanasi: Bharatiya Vidya Prakashan.

———, trans. (1934) 2003. *The Yoga-Darshana: Comprising the Sūtras of Patañjali with the Bhāṣya of Vyāsa*. 2nd revised edition. Translated into English with Notes. Fremont, California: Asian Humanities Press. ISBN: 9780895819512.

Jwaideh, Wadie Elias, trans. (1959) 1987. *The Introductory Chapters of Yāqūt's Muᶜ-jam al-Buldān*. Reprint edition. Leiden: E. J. Brill.

Kak, Subhash. 2000. 'Birth and Early Development of Indian Astronomy: The History of Non-Western Astronomy'. In Astronomy Across Cultures, edited by Helaine Selin and Sun Xiaochun, 303–340. Science Across Cultures: The History of Non-Western Science 1. Dordrecht: Springer. https://doi.org/10.1007/978-94-011-4179-6_10.

Keller, Agathe. 2012. 'Dispelling Mathematical Doubts: Assessing Mathematical Correctness of Algorithms in Bhāskara's Commentary on the Mathematical Chapter of the Āryabhaṭiya'. In *The History of Mathematical Proof in Ancient Traditions*, edited by Karine Chemla, 487–508. Cambridge: Cambridge University Press. https://halshs.archives-ouvertes.fr/halshs-00151013v4.

Keller, Agathe, and Alexei Volkov. 2014. 'Mathematics Education in Oriental Antiquity and Middle Ages'. Chap. 4 in *Handbook on the History of Mathematics Education*, edited by Alexander Karp and Gert Schubring, 55–83. New York: Springer. https://doi.org/10.1007/978-1-4614-9155-2_4.

Kennedy, Edward Stewart. 1985. 'Spherical Astronomy in Kāshī's Khāqānī Zīj'. *Zeitschrift für Geschichte der Arabisch-Islamischen Wissenschafte* 2:1–46.

Kennedy, Edward Stewart, and Fuʾād Ḥaddād, trans. 1981. *The Book of the Reasons Behind Astronomical Tables (Kitāb fī ᶜilal al-zījāt) of ᶜAlī ibn Sulaymān al-Hāshimī: Facsimile reproduction of the unique Arabic text contained in the Bodleian MS Arch. Seld. A.11*. With annotations by Edward Stewart Kennedy and David Pingree. Studies in Islamic Philosophy and Science. Delmar, NY: Scholars' Facsimiles and Reprints. ISBN: 9780820112985.

Kennedy, Edward Stewart, and M. H. Kennedy. 1987. 'Al-Kāshī's Geographical Table'. *Transactions of the American Philosophical Society Held at Philadelphia for Promoting Useful Knowledge* 77 (7): 1–45. https://doi.org/10.2307/1006581.

King, David. 1986. *Islamic Mathematical Astronomy*. Collected Studies. London: Variorum Reprints. ISBN: 9780860781783.

King, Richard. 1995. *Early Advaita Vedanta and Buddhism: The Mahayāna Context of the Gauḍapādīya-Kārikā*. SUNY Series in Religious Studies. New York: State University of New York. ISBN: 9780791425138.

Kirfel, Willibald. 1920. *Die Kosmographie der Inder, nach den Quellen dargestellt*. In German. Bonn and Leipzig: Kurt Schroeder.

Knobel, Edward Ball. 1917. *Ulugh Beg's Catalogue of Stars: Revised from all Persian Manuscripts Existing in Great Britain, with a Vocabulary of Persian and Arabic Words*. Carnegie Institution of Washington Publication 250. Washington, DC: The Carnegie Institution of Washington.

Knudsen, Toke. 2021. 'Three Purāṇic Statements on the Shape of the Earth'. *History of Science in South Asia* 9:128–166. https://doi.org/10.18732/hssa55.

Kochhar, Rajesh, and Jayant Narlikar. 1995. *Astronomy in India, A Perspective*. New Delhi: Indian National Science Academy.

Krupp, Edwin Charles. (1983) 2003. *Echoes of the Ancient Skies: The Astronomy of Lost Civilizations*. Reprint edition. Dover Books on Astronomy. New York: Dover Publications. ISBN: 9780486428826.

Kunitzsch, Paul. 2008. '*Almagest*: Its Reception and Transmission in the Islamic World'. In *Encyclopaedia of the History of Science, Technology, and Medicine in Non-Western Cultures*, edited by Helaine Selin, 140–141. Dordrecht: Springer Netherlands. https://doi.org/10.1007/978-1-4020-4425-0_8988.

Mak, Bill. 2013. 'The Date and Nature of Sphujidhvaja's Yavanajātaka Reconsidered in the Light of Some Newly Discovered Materials'. *History of Science in South Asia* 1:1–20. https://doi.org/10.18732/H2RP4T.

Meliá, Juan Tous. 2001. 'La Isla de El Hierro y el meridiano origen'. *Anuario del Instituto de Estudios Canarios* 46:249–288. ISSN: 0423-4804.

Mercier, Raymond. 1992. 'Geodesy. Cartography in the Traditional Islamic and South Asian Societies'. Chap. 8 in *The History of Cartography*, edited by John Brian Harley and David Woodward, vol. 2, bk. 1, 175–188. The History of Cartography Series. Chicago: The University of Chicago Press. ISBN: 9780226316352.

———. 2004. *Studies on the Transmission of Medieval Mathematical Astronomy*. 1st ed. Variorum Collected Studies 787. Aldershot, Hampshire: Ashgate Publishing Company. ISBN: 9780860789499.

———. 2018. 'On the Originality of Indian Mathematical Astronomy'. In *The Interactions of Ancient Astral Science*, edited by David Brown, 734–776. With contributions by Jonathan Ben-Dov, Harry Falk, Geoffrey Ernest, Raymond Mercier, Antonio Panaino, Joachim Quack, Alexandra van Lieven, Michio Yano. Bremen: Hempen Verlag. ISBN: 9783944312552.

Minkowski, Christopher. 2002. 'Astronomers and their Reasons: Working paper on Jyotiḥśāstra'. *Journal of Indian Philosophy* 30 (5): 495–514. https://doi.org/10.1023/A:1022870103341.

———. 2004. 'Completing Cosmologies and the Problems of Contradiction in Sanskrit Knowledge Systems'. In *Ketuprakāśa: Studies in the History of the Exact Sciences in Honour of David Pingree*, edited by Charles Burnett, Jan P. Hogendijk, Kim Plofker and Michio Yano, 349–385. Islamic Philosophy, Theology and Science. Texts and Studies 54. Leiden: Brill. https://doi.org/10.1163/9789047412441_016.

———. 2008. 'Why Should We Read the Maṅgala Verses?' In *Śāstrārambha: Inquiries into the Preamble in Sanskrit*, edited by Walter Slaje, 1–24. Abhandlungen für die Kunde des Morgenlandes 62. Wiesbaden: Otto Harrassowitz Verlag. ISBN: 9783447056458.

———. 2014. 'Learned Brahmins and the Mughal Court: The Jyotiṣas'. In *Religious Interactions in Mughal India*, edited by Vasudha Dalmia and Munis Daniyal Faruqui, 102–134. New Delhi: Oxford University Press. https://doi.org/10.1093/acprof:oso/9780198081678.003.0004.

Minorsky, Vladimir Fedorovich, and Clifford Edmund Bosworth. 1982. *Ḥudūd al-ᶜĀlam 'The Regions of the World': A Persian Geography, 372 A.H.–982 A.D.* 2nd reprint edition. E. J. W. Gibb Memorial Series, New Series XI. Cambridge: Cambridge University Press.

Mishra, Krishna. 2016. *The Rise of Wisdom Moon.* Translated by Matthew Kapstein and J. N. Mohanty. Clay Sanskrit Library 1. New York: New York University Press. ISBN: 9781479852642.

Misra, Anuj. 2021. 'Persian Astronomy in Sanskrit: A Comparative Study of Mullā Farīd's *Zīj-i Shāh Jahānī* and Its Sanskrit Translation in Nityānanda's *Siddhāntasindhu*'. *History of Science in South Asia* 9:30–127. https://doi.org/10.18732/hssa64.

———. 2022. 'Sanskrit Recension of Persian Astronomy: The Computation of True Declination in Nityānanda's *Sarvasiddhāntaraja*'. *History of Science in South Asia* 10:68–168. https://doi.org/10.18732/hssa75.

Misra, Anuj, Clemency Montelle and Krishnamurthi Ramasubramanian. 2016. 'The Poetic Features in the *golādhyāya* of Nityānanda's *Sarvasiddhāntarāja*.' Gaṇita Bhāratī 38 (2): 141–156.

Monier-Williams, Monier, Ernst Leumann and Carl Cappeller. (1899) 1960. *A Sanskrit-English Dictionary: Etymologically and Philologically Arranged, with Special Reference to Cognate Indo-European Languages.* Reprint edition. Oxford: Clarendon Press.

Montelle, Clemency, and Kim Plofker. 2018. *Sanskrit Astronomical Tables.* Sources and Studies in the History of Mathematics and Physical Sciences. Cham, Switzerland: Springer Nature Switzerland AG. https://doi.org/10.1007/978-3-319-97037-0.

Montelle, Clemency, and Krishnamurthi Ramasubramanian. 2018. 'Determining the Sine of One Degree in the Sarvasiddhāntarāja of Nityānanda'. SCIAMVS 19:1–52. ISSN: 1345-4617.

Montelle, Clemency, Krishnamurthi Ramasubramanian and Jambugahapitiye Dhammaloka. 2016. 'Computation of Sines in Nityānanda's *Sarvasiddhāntarāja*'. SCIAMVS 17:1–53. ISSN: 1345-4617.

Nair, Shankar. 2020. *Translating Wisdom: Hindu-Muslim Intellectual Interactions in Early Modern South Asia.* Oakland, California: University of California Press. https://doi.org/10.1525/luminos.87.

Narasimha, Roddam. 2007. 'Epistemology and Language in Indian Astronomy and Mathematics'. *Journal of Indian Philosophy* 35 (5–6): 521–541. https://doi.org/10.1007/s10781-007-9033-5.

———. 2013. 'Axiomatism and Computational Positivism: Two Mathematical Cultures in Pursuit of Exact Sciences'. *Economic and Political weekly* 38 (35): 3650–3656. https://www.jstor.org/stable/4413961.

Neugebauer, Otto Eduard. (1951) 1969. *The Exact Sciences in Antiquity.* 2nd reprint edition. New York: Dover Publications. ISBN: 9780486223322.

Neugebauer, Otto Eduard. 1975. *A History of Ancient Mathematical Astronomy*. Vol. 1. Studies in the History of Mathematics and Physical Sciences 1. Berlin and Heidelberg: Springer. https://doi.org/10.1007/978-3-642-61910-6.

North, John David. 1986. *Horoscopes and History*. Warburg Institute Surveys and Texts XIII. London: The Warburg Institute University of London. ISBN: 9780854810680.

Ôhashi, Yukio. 1993. 'Development of Astronomical Observation in Vedic and Post-Vedic India'. *Indian Journal of History of Science* 28 (3): 185–251.

———. 1994. 'Astronomical Instruments in Classical Siddhāntas'. *Indian Journal of History of Science* 29 (2): 155–313.

———. 2002. 'The Legends of Vasiṣṭhaha—A Note on the Vedāṅga Astronomy'. In *History of Oriental Astronomy*, edited by Shaikh Mohammad Razaullah Ansari, 75–82. Astrophysics and Space Science Library 275. Dordrecht: Kluwer Academic Publishers. https://doi.org/10.1007/978-94-015-9862-0_7.

Orsini, Francesca. 2012. 'How to Do Multilingual Literary History? Lessons from Fifteenth and Sixteenth-century North India'. *The Indian Economic and Social History Review* 49 (2): 225–246. https://doi.org/10.1177/001946461204900203.

Ossendrijver, Mathieu. 2012. *Babylonian Mathematical Astronomy: Procedure Texts*. 1st ed. Sources and Studies in the History of Mathematics and Physical Sciences. New York: Springer. https://doi.org/10.1007/978-1-4614-3782-6.

Pannekoek, Anton. (1961) 1989. *A History of Astronomy*. Unabridged reprint edition. Dover Books on Astronomy. New York: Dover Publications. ISBN: 9780486659947.

Pargiter, Frederick Eden, trans. 1904. *The Mārkaṇḍeya Purāṇa: Translated with Notes*. Bibliotheca Indica: A Collection of Oriental Works. Calcutta: The Asiatic Society of Bengal.

Pedersen, Olaf. (1974) 1993. *Early Physics and Astronomy: A Historical Introduction*. Revised edition. Cambridge: Cambridge University Press. ISBN: 9780521408998.

———. 2011. *A Survey of the Almagest: With Annotation and New Commentary by Alexander Jones*. Revised edition. Edited by Alexander Jones. Sources and Studies in the History of Mathematics and Physical Sciences. New York: Springer. https://doi.org/10.1007/978-0-387-84826-6.

Peters, Christian Heinrich Friedrich, and Edward Ball Knobel. 1915. *Ptolemy's Catalogue of Stars: A Revision of the Almagest*. Carnegie Institution of Washington Publication 86. Washington, District of Columbia: The Carnegie Institution of Washington.

Petrocchi, Alessandra. 2016. 'The Bhūtasaṃkhyā Notation: Numbers, Culture, and Language in Sanskrit Mathematical Literature'. In *On Meaning and Mantras: Essays in Honor of Frits Staal*, edited by Richard K. Payne and George Thompson, 477–502. Contemporary issues in Buddhist studies. Moraga, California: Institute of Buddhist Studies / BDK America Incorporated.

Pingree, David. 1963. 'Astronomy and Astrology in India and Iran'. *Isis* 54 (2): 229–246. https://www.jstor.org/stable/228540.

———. 1970b. 'On the Classification of Indian Planetary Tables'. *Journal for the History of Astronomy* 1:95–108. https://doi.org/10.1177/002182867000100201.

———. 1971. 'On the Greek Origin of the Indian Planetary Model Employing a Double Epicycle'. *Journal for the History of Astronomy* 2:80–85. https://doi.org/10.1177/002182867100200202.

———. 1972. 'Precession and Trepidation in Indian Astronomy before A.D. 1200'. *Journal for the History of Astronomy* 3:27–35. https://doi.org/10.1177/002182867200300104.

———. 1973a. 'The Greek Influence on Early Islamic Mathematical Astronomy'. *Journal of the American Oriental Society* 93 (1): 32–43. https://doi.org/10.2307/600515.

———. 1973b. 'The Mesopotamian Origin of Early Indian Mathematical Astronomy'. *Journal for the History of Astronomy* 4:1–12. https://doi.org/10.1177/002182867300400102.

———. 1976b. 'The Recovery of Early Greek Astronomy from India'. *Journal for the History of Astronomy* 7:109–123. https://doi.org/10.1177/002182867600700202.

———. 1978a. 'History of Mathematical Astronomy in India'. In *Dictionary of Scientific Biography*, vol. XV (supplement 1): 533–633. New York: Charles Scribner's Sons.

———. 1978b. 'Indian Astronomy'. *Proceedings of the American Philosophical Society* 122 (6): 361–364. https://www.jstor.org/stable/986451.

———. 1978c. 'Islamic Astronomy in Sanskrit'. *Journal for the History of Arabic Science* 2:315–330.

———, ed. and trans. 1978d. *The Yavanajataka of Sphujidhvaja*. 2 vols. Harvard Oriental Series 48. Cambridge, Massachusetts: Harvard University Press. ISBN: 9780674963733.

———. 1981b. 'Jyotiḥśāstra: Astral and Mathematical Literature'. In *A History of Indian literature*, edited by Jan Gonda, vol. VI, fasc. 4. Wiesbaden: Otto Harrassowitz Verlag.

———. 1990. 'The Purāṇas and Jyotiḥśāstra: Astronomy'. *Journal of the American Oriental Society* 110 (2): 274–280. https://doi.org/10.2307/604530.

———. 1996a. 'Indian Reception of Muslim Versions of Ptolemaic Astronomy'. In *Tradition, Transmission, Transformation*, Proceedings of Two Conferences on Premodern Science Held at the University of Oklahoma, edited by F. Jamil Ragep, Sally P. Ragep and Steven Livesey, 471–485. Collection de Travaux de l'Académie Internationale d'Histoire des Sciences 37. Leiden: Brill. ISBN: 9789004101197.

———. 1996b. 'Sanskrit Geographical Tables'. *Indian Journal of History of Science* 31 (2): 173–220.

———. 1997. 'Tājika: Persian Astrology in Sanskrit'. Chap. 7 in *From Astral Omens to Astrology: From Babylon to Bīkāner*, 79–90. Serie Orientale Roma 78. Rome: Istituto Italiano per l'Africa e l'Oriente. ISBN: 9788863230918.

———. 1999. 'An Astronomer's Progress'. Papers Delivered at a Joint Meeting with the Royal Swedish Academy of Sciences. Stockholm, 24–26 May 1998 (Mar., 1999), *Proceedings of the American Philosophical Society* 143 (1): 73–85. https://www.jstor.org/stable/3181975.

———. 2001. 'From Alexandria to Baghdād to Byzantium. The Transmission of Astrology'. *International Journal of the Classical Tradition* 8 (1): 3–37. https://doi.org/10.1007/BF02700227.

———. 2003b. 'The *Sarvasiddhāntarāja* of Nityānanda'. In *The Enterprise of Science in Islam: New Perspectives*, edited by Jan P. Hogendijk and Abdelhamid I. Sabra, 269–284. Dibner Institute Studies in the History of Science and Technology. Cambridge, Massachusetts: MIT Press. ISBN: 9780262519168.

Plofker, Kim. 2000. 'The Astrolabe and Spherical Trigonometry in Medieval India'. *Journal for the History of Astronomy* 31 (1): 37–54. https://doi.org/10.1177/002182860003100103.

———. 2005. 'Derivation and Revelation: The legitimacy of Mathematical Models in Indian Cosmology'. Chap. 2 in *Mathematics and the Divine: A Historical Study*, edited by Teun Koetsier and Luc Bergmans, 61–76. Amsterdam: Elsevier Science. https://doi.org/10.1016/B978-044450328-2/50004-7.

———. 2008. 'Sanskrit Mathematical Verse'. Chap. 6.2 in *The Oxford Handbook of the History of Mathematics*, edited by Eleanor Robson and Jacqueline Stedall, 519–536. Oxford: Oxford University Press. ISBN: 9780199213122.

———. 2009. *Mathematics in India*. Illustrated edition. Princeton: Princeton University Press. ISBN: 9781400834075.

Pollock, Sheldon. 2001. 'New Intellectuals in Seventeenth-century India'. *Indian Economic and Social History Review* 38 (1): 3–31. https://doi.org/10.1177/001946460103800101.

Prabhupāda, Abhaya Caraṇāravinda Bhaktivedānta Svāmī. 1975. *Śrimad Bhāgvatam, Fifth Canto: The Creative Impetus*. 1st ed. two vols. California: The Bhaktivedanta Book Trust. ISBN: 9789171496386.

Ragep, Sally P. 2016. *Jaghmīnī's* Mulakhkhaṣ: *An Islamic Introduction to Ptolemaic Astronomy*. 1st ed. Sources and Studies in the History of Mathematics and Physical Sciences. Cham, Switzerland: Springer Nature Switzerland AG. https://doi.org/10.1007/978-3-319-31993-3.

Rahman, Abdur, ed. 1998. *History of Indian Science, Technology and Culture AD 1000–1800*. Vol. 3, bk. 1. History of Science, Philosophy and Culture in Indian Civilization. New Delhi: Oxford University Press. ISBN: 9780195646528.

Raja, Kumarapuram Kunjunni. (1963) 1977. *Indian Theories of Meaning*. Reprint edition. Madras: Adyar Library / Research Centre.

Ramasubramanian, Krishnamurthi. 1998. 'Model of Planetary Motion in the Works of Kerala astronomers'. *Bulletin of the Astronomical Society of India* 26:11–31. https://adsabs.harvard.edu/full/1998BASI...26...11R.

Ramasubramanian, Krishnamurthi, Mandyam D. Srinivas and M. S. Sriram. 1994. 'Modification of the Earlier Indian Planetary Theory by the Kerala Astronomers (c. 1500 AD) and the Implied Heliocentric Picture of Planetary Motion'. *Current Science* 66 (10): 784–790. https://www.jstor.org/stable/24098820.

Rao, S. Balachandra. 2000. *Indian Astronomy: An Introduction*. Hyderabad: Universities Press. ISBN: 9788173712050.

Reinaud, Joseph Toussaint, and Baron William MacGuckin de Slane), eds. 1840. *Kitāb Taqwīm al-buldān. Géographie d'Aboulféda, texte arabe publié d'après les manuscrits de Paris et de Leyde aux frais de la Société Asiatique*. Paris: Imprimerie Royale.

Rochberg, Francesca. 1999. 'Empiricism in Babylonian Omen Texts and the Classification of Mesopotamian Divination as Science'. Journal of the American Oriental Society 119 (4): 559–569. https://doi.org/10.2307/604834.

———. 2002. 'A Consideration of Babylonian Astronomy within the Historiography of Science'. *Studies in History and Philosophy of Science Part A* 33 (4): 661–684. https://doi.org/10.1016/S0039-3681(02)00022-5.

Rocher, Ludo. 1986. 'The Purāṇas'. In *A History of Indian Literature*, edited by Jan Gonda, vol. II: Epics and Sanskrit Religious Literature, fasc. 3. Wiesbaden: Otto Harrassowitz Verlag.

Rosenfeld, Boris A., and Ekmeleddin İhsanoğlu. 2003. *Mathematicians, Astronomers and Other Scholars of Islamic Civilisation and Their Works (7th–19th c.)* Istanbul: Research Centre for Islamic History, Art / Culture. ISBN: 9789290631279.

Sachau, Edward C, ed. and trans. (1910) 2013. *Alberuni's India. An Account of the Religion, Philosophy, Literature, Geography, Chronology, Astronomy, Customs, Laws and Astrology of India About A.D. 1080: An English Edition, with Notes and Indices*. Reprint edition. Trübner's Oriental Series I. London: Routledge.

Saliba, George. 1995. *A History of Arabic Astronomy*. New York University Studies in Near Eastern Civilization 19. New York: NYU Press. ISBN: 9780814780237.

———. 2007. *Islamic Science and the Making of the European Renaissance*. Transformations: Studies in the History of Science and Technology. Cambridge, Massachusetts: MIT Press. ISBN: 9780262195577.

Sarasvati Amma, Tekkath Amayankottukurussi Kalathil. (1979) 2007. *Geometry in Ancient and Medieval India*. 2nd revised and reprint edition. Varanasi: Motilal Banarsidass Publishers. ISBN: 9788120813441.

Sarma, Krishna Venkateswara, ed. 1976. *Āryabhaṭīya of Āryabhaṭa I with Commentary of Sūryadeva Yajvan: Critically Edited with Introduction and Appendices.* New Delhi: Indian National Science Academy.

———, ed. 1977. *Jyotirmīmāṃsā of Nilakaṇṭha Somayājī: Edited with Critical Introduction and Appendices.* Panjab University Indological Series 11. Hoshiarpur: Vishveshvaranand Vishva Bandhu Institute of Sanskrit and Indological Studies.

———. 2008. 'Armillary Spheres in India'. In *Encyclopaedia of the History of Science, Technology, and Medicine in Non-Western Cultures*, edited by Helaine Selin, 243. Dordrecht: Springer Netherlands. https://doi.org/10.1007/978-1-4020-4425-0_9552.

Sarma, Sreeramula Rajeswara. 1999. 'Yantrarāja: The Astrolabe in Sanskrit'. *Indian Journal of History of Science* 34 (2): 145–158.

———. 2002. 'The Rule of Three and Its Variations in India'. In *From China to Paris: 2000 Years Transmission of Mathematical Ideas*, edited by Yvonne Dold-Samplonius, Joseph Warren Dauben, Menso Folkerts and Benno van Dalen, 133–156. Boethius Series 46. Stuttgart: Franz Steiner Verlag. ISBN: 9783515082235.

———. 2011. 'A Bilingual Astrolabe from the Court of Jahāngīr'. *Indian Historical Review* 38 (1): 77–117. https://doi.org/10.1177/037698361103800105.

———. 2018. 'A Monumental Astrolabe Made for Shāh Jahān and Later Reworked with Sanskrit Legends'. Chap. 6 in *Astrolabes in Medieval Cultures*, edited by Josefina Rodríguez-Arribas, Charles Burnett, Silke Ackermann and Ryan Szpiech, 198–262. Leiden: Brill. https://doi.org/10.1163/9789004387867_009.

———. 2019. *A Descriptive Catalogue of Indian Astronomical Instruments.* 2nd revised edition. Electronic Open Access version. Duesseldorf: Sreeramula Rajeswara Sarma. ISBN: 9783946742579. https://doi.org/10.11588/xarep.00004167. http://crossasia-repository.ub.uni-heidelberg.de/4167.

Schubring, Walther. (1962) 2000. *The Doctrine of the Jainas: Described after the Old Sources.* 2nd revised edition. Translated from the revised German edition by Wolfgang Beurlen. With the three indices enlarged and added by William Bollée and Jayandra Soni. New Delhi: Motilal Banarsidass Publishers. ISBN: 9788120809338.

Schwartzberg, Joseph E. 2016. 'Maps and Mapmaking in India'. In *Encyclopaedia of the History of Science, Technology, and Medicine in Non-Western Cultures*, edited by Helaine Selin, 2662–2664. Dordrecht: Springer Netherlands. https://doi.org/10.1007/978-94-007-7747-7_9780.

Sen, Samarendra Nath, and Kripa Shankar Shukla. (1985) 2000. *History of Astronomy in India.* 2nd revised edition. New Delhi: Indian National Science Academy.

Shastri, T. S. Kuppanna. 1969. 'The School of Āryabhaṭa and the Peculiarities thereof'. *Indian Journal of History of Science* 4 (1 and 2): 126–134.

Shukla, Kripa Shankar. 1967. 'Āryabhaṭa I's Astronomy with Midnight Day-Reckoning'. *Gaṇita Bhāratī* 18 (1): 83–105.

———. 1973. 'Use of Hypotenuse in the Computation of the Equation of Center under the Epicyclic Theory in the School of Āryabhaṭa???' *Indian Journal of History of Science* 8:43–57.

———, ed. 1976. *Āryabhaṭīya of Āryabhaṭa I with Commentary of Bhāskara I and Someśvara: Critically edited with Introduction and Appendices*. New Delhi: Indian National Science Academy.

———. 1977. 'Glimpses from the *Āryabhaṭa-Siddhānta*'. *Indian Journal of History of Science* 12 (2): 181–186.

Sircar, Dinesh Chandra. 1990. *Studies in the Geography of Ancient and Medieval India*. 2nd ed. History and Culture Series 11. Varanasi: Motilal Banarsidass Publishers. ISBN: 9788120806900.

Sivapriyananda, Swami. 1990. *Astrology and Religion in Indian Art*. New Delhi: Abhinav Publications. ISBN: 9788170172314.

Somayaji, Dhulipala Arka. 1971. *A Critical Study of the Ancient Hindu Astronomy in the Light and Language of the Modern*. Karnatak University; Research Publications 11. Dharwar: Karnatak University.

Srinivas, Mandyam D. 2008. 'Proofs in Indian Mathematics'. In *Encyclopaedia of the History of Science, Technology, and Medicine in Non-Western Cultures*, edited by Helaine Selin, 1831–1836. Dordrecht: Springer Netherlands. https://doi.org/10.1007/978-1-4020-4425-0_9347.

Subbarayappa, Bidare Venkatasubbaiah, and Krishna Venkateswara Sarma. 1985. *Indian Astronomy: A Source Book (Based primarily on Sanskrit Texts)*. Bombay: Nehru Centre.

Tagare, Ganesh Vasudeo, trans. (1960) 1987–88. *The Vāyu Purāṇa*. Reprint edition. 2 vols. Ancient Indian Tradition and Mythology, 37 and 38. New Delhi: Motilal Banarsidass Publishers.

Trovato, Paolo. 2014. *Everything You Always Wanted to Know about Lachmann's Method: A Non-Standard Handbook on Genealogical Textual Criticism in the Age of Post-Structuralism, Cladistics, and Copy Editing*. Storie e Linguaggi. Padova: Libreria Universiteria. ISBN: 9788862925280.

Truschke, Audrey. 2012. 'Defining the Other: An Intellectual History of Sanskrit Lexicons and Grammars of Persian'. *Journal of the Indian Philosophy* 40:635–668. https://doi.org/10.1007/s10781-012-9163-2.

———. 2015. 'Contested History: Brahmanical Memories of Relations with the Mughals'. *Journal of the Economic and Social History of the Orient* 58:419–452. https://doi.org/10.1163/15685209-12341379.

———. 2016. *Culture of Encounters: Sanskrit at the Mughal Court*. Edited by Muzaffar Alam, Robert Goldman and Gauri Viswanathan. South Asia Across The Disciplines. New York: Columbia University Press. ISBN: 9780231173629.

Vahia, Mayank, and Srikumar M. Menon. 2011. 'Theoretical Framework of Harappan Astronomy'. In *Mapping the Oriental Sky*, Proceedings of the Seventh International Conference on Oriental Astronomy, edited by Tsukō Nakamura, Wayne Orchiston, Mitsuru Sôma and Richard Strom, 27–36. Tokyo: National Astronomical Observatory of Japan.

van Vloten, Gerlof, ed. (1895) 1968. *Liber Mafâtîh Al-Olûm: Explicans Vocabula Technica Scientiarum Tam Arabum Quam Peregrinorum. Auctore Abû Abdallah Muhammed ibn Ahmed ibn Jûsof (Al-Kâtib Al-Khowarezmi). Editit, Indices Adjecit.* Reprint edition. Leiden: E. J. Brill.

Vasunia, Phiroze. 2013. *The Classics and Colonial India*. Classical Presences. Oxford: Oxford University Press. https://doi.org/10.1093/acprof:oso/9780199203239.001.0001.

Velankar, Hari Damodar, ed. 1949. *Jayadāman: A Collection of Ancient Texts on Sanskrit Prosody and a Classified List of Sanskrit Metres with an Alphabetical Index.* Haritoṣamālā 1. Bombay: Haritosha Samiti.

Wilson, Horace Hayman, trans. 1864–77. *The Vishńu Puráńa: A System of Hindu Mythology and Tradition*. Edited by Fitzedward Hall. 5 vols. Translated from the original Sanskrit, and illustrated by notes derived chiefly from other Puráńas. London: Trübner and Company.

Wright, J. Edward. 2000. *The Early History of Heaven*. 1st ed. New York: Oxford University Press. ISBN: 9780195130096.

Wujastyk, Dominik. 2004. '*Jambudvīpa*: Apples or Plums?' In *Ketuprakāśa: Studies in the History of the Exact Sciences in Honour of David Pingree*, edited by Charles Burnett, Jan P. Hogendijk, Kim Plofker and Michio Yano, 287–301. Islamic Philosophy, Theology and Science. Texts and Studies 54. Leiden: Brill. https://doi.org/10.1163/9789047412441_013.

Yule, Henry, and Arthur Coke Burnell. 2013. *Hobson-Jobson: The Definitive Glossary of British India*. Edited by Kate Teltscher. Oxford World's Classics. Oxford: Oxford University Press. ISBN: 9780199601134.

Index

Note: **Bold** page numbers refer to tables; *italic* page numbers refer to figures and page numbers followed by "n" denote endnotes.

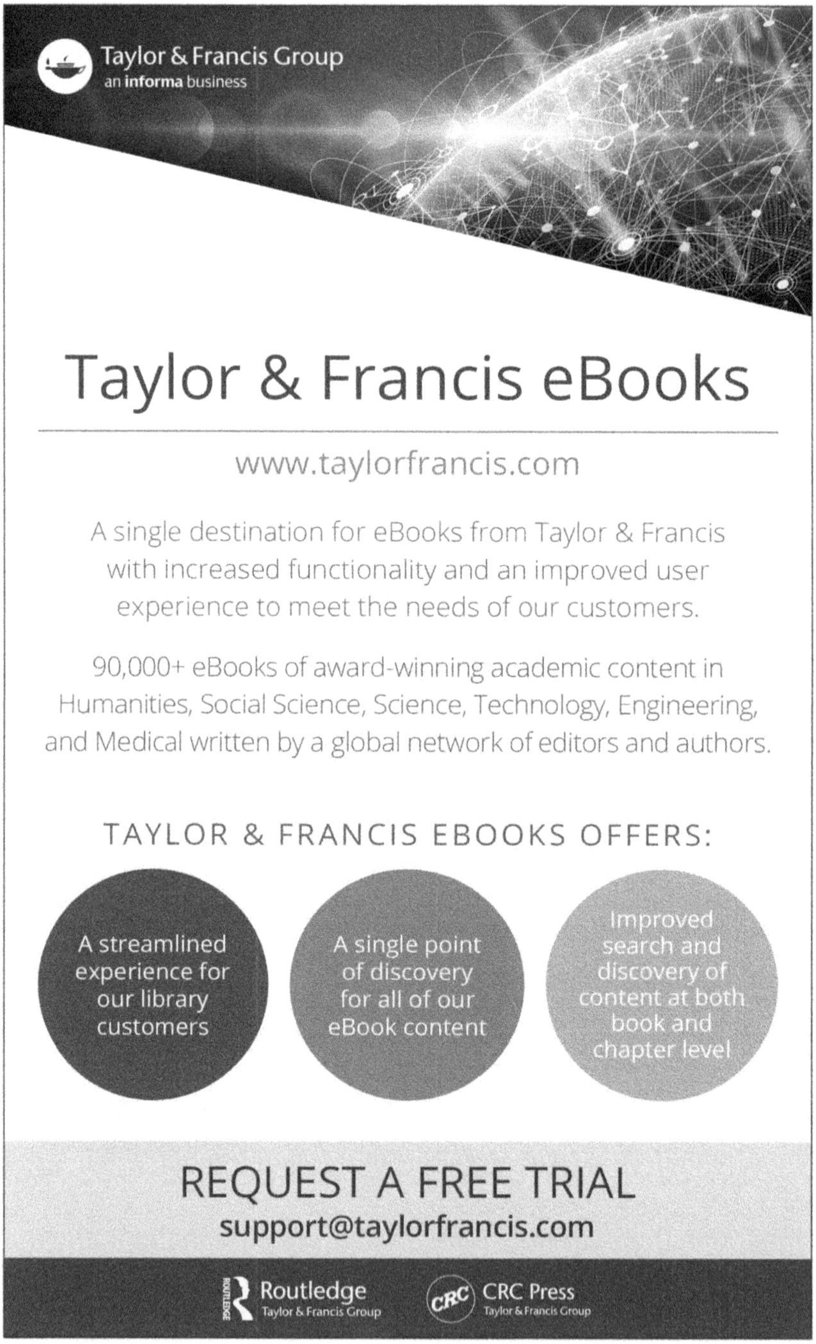

Taylor & Francis Group
an informa business
Taylor & Francis eBooks
www.taylorfrancis.com
A single destination for eBooks from Taylor & Francis with increased functionality and an improved user experience to meet the needs of our customers.
90,000+ eBooks of award-winning academic content in Humanities, Social Science, Science, Technology, Engineering, and Medical written by a global network of editors and authors.
TAYLOR & FRANCIS EBOOKS OFFERS:
A streamlined experience for our library customers
A single point of discovery for all of our eBook content
Improved search and discovery of content at both book and chapter level
REQUEST A FREE TRIAL
support@taylorfrancis.com
Routledge
Taylor & Francis Group
CRC Press
Taylor & Francis Group

For Product Safety Concerns and Information please contact our EU representative GPSR@taylorandfrancis.com
Taylor & Francis Verlag GmbH, Kaufingerstraße 24, 80331 München, Germany

www.ingramcontent.com/pod-product-compliance
Ingram Content Group UK Ltd.
Pitfield, Milton Keynes, MK11 3LW, UK
UKHW020317270926
475702UK00050B/941

* 9 7 8 1 1 3 8 5 8 3 5 7 3 *